AF531411

COMPREHENSIVE ENVIRONMENTAL IMPACT ASSESSMENT OF WATER RESOURCES PROJECTS

WITH SPECIAL REFERENCE TO SATHANUR RESERVOIR PROJECT (TAMIL NADU)

COMPREHENSIVE ENVIRONMENTAL IMPACT ASSESSMENT OF WATER RESOURCES PROJECTS

WITH SPECIAL REFERENCE TO SATHANUR RESERVOIR PROJECT (TAMIL NADU)

VOLUME—I

By

Dr. K.B. Chari

Dr. Richa Sharma

Prof. S.A. Abbasi

Centre for Pollution Control & Energy Technology
Pondicherry University
Pondicherry – 605 014

DISCOVERY PUBLISHING HOUSE
NEW DELHI-110 002

First Published - 2005

ISBN: 81-8356-044-X (Set)

© **Authors**

Published by:

DISCOVERY PUBLISHING HOUSE PVT. LTD.
4383/4B, Ansari Road, Darya Ganj
New Delhi-110 002 (India)
Phone: +91-11-23279245, 23253475, 43596065
E-mail: discoverypublishinghouse@gmail.com
sales@discoverypublishinggroup.com
web: www.discoverypublishinggroup.com

Printed at:
Amit Enterprises

Dedicated

To

Shri J .P. Singh IPS (Ret)

The Police Chief with a Healing touch

—S. A. Abbasi

My friends—Gajalakshmi, Mallick

And

Ravi—for being with me through the thick and thin

—Chari

Dedication

To

My Parents

—Richa Sharma

Preface

There was a phase during which, in the two decades following the independence of India, creation of reservoirs by the damming of rivers was an activity blessed by the government as well as the public.

But most reservoirs thus created, and the watersheds that fed them, did not happen to be managed in an environment-friendly manner. This wasn't as much due to the inefficiency of the governmental machinery as it was due to the then prevailing lack of environmental consciousness among the executives as well as the lay public. This led to denudation of the catchments, erosion of the watershed soils, silting of the reservoirs, eutrophication, loss of fisheries etc etc. The canal networks in the command areas were also not maintained properly, leading to breeches, weed infestation, puddle formation, and associated evils. Then, again, the available reservoir water was often appropriated inequitably by the more influential farmers to grow too many water-guzzling crops too soon which impoverished the command area croplands leading to diminishing returns. The quicker-than-anticipated rate of siltation of reservoirs meant lesser storage capacity, poorer water quality, and weaker flood control. Most of these problems were caused by improper management rather than any treachery inherent in large dams. But the public ire began to direct itself towards large dams instead of the forces which had contributed to their ill-management.

By the beginning of the 1970s, large dams had acquired the reputation of environment-plundering monsters. This image continued to darken through the 1980s and 1990s during which mini-hydel and micro-hydel projects were increasingly touted as the environment-friendly alternatives to large dams.

But, half-a-decade into the new millennium, we know that mini and micro hydel projects can at best service a small fraction of the requirements of a giant, fast emerging, economic superpower that India is. Further, thousands of dams, constructed at the cost of billions of rupees, are very much in existence. If only we can develop appropriate guidelines to manage these dams in an environmentally sound manner, we can minimize their harmful impact on the environment and maximize the benefits derivable from them. With this objective in view, the Ministry of Water Resources, Government of India, had asked us to conduct an elaborate study of the Sathanur Reservoir Project System (SRPS), situated near Thiruvannamalai, northwestern Tamil Nadu. The present book documents the gist of our findings which aim at assessing the present status of the environmental impact and its management at SRP. In the backdrop of how things are, and what they ought to be, we have tried to develop a body of knowledge that may be helpful in the better management of not only SRP but other dam-based projects as well.

We are grateful to the Ministry of Water Resources, Government of India, for sponsoring the study through its Indian National Committee on Hydrology (INCOH). We were helped immensely by the support and guidance of Dr K. K. S. Bhatia who was the Member-Secretary of INCOH during the tenure of the project. Shri Chetan Pandit who was in that period directing the R&D Division of the Ministry of Water Resources was a consistant source of strength and inspiration. Our grateful thanks to Dr Bhatia and Shri Pandit. We also express thanks for the patronage extended to us by Dr K. D. Sharma, Director, NIH, and for the swift follow-up to our requests by Shri S. Masood Husain (Director & Head, R&D Division, MoWR) and Dr Ramakar Jha (Member-Secretary, INCOH).

S. A. Abbasi

K. Brahmananda Chari

Acknowledgement

We gratefully acknowledge the support of Ministry of Water Resources (Government of India), through its Indian National Committee on Hydrology, for the project which made possible this study.

We fondly thank Prof A. Gnanam, during whose tenure as the Vice Chancellor of Pondicherry University, the project had taken off.

We gratefully acknowledge the help extended to us by the government departments:

Office of the Assistant Executive Engineer
SRBC,
PWD, Sankarapuram
Villupuram

Office of the Assistant Executive Engineer
SRBC,
PWD, Sankarapuram
Sathanur Dam

Office of the Agriculture Officer
Officer of Joint Director
Thiruvannamalai

Office of the Agriculture Officer
Office of the Joint Director
Villupuram

Office of the Assistant Executive Engineer
PWD, Moongilthuraipattu

Office of the District Forest Officer
Kallakurichi
Villupuram

Office of the Director of Public Health Service
Chennai

Office of the District Entomologist
Office of Deputy Director of Public Health Services (DDHS)
Thiruvannamalai

Office of the Divisional Forest Officer
Thiruvannamalai

Office of the Director
Watershed Management Board
PWD, Pollachi

Office of the Director
IHH, Poondi

Office of the Divisional Forest Officer
Kallakurichi
Villupuram

Office of the Directorate of Statistics and Economics
Thiruvannamalai

Office of the Directorate of Statistics and Economics
Villupuram

Deputy Director
Office of the Deputy Director of Health Service
Kallakurichi

Office of the Executive Engineer
Tamil Nadu Electricity Board
Sathanur Dam Hydro-electricity Board
Sathanur Dam

Office of the Executive Engineer
SM and R Division, PWD
Chennai

Office of the Executive Engineer
(Middle Ponnaiyar Region
Thiruvannamalai

Office of the Forest Ranger
Sathanur Dam

Office of the Forest Ranger
Thiruvannamalai

Office of the Health Inspector
D.D.H.S., Kallakurichi

Office of the Health Inspector
Primary Health Centre
Moongilthuraipattu

Office of the Joint Chief Engineer
Water Resources Commission
Tamil Nadu PWD
Chennai

Office of the Medical Officer
Primary Health Centre (PHC)
Sathanur Dam

Office of the Malaria Inspector
Thandarampet

Office of the Manager
Tamil Nadu Fisheries Development Corporation
Sathanur Dam

Office of the Manager
Kallakurichi Cooperative Sugar Mill
Kallakurichi
Villupuram

We thank the support of the following Doctoral, M. Phil, and MCA scholars who helped us in the development of SRPIS:

Rameswaran S, MCA
Development of SRPIS

Girija K
M. Sc, M. Phil
Digital Mapping of SRP maps

Richa Sharma
M. Sc, M. Phil, PhD

Karunakaran N
M. Sc, M. Phil;

Alexander R
MS, M. Phil
SRPIS data collection

Rajalakshmi K
M. Sc, M. Phil
SRPIS Data Organization and Data Entry Coordination

List of Abbreviations Used

AIU	Agriculture Impact Unit
API	Annual Parasite Index
EIU	Environmental Impact Unit
IIU	Irrigation Impact Unit
LBC	Left Bank Canal
RBC	Right Bank Canal
SLBC	Sathanur Left Bank Canal
SRBC	Sathanur Right Bank Canal
SRP	Sathanur Reservoir Project
SCA	Sathanur Command Area
SIN	Socio-economic Impact Unit
TRA	Total Reported Area
PHC	Primary Health Centre

Contents

PART—A

Developing Guidelines for the EIA of Water Resources Projects with Reference to Sathanur Reservoir Project (Tamil Nadu)

Section—I: Introduction

1

Environmental Impact of Water Resources Projects

A General Introduction

ENVIRONMENTAL IMPACT OF WATER RESOURCES PROJECTS BASED ON RIVER DEVELOPMENT

The effects of river development schemes on the environment manifest themselves in a number of ways. They are felt upstream, in the storage reservoir, downstream and often over the whole region. They may be beneficial or detrimental. Dealing with detrimental effects means evaluating the degree to which they are harmful and finding ways of controlling them.

The physical and biological effects of water resources projects based on damming of a river arise from the obstacle the dam causes to the natural flow of the river, the climate changes caused by the reservoir interacting with the overlying atmosphere, the effects of the structures on the water in and near the reservoir, and the seismicity induced or compounded by the stored water.

EFFECT OF OBSTACLE CAUSED BY THE DAM

Floating Debris, Fish, Boats

A dam is an obstacle to the passage of trees, ice and other floating debris, wildlife including fish, and boats.

The adverse effects can be overcome by providing timber chutes, fish ladders or lifts, and shipping locks. Overflow spillways also permit the passage of debris past a dam.

Sediment Load

A dam reduces or completely blocks the passage of the sediments conveyed by the river creating varying degrees of disturbance in the natural conditions. Blocking the bed load may disturb the balance of delta areas if natural erosion processes are no longer offset by the arrival of new material. River training works may be necessary to keep the banks stable as was done on the Damieta branch of the Nile.

Reducing the suspended solids load may deprive arable land of the silt brought down by river floods, an impact brought to sharp focus by the Aswan dam. This is the cost one may have to pay for extended irrigation, but the harm it does may in fact be illusory—silt is not such a good fertilizer as imagined (it contains only 4% by weight of nitrogen available to plants) and chemical fertilizers can be substituted. In addition land formerly fertilized by silt can be cultivated for a greater proportion of the year with controlled and comparatively silt-free applications of irrigation water.

The construction of a dam or barrage across a river channel creates a reservoir, which can hold back and retain most of the suspended material normally carried in the water flowing in the river. The water freed from its sediments tends to reacquire the sediment load. This leads to erosion in the channels in the zone immediately downstream. The zone gradually extends downstream until a new equilibrium is reached, when the slope of channel is once more balanced by the reduced suspended load. Such erosion can be corrected (or reduced to some degree) by reconstructing the slope of the channel and/or by changing the discharge from the dam.

Various methods have been used such as building weirs (sometimes also used incidentally for power purposes), spreading flood spillage over a longer period of time, or diverting part of the river flood into an offshoot in times of flood, as is done on the Nile at Aswan.

Sedimentation

A reservoir is a veritable sedimentation tank. The problem of siltation due to water resources development is a global phenomenon. Countries experiencing difficulties include Nepal, Pakistan, Phillipines, Burma, Columbia, Egypt, India, Indonesia, Tanzania and Thailand. Twenty one reservoirs in India have experienced sediment inflows at least 200 per cent higher than anticipated (Ahmad and Singh, 1991). The sediments brought in by the feeder streams and the run-in from the reservoir catchment get an opportunity to settle down in the quiescent waters of the reservoir.

This results in two impacts of far-reaching consequences. The first is that the settled sediments reduce the storage capacity of the reservoir. The impact can be strong enough to drastically reduce the reservoir life. For example, the capacity of Nizamsagar reservoir has been reduced to less than half due to sedimentation and sufficient water is not stored any more to irrigate 1100 Km^2 of sugarcane and paddy for which the reservoir was mainly built. The second impact is caused when the desedimented water from the reservoir is released into the river downstream and in the irrigation canals. The water tends to reacquire its sediment load and erodes the banks of the river or the canal carrying it. The erosion can play havoc with the canal stability.

The catchment area of Chittaurgarh irrigation project which consisted of dense natural forests in its upper reaches has been denuded during dam construction. The expected rate of sedimentation is 0.14 million $m^3 yr^{-1}$ and the life span of the dam is estimated to be 50 years. The vegetation cover upstream of the dam is poor and it is estimated that after 5 to 10 years a large quantity of silt and sand will be deposited at the head of the dam and in the canal system affecting the fish fauna seriously. A part of this is slowly transferred to the farmers' fields where it will cause congestion of irrigation canals and ditches and a 50 per cent reduction in the life of dam. The canal embankment is devoid of any vegetation cover; therefore it is highly vulnerable to erosion during heavy rainfalls and subject to increased sedimentation (Ahmad and Singh, 1991).

Due to geological and climatic peculiarities, the Himalayan rivers carry some of the highest sediment loads in the World (Bandhopadhyay and Gyawali 1994). The major Himalayan dams like Pong in Himachal Pradesh or Ramganga in Uttar Pradesh have been silted up at a rate four or five times higher than the assumed ones (Bandhopadhyay, 1995).

A turbid layer in the otherwise clear reservoir may flow along the bottom to form a mud lake just in front of the dam capable of blocking the bottom outlets as it hardens, or may remain at mid-depth until it rises to surface near the dam, sometimes entering the intakes or spillway. In some reservoirs, turbidity may extend throughout the reservoir because of the fineness of the material. India will reap the benefit of the Indo-Bhutan Sankosh hydel project by diverting a large volume of the Sankosh river water to the Farakka barrage and consequently clean out siltation of Calcutta port (Down to Earth, 1996).

Nutrient Transport

Deposition of suspended nutrients in the reservoir may have consequences for the aquatic biota downstream. It has been blamed for the smaller sardine catches in the Mediterranean off the Nile delta after construction of the High Aswan dam.

On the other hand, however, shallow reservoirs may increase the plankton yield, as in the pool on the Seine river near the town of Troyes.

Retention of Waters from Small and Moderate Floods

Small floods have a beneficial effect in that:

(a) they provide ready access to spawning grounds (ponds and lakes) and renew the water in them;

(b) they form small islands used by migrating birds to escape from the predators on the shore;

(c) they prevent the river banks becoming overgrown with trees, and stop mammals destroying the river-side grasses needed to sustain migrating birds; and

(d) they bring nutrients into lakes and ponds.

Water can be released through the dam to stimulate such floods by a proper choice of time, duration and volume released.

Overlarge or untimely releases can cause harm, and the gate operation guidelines drawn up with reference to ecological requirements downstream must be strictly followed to prevent it.

Tidal Barriers and Barrages

Barriers and barrages constructed in the tidal zone of a river estuary or coastal embankment may be considered as a special case of a barricade.

The characteristics of an estuary may be changed in a variety of ways by a barrage scheme. The inflow of tidal waters may be restricted, leading to effective loss of volume and thereby reduced scouring effect of ebb tides. If this problem is not considered and dealt with, bars may form which handicap navigation. The estuarial regime of salt and fresh water may change from stratified to mixed or vice versa, and this may affect siltation and aquatic life. Wave patterns may also be significantly modified. Upland drainage in terms of both freshwater river run-off and riparian land drainage in the estuary area have to be dealt with. The latter could involve pump schemes. Sewage previously allowed to flow to the sea with limited treatment could require full treatment if liable to be trapped behind a barrage. Docks may be required for shipping access to up-river ports. Fish passes may be required if the estuary is a route for migratory fish, and ice passes if the river carries ice floes in winter. Thought should be given to the effects on flora and fauna previously inhabiting the tidal range which may become permanently inundated.

A barrier differs from a barrage in that it may not be closed permanently or regularly. In the case of the Thames Barrier at Woolwich, the intention is to block only major tidal flood surges from penetrating upstream and flooding London. This may involve closure for a few hours on a few occasions per year. Thus effects on regimes and the environment are only marginal. In other cases of barrier, the environmental effects will approach those of barrages as the frequency of closure increases.

Both barrages and barriers have the effect of preventing some upstream tidal flow. This volume of water rebounds downstream. Coastal protection works need to be raised against this effect.

Effect of Dam on Sea

The large number of dams in operation around the world today could affect the food web structure and bio-geo-chemical cycling of materials in coastal seas. This is the finding of research conducted by scientists at the Hamburg University who studied the effect of Danube river dam on the Black Sea ecosystem recently (Science and Technology, *The Hindu*, 1997).

The research supports the fact that man-made barriers are trapping vital nutrients suspended in the river water and preventing them from reaching the oceans. Scientists warn that fishes will disappear and rising tides of toxic algae would float across ocean waters if what happened in the Black Sea is repeated in the other major oceans. The long term data on water and nutrient discharge from the river Danube to the Black Sea revealed a reduction in the dissolved silicate load of the river by about two-thirds since dam constructions in the early 1970s. A concomitant decrease in dissolved silicate concentrations by more than 60 per cent during winter was observed in the Central Black Sea surface waters. The consequent changes in silicon to nitrogen ratio of the Black Sea nutrient load appear to be larger than those caused by eutrophication alone and seem to be responsible for dramatic shifts in phytoplankton species composition from diatoms (siliceous) to coccolithophores and flagellates (non-siliceous). Human interventions have caused a worldwide increase in river inputs of nitrogen (N) and phosphorus (P) to the coastal seas by more than a factor of four leading to considerable eutrophication and to an increase in the frequency of unusual and/or noxious algal blooms.

The construction of dams in rivers can also cause considerable reductions in nutrient loads owing to the removal of these nutrients in reservoir sediments (the artificial lake effect) whereas this removal might be over compensated by anthropogenic nitrogen and phosphorus inputs downstream of the reservoir, no such compensation has been observed for silicate, for example in the Nile, Dr. Venugopalan pointed out. According to him the Iron Gates hydroelectric dam on the river Danube had captured tiny specks of silicate eroded from soils upstream and carried in the water. For the past 25 years that silicate has settled in sediments on the reservoir floor. He estimates that the dam has so far captured 12 million tonnes of silicate for 15 years' natural supply to the Black Sea turning the sea's biochemistry upside down.

Results published in 'Nature' revealed that water and sediment storage in reservoirs behind the 'Iron Gates'. have altered the bio-geochemistry not just of the river and the adjacent coastal waters but also of the entire Black Sea basin (Venkiteswaran, 1997). The observed species shift towards carbonate producing coccolithophores in the coastal waters exert significant control over seawater chemistry. Furthermore, the occurrence of potentially toxic flagellate blooms may become more frequent. Similar effects may be expected for the Central Black Sea where, in contrast to pre-dam conditions silicate depletion appears now to be more frequent.

Today more than 36,000 dams are in operation around the world and more are being constructed at an appreciable rate. Though the observed effects of these dams can be more severe in closed seas (such as the Black Sea) than in rivers feeding into open systems, these effects could be of global consequence.

EFFECT OF FLOODING ON FAUNA

A new reservoir directly affects the upstream fauna in a number of ways:

(a) many animals die although there may be ways of saving them;

(b) some animals migrate to new areas;

(c) a few animals accommodate to the new environment, particularly amphibians and riparian fauna;

(d) birds such as water-fowl and waders move into the new water habitat.

The proposed 141 km long canal to be built across the northern part of West Bengal in the Tista river as a part of the Indo-Bhutan Hydel project would spell total doom to the wild life of the area. The canal 60 m wide and 6 m deep with a bed width of

26 m will cut off population of all wild life species from one side to another and will cut off the gene flow and migratory routes of various endangered species, especially the elephant (Down to Earth, 1995).

The Three Gorges dam hydro-electric project in China once promoted as a major money spinner for MNCs is has turned out to be a liability for Beijing as the project, scream the detractors, will spell the doom for the fragile eco-system of the Three Gorges lake (Down to Earth 1995).

The removal of trees and shrubs upstream of the Chittaurgarh dam has led to wildlife loss. Disturbances caused by heavy machinery and vehicles, used in dam construction have forced wild life such as sambhar *(Cervus duvaceli)*, cheetal *(Axis axis)*, Swamp deer *(Cervus duvauceli)*, tiger *(Panthera tigris)* to emigrate from the area. There is concern that the forest might lose its deer population entirely. The erection of an 11 km. long and 15m high dam for water storage in the reservoir is now acting as a major barrier to wild fauna migration (Ahmad and Singh, 1991).

The proposed hydro-electric project across the Rathong Chu river which meanders through Yuksam in West Sikkim, envisages the channelling of the river near the base of Mount Kanchenjunga. Struck by Sikkim's rich flora and several endangered animals like the snow leopard, Himalayan black bear and red panda, the World Wide Fund for nature classified it a priority reservation 'biodiversity hotspot'(Balakrishnan, 1995).

It is sometimes possible to carry out rescue and salvage operations, along the Noah's Ark principle. For instance more than ten thousand tortoises, tapirs, snakes, sloths, armadilloes, etc. were rescued from the Afobaka lake on the Brokopondo in Surinam, whereas zoologists had estimated that there were no large animals left in the area.

An important secondary effect which can arise is that animals displaced from the reservoir area must fit into a new habitat which was previously in equilibrium, a situation which tends to favour short-lived species with high reproduction rates and which are not always the most useful, desirable or attractive. This problem can be common to both forced migration caused by the rising waters of the lake and to the re-establishment of the "rescued" animals in the new habitats.

The creation of a new water body can establish conditions conducive to a large increase in the variety and numbers of birds, particularly water fowl, waders and their predators. The loss of primarily terrestrial habitat is often more than compensated for, by the large increase in water riparian habitat and the restoration and protection of endangered species of water fowl. The effect of flooding on fauna is generally more pronounced in tropical regions.

EFFECT ON CLIMATE

Whether, and to what degree climate can be changed by large reservoirs is the subject of controversy. The formation of fog by evaporation from a river or lake depends on air humidity, the temperature difference between the water surface and the ambient air, and the salinity of the water.

As a very rough rule, the air must be cooler than the water and its relative humidity must be greater than 90 per cent, to create a fog. This means that fog occurs in temperate climates, mainly on winter nights and the early morning. In other climates, shallow reservoirs may make the mist thicker on cool days. It has been demonstrated experimentally that large reservoirs produce a new micro-climate. A change in the rainfall pattern has been detected in Ghana around lake Volta, the peak having shifted from October to July and August, and rain fell for the first time at Aswan when lake Nasser was filled. It is generally accepted that the capacity of large natural lakes in temperate regions for storing heat and cold is the prime cause of the particular micro-climates in those areas. Examples are lake Leman, le Bourget lake, Lake Annecy, the Italian lakes and the five Great Lakes.

EFFECTS ON RESERVOIR WATER

Water Temperature

In deep reservoirs, a cold layer underlies a warmer layer of water in summer and knowledge of such thermal layering is essential for an understanding of the effects that deep reservoirs might have on water quality.

In warm seasons, the upper layer of the reservoir (the epilimnion) is heated by sunlight and the warmer inflow, whereas the lower layer (the hypolimnion) remains at around 4°C, representing the maximum density of water. The two layers are separated by a sharp thermal gradient zone, the metalimnion.

Temperature changes due to the wind and changes in air temperature are confined to the epilimnion. In colder weather, wind and the lower air and inflow temperature cool down the epilimnion until it is colder than the hypolimnion; vertical currents appear, causing the metalimnion to disappear and the temperature to even out in the body of water (overturn).

At times when there is no stratification, water drawn off from the bottom of the reservoir is provided from the whole vertical slice just in front of the intake, but at other times, the hypolimnion is drawn off first while new inflow into the reservoir stays in the epilimnion. The abstracted water is cold.

A lower water temperature in warm weather is beneficial as regards supply to homes and industry, for thermal power station cooling and fish spawning at the end of summer, but harmful when spawning takes place earlier.

Dissolved Gas Content

Oxygen Deficit

Oxygen is necessary to sustain aquatic biota and provide a self-purification capacity for the water. Dissolved oxygen penetrates the surface layer (detergent or oil films are obstacles to this process) or is produced by photosynthesis in the phytoplankton from dissolved mineral salts and carbon dioxide. Biodegradation of dissolved, suspended and deposited organic material depends on oxygen, as does of course respiration of the aquatic biota. Warm water fish needs 4 mg/l to sustain life and reproduce.

If the river is heavily loaded with organic material, the amount of oxygen consumed to degrade it may be more than can be absorbed through the water/air interface, so that the oxygen content quickly falls. Once all the oxygen has been removed in this way, the river reaches an advanced stage of mal-odorous pollution with a scum over the surface, and all aquatic life ceases.

Storing of water in deep reservoirs produces a considerable oxygen deficit in the deeper layers in some periods of the year. No harm is done if little water is abstracted or spilled in such periods and the oxygen demand downstream is small. The water recovers its normal oxygen content by contact with the atmosphere very quickly. Only large amounts of water released through the dam have detrimental effects because of oxygen depletion.

The oxygen content of water released from deep reservoirs depends on four factors:

1. The extent of thermal layering in the reservoir;
2. The depth of the intakes under the water surface;
3. The amount of organic material present;
4. The amount of decomposition products in the reservoir.

The current practice is to build the intakes with several inlets at different heights so that the water with the highest oxygen content is drawn off. Blowing a stream of oxygen or air through the abstracted water has also been tried.

Nitrogen and Oxygen Super Saturation

Water below a spillway may become supersaturated with nitrogen and oxygen. This effect decreases with depth but nitrogen contents of up to 130 per cent have been recorded below high spillways.

The mortality threshold for some fish species appears to be about 120 per cent. The effect was demonstrated on the Columbia and Snake rivers where in 1970, it was estimated that 90 per cent of the young salmon (fry) making their first passage to the sea, died.

Fries moving close to the surface were noticeably more affected. Numbers of salmon moving upstream to spawn in the following year were apparently unaffected.

Eutrophication

The condition known as "eutrophication" (from the Greek "*trephein*" to nourish) involves the water and bottom sediments becoming enriched with nutrients to a point where the water quality deteriorates; the nutrient content of the lake rises to a dangerous level, sometimes known as the "trophic" level.

Until recent times, eutrophication was a slow process, comparable to ageing, affecting natural as well as man-made lakes, but in the last few decades, more widespread use of fertilizers and detergents and growing quantities of waste water have speeded it up. Phosphorus and nitrogen are the main factors.

The first symptoms of eutrophication are:

(i) Excessive development of bed plankton and macroscopic plants and weeds near the banks;

(ii) Displacement of Salmonidae by Cyprinidae;

(iii) Reduced water clarity caused by increase in the number of swimming micro organisms, and changes in the colour of the water due to the red algae;

(iv) Reduced oxygen content near the bottom, accompanied by an increase near the surface;

(v) Swimming micro-organisms, and changes in the colour of water due to the red algae;

(vi) Reduced oxygen content near the bottom, accompanied by an increase near the surface;

(vii) Development of luxuriant weed growth;

(viii) Formation of massive amounts of rotting matter on the bed;

(ix) Complete disappearance of oxygen in the deeper layers in summer;

(x) Appearance of sulphuretted hydrogen, free iron, manganese and ammonium ions;

(xi) Appearance of gas bubbles.

By this stage, the lake becomes a 'breeder', causing more pollution from what it has received rather than digesting it.

Impacts of Eutrophication

The impacts of Eutrophication are listed below (Rast and Holland, 1988):

(i) Taste and odour problems can develop in drinking water taken from a eutrophic reservoir even though water is treated prior to its use;

(ii) The water treatment process itself can also become more expensive and time consuming for waters taken from eutrophic water bodies;

(iii) Excessive algae and aquatic plant growths are highly visible and can detract significantly from the aesthetic quality of a water body;

(iv) With dense quantities of algae in a reservoir, the transparency of the water is greatly reduced and the water body can acquire an undesirable "pea-soup" green colour;

(v) Excessive algal densities can interfere significantly with swimming and other recreational activities;

(vi) Large masses of dead algae can accumulate on beach with negative aesthetic impacts;

(vii) As algae and aquatic plants sink to the bottom of a water body their decay by bacteria can reduce oxygen concentration in bottom waters making it too low to support fish life, resulting in fish kills;

(viii) Excessive levels of iron, manganese and foul smelling hydrogen sulphide in the bottom waters may also occur as a result of oxygen depletion;

(ix) There are also potential negative health impacts, especially in tropical and sub-tropical regions. Cultural eutrophication can aggravate the occurrence of parasitic diseases such as schistosomiasis, onchocerciasis and malaria by enhancing the habitats of the organisms responsible for their transmission to humans;

(x) Eutrophication of water bodies does have some positive aspects. For example in some countries, control cultural eutrophication is used to enhance fish production and other forms of aquaculture for the purpose of producing food.

In such cases, the management goal is to maximize this productivity with a minimum of cost and effort.

Development of Eutrophication Management Strategies

A basic practical framework for development of a eutrophication management strategy consists of the following steps (Rast and Holland, 1988):

(i) Identify eutrophication problem and establish management goals;

(ii) Assess the extent of available information about the reservoir;

(iii) Identify available, feasible eutrophication control methods;

(iv) Analyse all costs and expected benefits of alternative management strategies;

(v) Analyse adequacy of existing institutional and regulatory framework for implementing alternative management strategies;

(vi) Select desired control strategy and disseminate plan to interested parties;

(vii) Use institutional mechanisms to minimize future eutrophication problems.

Remedy

1. Purifying the inflow and reduction of nutrient inputs.
2. Agricultural non-point source control measures include:
   (i) Conservation tillage
   (ii) Vegetative buffer strips
   (iii) Contour cultivation
   (iv) Correct fertilizer application practices
   (v) Management of livestock manure
3. In Reservoir control measures:

   Some treatment procedures can be applied directly to a reservoir,

   (i) The harvesting of aquatic plants
   (ii) Use of algicides

(iii) Nutrient inactivation in the reservoir

(iv) Artificial oxygenation of bottom waters

(v) The dredging or covering of bottom sediments

(vi) Increasing the water flushing or circulation rates

(vii) Artificial manipulation of in-lake biological communities

4. Use of nutrient rich waters for irrigation or aquaculture purposes

5. Building intercepting ditches:

Purification of inflow oxygen injection and building interception ditches have been used for Nantia, Bourget and Annery lakes in France (Rast & Holland, 1988).

Dissolved Solids Content

A reservoir can have a complex effect on dissolved solids content of waters, by such mechanisms as oxidation, reduction and ion exchange, but this can often be turned to advantage. In the simple case of constant high salinity inflow, there is some concentration of salts in the reservoir as evaporation takes place from an increased surface area. However, high salinity inflows are normally in the form of concentrated 'slugs' at a particular time of year. In this case the 'slugs' are diluted by fresher water in the reservoir, and even after allowing for evaporation, effluents are generally at lower salinity levels than inflows. The same applies when one stream forming part of a multiple-tributary system has a high salts concentration.

It has been pointed out that on a global scale, at least 200,000 to 300,000 ha. of irrigated land are lost every year due to salinization and water-logging. In Pakistan as many as 11 million ha. to 15 million ha. of irrigated land are already suffering from salinization and water logging. Afghanisatan, Egypt, Iran, Iraq, Sudan and Syria also have similar problems.

The Chittaurgarh area under Chittaurgarh irrigation project shows high water table indicating that the area is vulnerable to water-logging and salinization because of clay-dominated soils. A rise in the water table under similar conditions, caused water logging of 45,000 ha of the Sarda canal in Sarda Sahayak Irrigation Project of Uttar Pradesh and of 30,000 ha area in the Gandak canal in Gandak canal irrigation project (Ahmad and Singh, 1991).

A further complicating effect exists where reservoir area ground waters are saline. At Chowilla Dam site in South Australia substantial seepage flows of highly saline groundwater were predicted, flushed out by the head of fresh impounded reservoir water. The designs incorporated a comprehensive system of relief wells to intercept this flow and prevent it from polluting the river downstream. The seepage was disposed of by pumping to large evaporation ponds.

Health Effects for Riparian Dwellers

New storage reservoirs and their related water systems have repercussions on the main endemic parasites. In hot, damp regions, certain of the more chronic and widespread endemic diseases are transmitted through pathogens whose vectors live in fresh water. Such diseases are endemic even before construction of the dam but the reservoir increases the size of the hosts' habitat, thus promoting their expansion. The main diseases spread in this way are malaria (transmitted through the Anopheles mosquito), bilharziasis caused by blood flukes of the genus Bilharzia transmitted through a snail, and onchocerciasis, a very widespread disease of Africa and tropical America, transmitted by the bite of a small fly of the Simuliidae family.

Anopheles mosquitoes breed in small puddles and swampy areas rather than large lakes, so it is not that the full storage reservoirs behind dams that are responsible for the spread of malaria, but rather the temporary or permanent marginal systems created by seasonal water level fluctuations and human colonisation around the edge of the lake promote malaria disease. Irrigation malaria breaks out in irrigated agricultural areas with large, medium and small dams or tank irrigation projects. High malaria endemicity is recorded in Upper Krishna Irrigation Project in Karnataka and Sardar Sarovar Project on the Narmada in Gujarat (Jain,1994) A good monsoon and water logging from the Indira Gandhi canal combined to create an outbreak of cerebral malaria which quickly spread to epidemic proportions in 1994 and 4000 people were feared dead. Flood irrigation and stagnant water in the fields breed mosquitoes. Parts of the canal itself have stagnant water as construction is yet not complete. Seepage from the canal has caused massive water logging (Jain, 1994).

The Chittaurgrah project is situated in the Tarai region (foot-hills) which is humid and has a high water table, 0.43 m below ground level. The area is already suffering from waterlogging and flood. The construction of dam and the seepage around it will provide a good breeding ground for mosquitoes and cause malaria, filariasis and poliomyelites epidemics in the area (Ahmad and Singh, 1991).

Bilharziasis, which flourishes in America, Africa and tropical Asia, affects more than 300 million individuals. Conditions are particularly favourable near storage dams and reservoirs as the still or slow moving water promotes the growth of swamp plants on which the snails live. The snails' entire life cycle is aquatic, and they, and through them man, are contaminated in the water. The snails attach themselves to aquatic plants, and are infected through excrement containing eggs, which hatch into larvae living several days inside the snail, after which they emerge into the water and enter human feet, hands or other exposed parts of the body through the skin. Once in the body, the larvae develop into adults, hooking themselves into the liver, intestine and bladder. They lay eggs which are excreted, thus returning to the water for the life cycle to repeat itself. The larvae and pupae of the fly vector for onchocerciasis live in flowing water, which is better aerated and cooler than stagnant water. As with malaria, it is not the main lake that constitutes the health risk, but the marginal lands and auxiliary water systems.

Ghana between 1958 and 1960, constructed 104 small dams in the north-east part of the country in response to population demand and seasonal hunger. This led to the prevalence of Schistoma haematobium infection in the local population tripling from 17 per cent to 51 per cent in 38 surveyed areas with some areas reaching 78 per cent (WHO 1993).

Odei (1982) reported a higher prevalence of schistosomiasis following the creation of new reservoirs with some of the riparian communities recording rates as high as 100 per cent. The Weija lake was created by the construction of a dam on the river Densu in 1977. The creation of the lake to supply piped water and to support irrigation and fisheries programmes provided ideal conditions for disease transmission. Water related diseases such as guinea worm *(dracontiasis)* and schistosomiasis were reportedly endemic in and around the river Densu catchment area. Odei (1975) reported that *Bulinus globosus* and *Bulinus truncatus* were the proven host snails of the urinary schistosomiasis, in the river Densu and its tributaries, which existed at a prevalence rate of between 8.3 per cent and 43.3 per cent (Grant 1969). Migration of disease infected fishermen and peasent farmers, flaws in resettlement programmes, changes in the flow rate of water, proliferation of water weeds and conditions favouring the growth and development of host snails were found to be responsible for the increasing incidence of schistosomiasis in the lake basin (Ampofo and Zuta, 1995).

Man induced alterations of many South African rivers have created an environment which has abundant food supplies and sufficient flow to allow year-round colonisation by filter feeding invertebrates, of which black flies *(Simuliidae)* are of particular concern (Moor, 1997). The majority of the 39 species of black flies recorded from South Africa (Palmer, 1991) are completely harmless and play an important functional role in the ecology of streams and rivers. There are, however, four species *(Simulium agersi, Simulium nigritarse, Simulium damnosum* and *Simulium chutteri)* that cause problems of sufficient magnitude to man and his livestock to require remedial intervention. Only one species, S. *chutteri*, under certain circumstances, can attain densities which can be considered to be of plague proportional size.

It is the female blackfly that is in all instances, the blood sucking culprit. It feeds creating large wounds in the victims which often lead to secondary bacterial infection. In sheep, the loss of parts of the ears and, in cattle, badly damaged teats are often encountered following severe outbreaks of blackfly attacks.

Blackflies have also been implicated in the mechanical spreading of certain disease bearing organisms. These include Rift-Valley fever virus, Wessels Bron disease and the protozoan

Chlamydia, which causes blindness in sheep and enzootic abortion in cattle (Bath, 1978) Black flies first became a problem along the Vaal river in the early 1960.

The modified approach taken to keep the black fly population, particularly *S. chutteri*, down to a level where it is no longer a serious threat to riparian livestock is achieved by regulating the flow of the river to almost cease at certain times during spring. This approach of drying out the river bed, by stopping the flow for short intervals and then resuming normal flow again, can, unfortunately, only be carried out within a limited distance downstream of a dam.

The bacterium *Bacillus thuringiensis* var. *israclensis* ("B.t.i") has proved successful in blackfly population size control along the Vaal and Orange rivers (Car and de moor, 1984; de moor and Car 1986; Palmer, 1993).

The reservoirs created by large dams also have effects on parasitic illnesses not requiring intermediate hosts, such as various types of amoebiasis, and on viral and bacterial diseases, such as typhoid and intestinal infections. The preventive steps are:

(a) before the first filling of the reservoir, clear the area of vegetation; backfill, deepen, impound or drain any areas liable to become swampy through fluctuations in the lake level;

(b) during operation of the scheme, increase and lower the lake level a few centimetres weekly in the egg-laying season;

(c) use pesticides judicially.

The Tennessee Valley Authority used these measures and almost completely eradicated mosquitoes around their dams. The spraying and spreading of larvicides and pesticides must be repeated at frequent intervals, and the cost may be prohibitive in developing countries. The chemicals must not be toxic for man or the fish forming the staple diet of riparian dwellers.

It might seem cheaper and simpler to treat the disease rather than attempt to destroy the vectors, but unfortunately the available drugs have only a limited effect and in fact are so toxic that they cannot be considered for mass treatment of riverside population. It is interesting to record that China has now completely stamped out bilharziasis through health education, discipline and increased community awareness.

Effects on Aquatic Flora and Fauna

The uncontrolled growth of aquatic plant life in tropical regions may seriously hamper navigation and fishing, provide breeding places for disease-carrying insects, change the chemical composition of the water and block water intakes. The three most troublesome plants are the water lettuce *Pistia stratiotes*, a free floating aquatic herb of the Arum family; a fern *Salvinia molesta* and the worst, the water hyacinth *Eichhornia crassipes.*

In the more temperate regions, the Eurasian aquatic milfoil *Myrcophyllum spicatum L.*, an ornamental aquarium plant, is a great nuisance in shallow lakes. In the absence of suitable means of eradicating it, it is contained by lowering the reservoir level to dry the roots, and spraying selective herbicides by boat or helicopter.

Rushes overrun shallow reservoirs in France. But the most serious difficulties from aquatic flora arise from weeds which invade canals downstream of some large reservoirs—the Durance canals below Serre Poncon dam, and the Drac canals below Monteynard for example. The reduction in flow capacity may be large, and there is no known way of overcoming the problem to date.

Reservoirs may also promote considerable development in the aquatic fauna. For fish, the species composition often changes after a river is dammed. The indigenous fish populations of both North and South Indian man-made reservoirs have changed significantly after their impoundment. Transplanted species have established themselves which were almost nil in 1957-58 in Sathanur reservoir, formed a fishery constituting 25 per cent of the fish landed in 1964-65 and later gradually increased in the catches to occupy a predominant position, reaching 91 per cent in 1975-76. In Stanley reservoir prawns, once abundant on the sandy beds of Cauvery, have disappeared. River Tapti was devoid of major carps prior to its damming but accidental transplantation has resulted in the establishment of *Catla catla* and *Labeo rohita* in the lower stretches of the river. The fish community of Ranapratap Sagar also appears to be in a state of transition after dam formation. There is a decreasing trend of *Cirrhinus mrigala* year after year and the smaller size group is missing. Indications are such that the proper recruitment of *Cirrhinus mrigala* is not taking place.

Migratory fishes are of great significance economically and for sporting purposes on many rivers. For these reasons and to minimise ecological changes, fish passes are needed in dams constructed on such rivers. These may take the form of ladders in which the fish swim up steps through a series of connected pools, much as in nature. In the fish lift alternative, the fish enter a bottom chamber and after an interval of time the entrance is shut. The bottom chamber is then filled with water, as in a conduit connecting to an upper chamber at reservoir top water level. The fish swim up and out of the top chamber to the reservoir. Both these systems are efficient for fish migrating up-stream to spawn but there is more difficulty in passing spent fish downstream on the way back to the sea.

In association with fish protection works, screens are required to prevent fish entering water passages harmful to themselves and where they could damage or block machinery. The screens may be of steel bars or an electrical system can be used. Migratory fish can be directed to preferred streams by erecting fish breaks or barriers on other streams. Traps can be incorporated to catch fish temporarily and strip them of spawn for fish hatcheries. Hatcheries can be used to breed fry to stock new reservoirs for sporting purposes.

In the case of catchment diversions, the transfer of unwanted species from one catchment to another should be guarded against, and screens at diversion intakes may be required. A particularly difficult task is the prevention of eel migrations over land saddles close to new reservoirs. The following measures are usually adopted in water resources projects to maintain and rather increase and improve fish quantity and quality.

(i) Fish ladders, fish locks, fish lifts or other special means give positive results.

(ii) In case of certain diversion projects involving dams, tunnels or canal aqueducts, migration of fish species from one river to another can be permitted.

(iii) Continuation of fish resources may be ensured by stocking.with young fish.

(iv) Plantation of the upper surface of a reservoir with a flora similar to that existing outwards nearby at the same elevation, can help in preserving population of fish. Such a treatment was tried with success on the Marchlyn Manic dam in North Wales, U.K., to preserve population of salmon, seatrout, brown trout, eels, and minnows.

(v) Compensation water released through the dam helps in maintaining the fish life downstream.

(vi) Eutrophication can be controlled by clearance of new reservoir areas prior to impounding. Other remedial measures include aeration, purifying inflow and building interception ditches to carry away from the reservoir any land run-off liable to contain traces of fertiliser, particularly compounds of nitrogen and phosphorus.

(vii) In case of barrages, weirs and small diversion dams, under water stone revetments can be made which can create conditions of suitable flora and fauna so vital for the life and growth of fish.

According to Tyson Roberts, an expert with 20 years of research experience on fresh water fish in Thailand and the Mekong basin, the Mekong river rapids are the centres of biodiversity akin to coral reefs in the oceans. The river's ecosystem is described as one of the world's richest in fish biodiversity and productivity.

In the past two years alone, fish biologists have added 3 new species to the list of 1000 indigenous species already known. The region's experts warn that the dams will deprive biologists of the last chance to study this complex river system in its relatively undisturbed state. Any reduction in the Mekong's flow could affect the river rapids disrupting the riverine food chain and fish habitat. The dams will block fish migration routes cutting off spawning and feeding habitat.

Many Mekong species are believed to take annual migrations up and down the largest rivers, including the Mekong mainstream. Schools of Pangasius kempfii, a large economically important catfish reaching more than 1 metre in length and weighing more than 15 kg, are believed to travel over 1000 km. from the South China Sea and Mekong Delta (Rajesh, 1995).

The Three Gorges dam billed as the world's largest water control project on Yangtze river in China will cause damage on important fishing grounds at the mouth of the river Yangtze by cutting short their nitrogen and phosphorus supply because of the siltation (Down to Earth,1996).

It is suggested that physiographic and hydrologic complexity plays an important role in making rivers suitable for dolphins (Reeves and Leatherwood, 1994). Dams and other artificial constructions degrade dolphin habitat in so far as they reduce such complexity. Research is needed at project sites, both before and after construction of dams, barrages and dikes to document impacts. Specially-designed "Swimways" may allow upstream and downstream passage by dolphins and can mitigate at least one of the adverse effects of dam projects namely population fragmentation.

Indian Sub continent

(i) Dams and barrages have long been present on rivers of the Indian sub-continent and they have had a major impact on dolphins. The Indus dolphin *(Platanista minor;* also known as Indus susu or bhulan) now exists only as a metapopulation consisting of four or five artificially isolated sub populations.

(ii) Barrages built by India along the Nepalese border have left Nepal with only a few sub-populations of Ganges dolphins *(Platanista gangetica;* also known as Ganges susu), confined to upstream ends of Ganges tributaries such as Karnali, Narayani, Koshi and Mahakali.

(iii) Farakka Barrage, constructed near the Indian border with Bangladesh in the early 1970s, mainly to improve navigation in the Hooghly river, partitioned the dolphin population in the lower Ganges.

(iv) China's Yangtze dolphin L*ipotes vexillifer*; also known as baiji) may be the most endangered cetacean in the world, with an estimated population of less than two hundred.

(v) River bed scouring and altered flow caused by the planned high dam at Three Gorges will degrade about 10 per cent of the habitat currently used by Yangtze dolphins.

South America

South America has two kinds of river dolphins, the Amazon dolphin *Inia geoffrensis*; (also known as boto, tonina or *Bufeo colorado)* and *Sotalia fluviatilis* (known as tueuxi or *Bufeo gris*). Three large dams (Balbina, CuruaUna, and Tucurui) have already been built in the Brazilian Amazon and several others are planned or under construction.

How Dams Threaten Dolphins

A 1986 international workshop on river dolphins identified a number of potential problems caused by waterway obstruction.

(i) Fragmentation of populations into genetically isolated sub populations.

(ii) Reduction in prey due to blocked migratory routes.

(iii) Less diversity and smaller biomass of prey in impoundments upstreams of dams due to lowered nutrient availability.

(iv) Downstream effects on prey caused by changes in flow rate, sediment transport, and estuarine salinity.

(v) Limited dispersal of dolphins between river systems due to saline encroachment in estuaries.

Mitigative Measures

Dams should be concentrated as near headwaters as possible. Also a series (or "staircase") of dams on one tributary is usually preferable, from an ecological perspective, to having single dams constructed on several rivers.

Swimway Concept

There is a long history of using fishways to permit anatropous fish to maintain their upstream spawning migrations but river dolphins were neglected because of the lack of their economic value. But now for the sake of preserving biological diversity dolphin swimway concept has gained momentum which can be designed and built at a reasonable cost (Reeves and Leatherwood, 1994).

ACTION ON WATERS OTHER THAN IN THE RESERVOIR

Drying Up of Rivers

The diversion and drying up of mountain rivers is detrimental to fishing and is an eyesore. It can be remedied by allowing an adequate guaranteed compensation flow through the dam; the amount may be varied in daylight and night hours and from season to season, as required by tourist amenity considerations. The question is particularly acute when natural waterfalls, like Niagara, are bypassed.

Changes in Water Table

The depth of the water table is changed by storage schemes. It is a simple matter to prevent it rising nearer the ground surface generally at the toes of embankments, which may even be designed to set the level of the water table at the best depth for farming purposes.

River stretches in alluvial plains that are bypassed by the scheme, and deep tailrace canals may, on the other hand, lower the water table. Weirs keep it at an acceptable depth while safeguarding the countryside, water, sports and fishing, and construction of a watertight cutoff parallel to the tailrace prevents ground water draining into it.

Catchment Management

In order to protect a new reservoir it is often necessary or desirable to take certain measures in the catchment area. Following impounding of a new reservoir there is frequently additional access and development in the catchment upstream. Uncontrolled deforestation and incautious agricultural development can lead to land erosion problems, particularly on steep land. If not dealt with properly by such measures as controlled re-afforestation and contour ploughing such land erosion can result in rapid siltation of the reservoir as well as loss of good farming soil.

Pollution of reservoirs used for potable water supply must be avoided and controlled access for human beings and the exclusion of large animals may be necessary.

Landslides

Landslides typify the interaction between the river development scheme and the environment. A reservoir may cause a landslide on the banks, which may partially fill in the reservoir.

The most striking example was at Vajont in the Piava valley in Italy. On the second filling on 9 October 1963, 250 million cubic metres of rock slipped into the 168 million cubic metre reservoir and flung the water over the top of the dam, which stood up without damage. More than 3000 persons in the valley died.

There are other known examples of smaller slides, and adits are sometimes driven into the reservoir banks to keep a watch on stability. In new reservoir areas with steep sides, a geological and soil mechanics reconnaissance should precede reservoir filling. Any doubtful areas should be thoroughly investigated and analyzed taking account of wetting, new water table, and possible rapid drawdown. Zones of probable trouble should be stabilised, drained, or flattened. Slope indicators and other instrumentation such as piezometers should be used for monitoring behaviour of doubtful areas, and in important cases automatic tele-metered warning systems should be incorporated.

INDUCED EARTHQUAKES

The question of whether filling a reservoir will cause earth tremors was first raised in 1934 during construction of the Hoover dam in the USA, but interest was only really awakened among seismologists and dam builders after the earthquakes at Kremasta (1966) and Koyna (1967).

Kremasta dam in Greece is 147m high and impounds 4.8 km of water. The main earthquake in 1966 was of magnitude 6.2 with its focus 20 km under the reservoir. Koyna dam near Bombay is 103m high with a reservoir volume of 2.7 km^3. The magnitude of the 1967 earthquake was 6.0 with the focus 9km under the reservoir.

Extensive study during the filling (1959 to 1971) of the Kariba dam (Rhodesia), a 128m high dam with a 175 km reservoir, indicated that the coincidence between reservoir filling and ground motion at Kremasta and Koyna was not chance. Seismic activity under the Kariba reservoir was strikingly parallel to the rise in water level over the five-year filling period and the main earthquake of magnitude 6 occurred when it reached top level for the first time.

Concern about seismic safety of dams has been growing in recent years as it has become evident that the earlier pseudo-static design concept is inadequate (Malla and Weiland, 1993) The most intense public debate on the Tehri dam has centered around the issue of seismicity and dam safety.

The potential of earthquake disaster all along the Himalayan plate boundary has been well known and widely accepted (Khattri, 1987). A debate has been going on for years on the adequacy of present design of the Tehri dam to sustain the dam against the maximum credible earthquakes expected in the area. The occurrence of a major earthquake in 1991 in the upper catchment areas of dam further strengthens the environmental opposition. It may be the case that the feasibility of realising the hydrological dream of storage in the Himalayan rivers will be cut short by seismo logical realities. As cited by Khan and Miah (1983) to the hydrological possibilities of storing 400 MAF of water in Brahmaputra basin, they quote from the recommendations of the Geological Survey of India that the Technical Committee on construction of high dams in seismic zones found that only 30 MAF out of a total of 400 MAF i.e. barely 7 per cent would be available as potential storage in Brahmaputra basin. This is an important instance of geological factors limiting the availability of optimal storage volume, independent of the availability of hydrologically feasible stor-age sites.

By 1976 the correlation between seismic activity and first filling of the reservoir had been recorded in some twenty cases. There are already unambiguous examples in which sensitive instruments for recording earthquake activity were installed some considerable time before construction of the dam as a reference for subsequent events. At Talbingo dam in Australia for example, only one minor earthquake was detected in the 13 years before the reservoir was filled, whereas one hundred tremors of comparable intensity were recorded in the first 15 months afterwards. The strongest tremor occurred just when the reservoir reached its top level, with a magnitude of 3.5. All the foci were shallow, just near the dam.

Most large reservoirs impounded by high dams have not caused any significant earthquake activity on filling, even when they lie in areas of high seismic activity such as California or Mexico. Conditions conducive to induced earthquakes are not therefore clearly understood. However there is sufficient evidence to demonstrate that it is probably true that:

(i) the earthquakes are due to the release of pre-existing stresses which are triggered into instability by the relatively small stresses from the water load, or by weakening of rock by percolating water;

(ii) the earthquakes are shallow, generally less than 10 km deep, compared to many natural earthquakes;

(iii) it cannot be assumed that attempts to control induced seismicity by lowering the water level or moderating the rate of filling after the earthquake sequence has commenced, will be successful or even helpful;

(iv) there remains the possibility that load influenced seismicity can be reduced in some sense by maintaining as slow a rate of filling as practicable; and

(v) it should not be assumed without full study that every seismic swarm near a storage reservoir is a case of load induced seismicity.

Hence, designers of large (more than 100m high) dams or those impounding large (more than 1 km) storage reservoirs must examine the following points:

(a) What is the current state of tectonic stress in the whole region and within the shoreline contour of the proposed reservoir? How does this vary, especially with depth, and how has it varied over recent time, say during the Holocene? This can be approached by such means as stress measurements on outcrops, and by a detailed geological search for Holocene tectonic movements.

(b) Are there any recent (tertiary or quaternary) volcanic formations in the area?

(c) Are there any nearby active faults, which would be signs of recent tectonic action, especially where they exhibit large displacements (more than 1000m), where they are of great length (more than 100 km), where they are sources of hot springs, where they bring formations of different stiffness into contact with each other, or where they are subject to a non-compressive horizontal stress so that they tend to open?

SOCIAL EFFECTS

Beneficial Effects

With the very wide sense of the word "environment" used here, action on the environment includes all the purposes for which a multiple purpose dam is built in addition to induced effects. All river development schemes are built to produce beneficial effects for society, the main ones being irrigation, land reclamation in river flood plains and tidal areas, domestic and industrial water supply, flood control and improved dry weather flow in rivers, hydro-electricity, navigation, recreational activities, and so on. Many induced effects may be beneficial. Some are discussed below.

Effects on Countrysides, Tourism and Leisure

The beauty of the new countryside created by storage lakes, especially those in which the water level does not vary to any extent, is a favourable effect on the environment. The same applies where the surrounding area is developed for amenity purposes, and parks and camp sites are increasingly used as a barrier to uncontrolled or excessive building. The dam itself is often a great tourist attraction.

Storage reservoirs near populated areas are increasingly used for water sports and swimming when they can be kept nearly full in the summer season. Unattractive reservoir bank areas affected by drawdown can be limited by subsidiary dams. Chittaurgarh is a large man made dam situated in the northern region of Shivalik Himalaya with a volume of 100 million m^3, height 15.3 m from deep foundation level and gross catchment area of 194 km^2. So the area can be developed for water sports, tourist lodge, camping sites, huts and shopping centres. Practice of intensive and extensive farming will provide good employment possibilities (Ahmed and Singh, 1991). Besides this the agro-based rural industries, construction, manufacture and distribution of agriculture inputs and outputs, marketing, transport and land improvement etc. will provide additional employment. Development in the area of pharmaceutical, cotton, paper and sugar industries will favour the development of infrastructure facilities eg. banking, marketing, health care, warehouses, roads and schools (Ahmad and Singh, 1991).

Roads

Road relocation is often an opportunity for considerably improving the existing road system, while dams provide possibilities for cheap road crossings, often a major advantage on wide alluvial plain rivers. Roads may also provide access to otherwise remote areas for emergency activities such as firefighting. This is particularly important in the prevention and control of bush firs, forest fires and conservation of forest regions.

Fisheries Development

Large reservoirs generally have a fovourable effect on the development of the aquatic fauna. When the Tennessee Valley Authority began building its cascades of storage reservoirs, there was little hope for improved fishing. It was thought that stocks would be abundant for a short time but the reservoirs would afterwards turn into 'biological deserts'. Yet industrial and sport fishing have been continually growing after the fish tonnage increased fifty-fold, while the restocking stations built through fear of natural reproduction in the reservoirs being impossible, have been closed down.

Similar successes have been recorded at Kariba and Lake Nasser, to mention only the larger lakes. Lack of facilities, processing industry and transport do however limit fisheries development in some cases.

The only disappointing results concern the introduction of new species. This policy can however give good results when run by experienced biologists making suitable advance studies.

Fire Prevention

New bodies of water created by dams act as barriers to contain and reduce forest and bush fires. This is a particularly important effect in regions with prolonged hot dry summers.

Detrimental Effects

Population displacement, the drowning of arable land and archaeological sites, and danger to downstream population must be included on the debit side of river development schemes.

Deforestation

Often good storage sites for damming occur in hilly uplands which are thickly forested. Before a reservoir is filled up the entire area to be submerged has to be cleared and forests often become a casualty.

Despite the increased awareness about the need to protect forest cover, little attention is generally paid to this factor when a site is selected for a dam, basically because all forests are state property and no one has to be compensated for them. Even in cost-benefit calculations, forest loss is grossly undervalued and the recurring loss in terms of timber and firewood yield per year and social costs of degradation of environment are never taken into account. Undervaluation of standing forest helps in boosting the cost-benefit ratio for obtaining Planning Commission approval and finance from international agencies like the World Bank. Moreover, the contract for clearing the forest is also undervalued, thus leaving a bigger margin to be shared between the contractors and the forest department.

The forest cover bears the brunt of dam construction in more than one way. In the Himalayan valleys, India, contractors who come to clear the forest which is going to be submerged greedily cut trees even in unaffected area. The forest is also cleared for approach roads, offices, residential quarters and for storage of construction material and rehabilitation.

In September 1987, the Ministry of Environment and Forests (MEF) sanctioned the submergence of over 1,360 hectares of reserved forests for Sardar Sarovar Project under the strict condition that no forest land will be used for rehabilitation. But by June 1990, under pressure from Gujarat and Madhya Pradesh governments, MEF sanctioned the felling of another 2,769 hactares of reserved forest land in Taloda for rehabilitation. With the reduction in forest cover and the entry of people, the pressure on the remaining forest increases. The need for firewood leads, to further denudation. The construction of a dam, therefore, has a multiplier effect.

The 1983-84 annual report of the Department of Environment, Government of India, reveals that the construction of the Idukki dam over the Periyar river in Kerala, India, has hastened the degradation of vegetation and sharply reduced the forest cover.

It is estimated that during the period 1951-76, 0.49 million hectares of forest has been lost due to major river valley projects. This is roughly 10 per cent of the area that came under canal irrigation-5.5 million hectares. The construction of Chittaurgarh dam affected a vast area (405 hactare) of natural forests in the Himalayas. The removal of vegetation included the felling of valuable timber and fruits bearing trees and the removal of firewood.

The highest percentage of damaged trees upstream of the dam included Khair (Acacia catechu), Impli *(Tamarindus indica)*, Jamun *(Syzygium cumini)*, Teak *(Tectona grandis)* Shisham (*Dalbergia sissoo)*, Chandan *(Santalum album)* etc. The deforestation of these species has caused serious wood scarcity in the area. The construction of the Saryu canal project has also resulted in the denudation of 1250 hactare of a valuable timber yielding area (Ahmad and Singh, 1991).

The Three Gorges project of China will result in the drowning of 238 km^2 farmland and 50 km^2 orange groves (Chau Kwai-cheong, 1995). The Bakun dam to be built across the Balui in Malaysia will submerge 695 sq. km of virgin rain forests and will also inundate the farmlands (Iyer, 1996).

Environmental damages due to the construction of Auburn dam, the most expensive dam in the US history to be built on the American river in California, include the destruction of up to 80.45 km of deep, richly forested Canyons that provide habitat for wild life and support recreation for 50,000 people a year.

Degree of deforestation involved in the 4,060 MW Indo Bhutan Sankosh Hydel project will be great deal causing denudation of 700 hectare of prime forest land.

Population Displacement

Reservoirs displace a large number of people, the number depending on the reservoir area and the population density. In India, as elsewhere, rehabilitation is the most persistent problem encountered during the development of large dams. Development related displacement in India can be traced back to Raj days and the construction of the first state owned Yamuna Canal system in 1789.

(i) The construction of the Upper Anicut dam on the Cauvery in 1863 raised the issue of involuntary displacement.

(ii) Colonial administration came out with the Land Acquisition Act in 1894.

(iii) The Tata power company's building of a dam Mulshi Peta in 1920 near Pune displaced 11,000 people, and triggered off the first satyagraha against forcible displacement. (Mitra, 1995).

Since independence, over 1,600 major dams and tens of thousands of medium and smaller irrigation projects have been built with the attendant canal systems and the invariable consequences of water logging and soil salinisation. As a result, between 100-120 lakh people have been forcibly displaced(Paranjpye, 1990).

In the absence of firm project wise data the estimates of the total numbers displaced by planned development interventions from 1951-90 range from a conservative 110 lakhs to an overall figure of 185 lakhs (Fernandes and Thukral, 1989). These figures do not include the sizeable number of people who are not acknowledged as being project affected (i.e. by loss of livelihood caused by natural resource extraction or degradation), those displaced in urban areas and those victimised by the process of secondary displacement. If these are tallied, the number of those displaced since independence would be as high as four crores.

Walter Fernandes estimates that in the recent past the proportion of tribals among those displaced has been increasing(Fernandes,1991). For example, of the 11.6 lakh persons displaced by 20 representative dams above 50 metres either under construction or being planned in the 1990s 59 per cent are tribals. The figure will obviously increase for dams planned in predominantly tribal areas (Suvarnrekha, Pollavaram etc.)

The Central Water Commission's 1990 "Register of Large Dams" is also instructive. Of the 32 dams of more than 50 metres height completed between 1951 and 1970 only 9 (22.13 per cent) were in tribal areas. Between 1971 and 1990, 85 additional dams of similar sizes were either completed or were under construction. However, by now not only were they taller and more sophisticated around 60 per cent of them were in the tribal regions. A recent official report on the rehabilitation of tribals, based on a comprehensive study of 110 projects, concludes that of 16.94 lakh people displaced by these projects almost 50 per cent (8.14 lakhs) were tribals (GOI,1993).

Trauma of Displacement

The experience of the post independence period from projects across the country suggests that the long drawn out process of displacement has caused widespread traumatic psychological and socio-cultural consequences.

These include the dismantling of production systems, desecration of ancestral sacred zones or graves and temples, scattering of Kinship groups and family systems, disorganisation of informal social networks that provide mutual support, weakening of self management and social control, disruption of trade and market links etc.

Essentially, what is established in the accumulated evidence in the country suggests that except in the rarest of the rare cases, forced displacement has resulted in, what Michael Cernea calls "a spiral of impoverishments". In addition there is a loss of complex social relationships which provided avenues of representation mediation and conflict resolution. In essence, the very cultural identity of the community and the individual within it is disrupted causing immense physiological and psychological stress (Cernea,1991).

The process of displacement is also disempowering since it breaks up socio-political organizations articulating a critique of the project (and of development process itself). Activists of the Tehri Bandh Virodhi Samiti states that there was a conscious strategy of the project authorities and the state government to divide the united resistance to the Tehri project (Kothari,1996). The other neglected dimension of displacement is its adverse impact on women. Their trauma is compounded by the loss of access to fuel, fodder and food, the collection of which inevitably requires greater time and effort. Similarly, children are adversely affected since not only is schooling less accessible in most cases there is also a disruption in the traditional socialization processes.

Apathetic Planning

Planners and administrators invariably capitalize on and manipulate the relatively weaker socio-economic and political position of most of the people facing displacement. Their numbers are underestimated, they are treated indifferently and only minimal cash compensation, it at all, is paid. There is an extraordinary unwillingness to grant them clear rights, such as security of tenure on alternative developed land sites.

A graphic and painful example is that of the displaced of Singrauli who are part of the over two lakh people first displaced by the Rihand Dam in 1964 and now face displacement for the incomprehensible fifth time in a single generation. Some indicative data highlights the severity of the apathy and indifference of official agencies and the government to take responsibility for those who, in an overwhelming number of cases have been forced to forgo their ancestral habitats and have experienced social and cultural disruption in the past four and a half decades of planned development. For example 25 years after the building of the prestigious Bhakra-Nangal Project only 730 out of the 2,108 families displaced in the early 1950s from Bilaspur and Una districts of Himachal Pradesh had been resettled. A majority of those displaced by other renowned projects like the Hirakud dam in Orissa or the Rihand dam in Uttar Pradesh, have never been officially resettled. The oustees of the Pong dam in Himachal Pradesh who were displaced in the 1960s had worse in store. Out of the 30,000 families or more that were displaced, only 16,000 were found eligible for compensation and in the end only 3,756 were moved hundreds of miles to a completely different cultural, linguistic and ecological zone in Rajasthan (Bhanot and Singh,1992). Official indifference and callousness is also evident in the lack of data regarding the total number of persons displaced by different developmental interventions. Independent studies of the oustees of Hirakud placed the figure for those displaced at 1,80,000 while official figures were 1,10,000.

Official data for the controversial dam on the Narmada river, the Sardar Sarover Project (SSP) fluctuates between 10,758 families and 30,134 families(Morse et al, 1992). The latest figures of the oustees of the SSP show that since June 1994 the Gujarat government has been able to resettle just 94 families out of the over 41,000 families from all the three states of Gujarat, Madhya Pradesh and Maharashtra (Kothari, 1996). In case of Tehri dam, direct involuntary displacement will occur for about 80,000 people. Indirectly, another large number of villages would lose parts of land or access road to the nearest towns in the plains. Because of the Koel karo hydro electric project about 120 villages were to be submerged over 22,000 ha., displacing over one lakh people of the total and acquisition land, while 10,000 ha. comprise forests (Down To Earth, 1992).

The Tawa dam built on Tawa river in Madhya Pradesh was built for irrigation purposes. Forty one villages were submerged and almost all inhabitants were adivasis. These adivasi villages which were displaced due to dam have not been rehabilitated and they have hardly received any compensation. There was no attempt to give them alternative land suitable for agriculture, or give them suitable alternative employment. They were not given the minimum conveniences like drinking water, roads, grazing areas which were available to them before and to top all these they were forced to clear forests and rebuild their homes. Some of them who were entitled to compensation got amounts of Rs.150 per acre, Rs. 65 per acre, Rs. 35 per acre and Rs. 18 per acre. The case of Tawa is a unique example of the injustice inherent in displacement of people (Sunil and Smitha, 1996). After a long struggle and agitation the displaced of the Bargi dam have been awarded as part of their rehabilitation the fishing rights on a cooperative basis. The displaced villages formed 54 co-operative societies which were then linked to form a co-operative marketing federation which now looks over the work of the co-operatives. They have raised the wage rate for fishing, given employment to local people and at the end of the year have shown a saving worth Rs.25 lakhs (Sunil and Smitha,1996). The chief obstacle to the process of resettlement is the non-availability of land in Gujarat, where maximum resettlement is planned. The eleven dam projects on Mekong river in Thailand which will have a total capacity of 14,000 MW, will flood 1,900 sq.km and displace about 60,000 people (Rajesh, 1995).

Since 1949, some 86,000 water conservancy projects have been undertaken in China involving the resettlement of over 10 million people. Many of the displaced people were resettled in remote border areas far away from their home countries, resulting in widespread adaptation problems and conflicts with the minority groups. In addition, the relocatees were given one payment compensation with no other development oriented assistance. Roughly 30 per cent of the resettlement schemes failed, and because of this bitter experience many people are sceptical of China's competence to resettle the huge number of people who will be affected by the Three Gorges Project (Chan Kwai-Cheong, 1995). It is a standing practice for the bureaucracy to issue notices for evacuation the moment project is announced, though the actual shifting might not start for years.

A Centre for Social Studies, Surat study shows that most of the oustees in the 19 Gujarat villages facing submission due to SSP had been issued notices as early as 1980. The actual shifting is yet to take place from many villages. The 557 households who had been shifted by 1988 should have been given a total of 1,118 ha. of land, but only 526.5 ha. had actually been handed over (Mitra, 1995). In some cases, population movements are very extensive—30 000 persons had to be relocated from the Keban dam area in Turkey, 50000 from Kariba on the Zambesi, 70 000 from Tsimlian on the Don, and 90 000 for the development of the Volta in Ghana. Tehri dam is estimated to displace nearly a lakh people. An attempt was made to resettle the Ghanaian population near the reservoir by providing substitute jobs in tourism, irrigation and other new activities. At Kariba, a flourishing industrial fishing industry has replaced the former subsistence fishing. Some ground realities that continue to be neglected during resettlement are: In the pre-implementation phase.

(i) The desirability and justifiability of the project or developmental intervention itself.

(ii) Consultation with representatives of the local population, particularly and predominantly those who are more vulnerable and the institutionalization of this process.

(iii) The justifiability of reducing or winding down development programmes and resources for the area facing submergence.

(iv) The harassment, intimidation and repression of potential oustees.

(v) If displacement is inevitable, then the full participation of the affected communities in defining a comprehensive rehabilitation package. In the Implementation phases some of the neglected aspects are.

(vi) Effects to bribe community leaders with the intention of seeking the consent of the community.

(vii) A wide range of human rights violations including threats, harassment, police actions, firing on peaceful protesters, forced entry into villages and homes, destruction of private property etc.

(viii) The disruption of nomadic routes which are crucial to the survival of nomadic communities, and

(ix) Land acquisition that in turn displaces those dependent on the acquired land. The post-implementation stage which, for those displaced, is among the most traumatic and undocumented experience, since project authorities move on to the next assignment and government machineries are not oriented to undertake long-term monitoring.

(x) The condition of the displaced in the existing resettlement sites.

(xi) Tracing those who were paid cash compensation and who dispersed over a vast geographical area (as was the case with some of the Pong dam oustees).

(xii) Responsibility for those who were displaced by earlier projects but who continue to face a grim economic and social predicament.

(xiii) The displacement caused by the intensive development activity that inevitably follows an initial developmental intervention.

Drowning of Arable Land and Archaeological Sites

Reservoir basins are often rich in arable land and pasture, sometimes forming the only agricultural potential in the region. Just as some projects are abandoned because of the size of the population they would displace, other valleys are not developed because of essential agricultural interests. The decision to go ahead with the scheme is only justified if the benefits largely outweigh such upheavals. In some instances however, reclaimed land or land completely or partially protected by the scheme against former devastating floods represents a much larger area than that drowned by the reservoir. The Tennessee Valley Authority reservoirs cover 250,000 hectares but the area protected against river floods on the Ohio and Mississippi downstream is fifteen times larger.

On the question of the drowning of archaeological sites, three projects can be mentioned as examples: at lake Nasser impounded by the High Aswan dam, several monuments, including the Abu

Simbel temple, were moved higher up the banks. In Siberia, the Angara rock paintings which would have been drowned by the Bratsk and Oust-Illim reservoir were housed in the local museum. Valuable archaeological remains of the second century AD had to be moved out of the reservoir area of Nagarjunasagar dam. The proposed hydroelectric project across Rathong Chu river in Sikkim is a threat to Yuksam which is a sacred pilgrimage spot. Several Buddhist relics have been unearthed in the area. The Dubdi Gompa monastery which is also an archaeological treasure is close to the site (Balakrishnan, 1991).

DAM BUSTING

A spate of dam busting is sweeping the U.S. after the demolition of Newport Number II dam on the Clyde river in Newport, Vermont. For the last forty years, the dam has been preventing the salmon from reaching its spawning ground. After the dam disappeared with a loud roar and bang within a few weeks, the fish was back jumping in the free running water. Following this incident which is considered a turning point in the U.S. history of hydro-engineering in Maine, several dams on the Rogue river in Oregon and the Snake river in Idaho face a similar fate.

Plans are also on the anvil to dismantle two dams on the Elwha river in Washington state. Activists in Tasmania want the dam on the Franklin river dismantled so that the pink quartz beaches of Lake Pedder which were submerged by the reservoir be uncovered. And in France, the environmental group SOS Loire has convinced the government to remove the Saint Etienne du vigan dam on the Vienne river to restore passages for migratory fish through the upper Loire Valley. For dismantling a dam, engineers must consider the entire river system not just the concrete structure itself. Although smaller dams like the Newport Number II may pose relatively lesser problems, costing just about $1 million to be removed within six weeks, large ones would present several problems, and even cause degradation to the environment. Researches show that once liberated, the sediments in the water may prove to be toxic to fish spawning areas, damaging aquatic species habitats and altering water channels. Careful planned strategy for sediment removal and restoration can minimize the impact of the damage. In long term, environmental gains should outweigh any losses. For instance, report from the Bureau of Reclamation showed that removing the savage Rapids Dam on the Rogue river in Oregon could generate as much as $5 million a year from fishing alone.

INTERBASIN TRANSFER OF WATER

The feasibility plan, prepared by the experts of the National Water Development Agency for linking up the east flowing rivers of the south by a canal along the coast under the national water grid programme has categorized the Mahanadi in Orissa and the Godavari in Andhra Pradesh as surplus rivers to the extent off 280 tmc ft. and 530 tmc ft. respectively, and the Krishna and Pennar in A.P. and the Cauvery and the Vaigai in Tamil Nadu as deficit rivers. According to an assessment made by the Central Water Commission the country would face a shortage of 160 cubic km of water by 2050 AD with the requirement by then being 1,300 cubic km as against the present availability of 1,140 cubic km. While an estimated 1,800 cubic km of water is flowing in the rivers in a year on an average, it is being utilised only to the extent of 690 cubic km. Based on this the NWG or national perspective planning contemplated for equitable distribution and optimal utilization, envisages transfer of water from surplus basis to deficit ones, and has two components, the first Himalayan component, aiming at connecting the perennial Himalayan born rivers such as the Ganga, the Yamuna, and the Brahmaputra for the benefit of the drought prone areas in north and the second, Peninsular component, for lining up the east-flowing and west flowing rivers in the south by two separate canals.

Under the present inter-basin plan of the NWDA which was submitted to the Union Ministry of Water Resources for its approval and follow up, including a central legislation, 395 tmc ft. of water will be diverted from the Mahanadi by a 930 km long canal so that a quantum of 230 tmc will join the Godavari at Dowlais-waram. The Godavari is to be linked to the Krishna by three canals the first (300 km), connecting Ichamballi with Nagarjuna Sagar, carrying 580 tmc ft. but conveying 501 tmc ft. after wetting the fields en route, the second (270 km) connecting Ichampalli with Pulichintala carrying 154 tmc ft. but conveying only 57 tmc ft. into the Krishna after meeting the requirements midway, including that of Pulichintala Project proposed by the Andhra Pradesh government, and the third, connecting, Polavaram with Vijayawada (174 km), carrying 173 tmc ft but conveying 117 tmc ft. in the end.

The third segment consists of three canals connecting the Krishna with the Pennar. The first canal (564 km) is to divert 70 tmc ft from the Alamatti dam and convey 33 tmc ft after meeting the requirements en route, to the Maddileru stream. The second canal connecting Sri-sailam and the Pennar carries 80 tmc ft. but conveys only 70 tmc after meeting the needs en route, at Adinimmayapalli and the third, connecting Nagarjunasagar with Somasila to carry 430 tmc ft. The fourth segment has a link up canal (535 km) between Somasila (Pennar) and Grand Anicut (Cauvery) transferrring 302 tmc ft of water ending up as 135 tmc ft. after meeting the needs en route. Under the last segment (250 km) about 80 tmc ft of water of Cauvery will be diverted from the Kattalai regulator to the Vaigai in Tamil Nadu.

The NWDA has also prepared blueprints to connect the west-flowing rivers on the western coast according to which the Pamba in Kerala will be diverted from Achankovil to Vaippur in Tamil Nadu to the extent of 22 tmc ft. all to be used in the drought prone. Tirunelveli, Chidambaranath and Kamarajar districts in Tamil Nadu. The Netravati and Hemavati both in Karnataka are to be connected for 6 tmc for irrigation in Karnataka. The Bedthi and the Wardha, also in Karnataka are to be linked for transfer of 9 tmc into the Tungabhadra near Raichur.

RELEVANCE OF THE PRESENT STUDY

India being an agricultural economy, irrigation has always been the motive force behind the construction of dams. Sathanur reservoir was created with the aim of irrigating the water scarce regions of Thiruvannamalai and Villupuram. Built on an interstate river, Ponnaiyar, it is a large dam fulfilling the criteria laid down by the International Commission Of Larger Dams (ICOLD) with height more than 15m, storage capacity more than 229.4 Mm^3 and maximum flood discharge more than 2000 $m^3 sec^{-1}$. The Ponnaiyar river basin with its catchment falling between 2000-20,000 km^2 is classified as the medium river basin.

Sathanur reservoir project is also significant as it is situated in Peninsular India where large dams are needed due to the limited capacity of the valleys on account of the natural topography and configuration. The flow in most of the south Indian rivers is limited to 3-4 monsoon months of the years in contrast to the perennial rivers of north India. Ponnaiyar river, across which the Sathanur dam is constructed, is one such river. Tank irrigation has played a major role from time immemorial in the agriculture sector in many of the south Indian states, especially Tamil Nadu. Sathanur project irrigates a major portion of the land in the command area under indirect irrigation via 90 small and big tanks. The command region lying in the drought prone region of the country with the mean annual rainfall less than 1000 mm makes the role of Sathanur dam more crucial. The story of the Sathanur project has been that of a slow and steady evolution of a single purpose dam to a multipurpose project fulfilling the following objectives.

- Irrigation
- Hydropower generation
- Domestic Water supply
- Aquaculture
- Recreation

ORGANISATION OF THE BOOK

The book has been organized into two parts: Part A consists of four sections and 17 chapters. In each chapter we have presented a synthesis of the information we have obtained from the governmental and non-governmental agencies, and the information generated by us through extensive field work. The beneficial as well as the adverse impacts have been identified as perceived by the project authorities, by the end users, and by us. We have also made an attempt to identify the impacts of the eco-restoration measures undertaken by the project authorities and to develop guidelines for the management of this project, and other similar projects in future.

Part B of this book deals with the development and application of SRPIS (Sathanur Reservoir Project Information System).

2

Study Area

PONNAIYAR RIVER

History of Ponnaiyar River

Ponnaiyar, an interstate river, is one of the largest rivers of the state of Tamil Nadu (TN), India; often reverently called 'little Ganga* of the South'. The river has supported many a civilizations of peninsular India across the history and continues to play a vital role in supplying precious water for drinking, irrigation and industry to the people of the Indian states of Karnataka, Tamil Nadu and Pondicherry. Ponnaiyar originates from the south-eastern slope of Chennakesava hills in Mysore state and flows for 84.8 km in that state before entering Tamil Nadu at about 4.8 km north-west of Bagalur village, Dharmapuri district. The Krishnagiri dam was the first one to be constructed across the river, about 10 km from Krishnagiri town in the Dharmapuri district (TN). The river flows for a distance of 430 km from its point of origin to the sea 35 km in the Thiruvannamalai district (TN). The total drainage area of the river is 14,130 km^2 of which 3608 km^2 lies in Karnataka,

---

* *Ganga* or *Ganges, which* originates in the Himalayas and traverses large tracts of Northern India before crossing into Bangladesh, is considered a holy river by the Indians and is often called ' mother Ganga'. The river also signifies extreme purity though, in modern times, it has become gross by polluted.

95 km^2 in Andhra Pradesh and the remaining 10,427 km^2 falls in Tamil Nadu. Sathanur dam is about 113 km below the Krishnagiri dam i.e. 286 km from the origin of the river. The river below the dam runs in a southeasterly direction through Salem district. In this district, the river meanders through thick jungle of the Ponnaiyar reserve forest and flows through Cuddalore district, before finally merging with the Bay of Bengal near Cuddalore (TN).

The river flows for a major portion through wooded country, deep ravines and narrow gorges, with the country at the sides rising steeply to a height of about 90 m above the riverbed. Unlike the other major river of southern India-Cauvery-Ponniyar is not perennial. There is practically no flow in the river except during Northeast monsoon months (October-December) or the less precipitous southwest monsoon (June-August), during the rainy spells there are flash flows in Ponniyar. This has aptly given rise to a proverb in the Tamil language "the Ponnaiyar will rise and fall even before the butter melts".

SATHANUR RESERVOIR PROJECT (SRP)

Sathanur Dam

The Sathanur Reservoir Project (SRP) is constructed across the river Ponnaiyar in reserve forest area near Sathanur village (See Plate—2.1). It is situated about 32 km from the Thiruvannamalai town at longitude 78° 51′ 10″ and latitude 12° 4′ 48″ (See Figure 2.1).

A Brief Historical Background

Irrigation facilities available in Thiruvannamalai and Chengam sub-divisions were scanty and not quite dependable. There were no means of harnessing the flash flows of Ponnaiyar traversing through the district and the major portion of the water was going waste into the sea. Several options were explored from time to time but none was taken up owing to the drawbacks like the uneconomical return on the outlay, unreliability of flow data of Ponnaiyar etc.

After the Second World War, steps were taken to review all the past proposals, under Government's "grow more food" drive. The proposal, under which SRP has been established, was taken

Plate 2.1: A view of the Sathanur reservoir and the sluice gate of the dam constructed across it

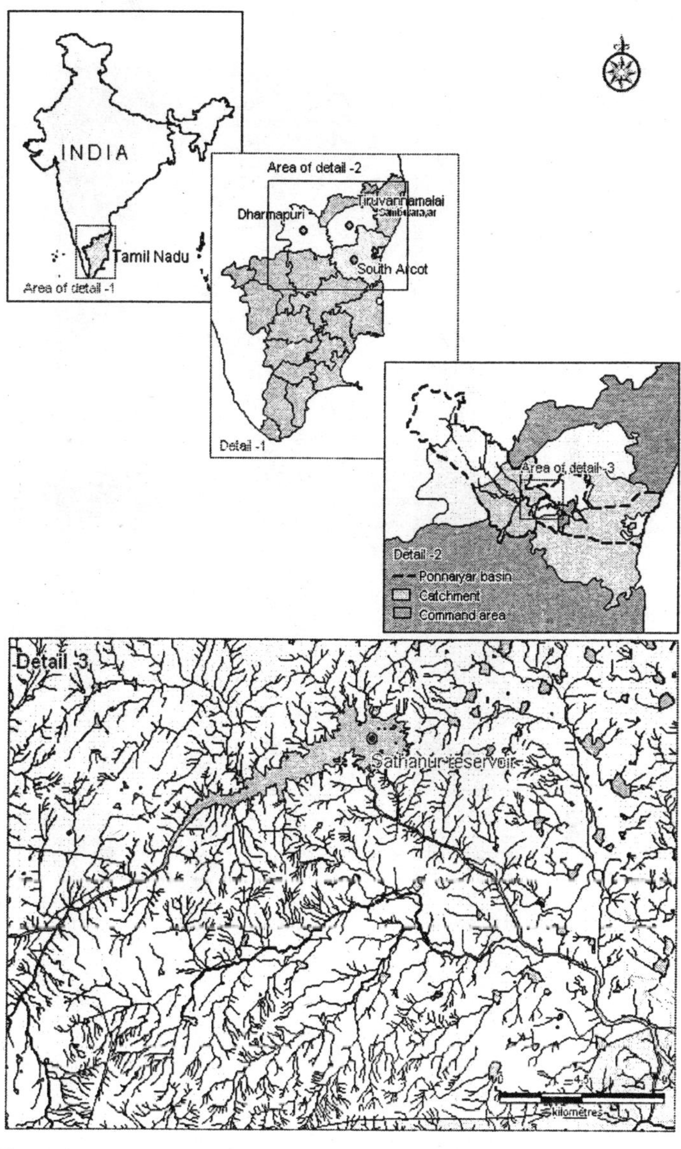

Fig. 2.1: Location map of Sathanur reservoir, detail—2 represents the Ponnaiyar basin, Sathanur catchment area and command area and Detail—3 represents the Sathanur reservoir

Plate 2.2: Pick-up dam of Sathanur reservoir

up for detailed investigation in 1954. The first stage was swiftly sanctioned at an estimated cost of Rs.29 millions. The project was inaugurated on 2.10.1954. The dam was constructed in two phases. A capacity of 2,89,940 cusecs was created in the first phase, which was dedicated to the nation on 10.11.1957.It had the potential to irrigate an area of 1714 acres. An extent of 14,159 acres was irrigated in 1958-59. The entire proposed area of 21,000 acres was brought under irrigation during the year 1959-60. It was further increased by an additional 2,978 acres in 1971.

The second stage of construction started in 1968 and the capacity of the reservoir was increased to 229.4 mm^3 by providing a 12.2 m x 6.1 m shutter. The current full reservoir depth is 366.3 m over the sill of river sluice. The right flank saddle, which was at low level, was converted into vented escapes. Thus, full reservoir level (FRL) of reservoir was raised to 220.9 m, which was previously maximum water level (MWL) of first stage. The extra capacity thus created was meant to stabilize 5000 acres of second crop under Thirukoilur barrage system.

Salient Features of the Dam

The total length of the dam is 779.8 m of which the masonry section is 418.5 m and rest is the earthen dam (See Table 2.1). The masonry dam is designed as a gravity structure, 40.9 m above the riverbed and 44.7 m above the deepest foundation level. The uncontrolled spillway consists of 9 vents of 12.2 m x 6.096 m each with a maximum discharge of 2,80,940 cusecs. ARCC bridge with 3.6 m clear roadway is provided over the spillway. The requirement of water, both for existing and new irrigation, is supplied though 5 river sluices of size 1.5 m x 1.5 m located at the left end of the masonry dam. The earthen dam is zoned, rolled fill type with front slopes of 2:1 and 3:1, rear slopes of 2:1 and 2:5:1 and top width of 6 m. A cut of trench is provided for section of more than 3.03 m wide beams at 6.1 m vertical intervals. For effective drainage, horizontal and transverse filters are provided.

Pick-up Dam and the Canal System

The margin of the river below the dam is rocky and rugged for a considerable distance, thus rendering a direct off-take canal from the reservoir for irrigation purpose impracticable. Water from reservoir is, therefore, let down in the river itself and picked up about 7.2. Km lower down by a dam (Plate—2.2). The length of the Pick-up dam is 121.9 m excluding the 6.1 m bye wash on the right flank. The left and right head sluices have been provided in the pick-up dam itself. They are capable of discharging 400 cusecs of water.

Table—2.1 Morphometric details of Sathanur reservoir

River at bed level	E1+182.5 m
Top of dam	El+222.6 m
Sill of river sluices	El+185.9 m
Crest of spillway and FRL 1st stage	El+216.1 m
FRL 2nd stage and MWL	El+222.2 m
Maximum rear water level	El+195.1 m
Capacity of reservoir 1st stage	129.3 Mm^3
2nd stage	229.4 Mm^3
Catchment area at dam site	10826 km^2
Water spread area 1st stage	12.5 km^2 (3100 acres)
Water spread area 2nd stage	18.2 km^2 (4500 acres)
Maximum flood discharge	5663.3 m^3 sec^{-1} (200000 cusecs)
Length of dam	
a. Length of masonry dam	418.5 m
b. Length of earthen dam	361.2 m
River sluices	
Vents	5 nos.
Size of vents	1.5 m x 1.8 m
Maximum discharge	240.69 m^3sec^{-1} (8500 cusecs)
Spill way	
Number of vents	9 nos.
Size of vents	12.2m x 6.1 m
Maximum discharge	3281.89m^3sec^{-1} (115900 cusecs)
Saddle	
Length of masonry between abutments	161.5m
Vents	11 nos.
Size of vents	12.2 m x 4.6 m
Sill level	El+217.6 m
Maximum height of masonry above foundation	16.8 m
FRL and MWL	El +222.2 m

(Table Contd...)

Flanking bund	
Width of the road-way between kerbs	3.7 m
Top of road-way	El+222.6 m
Length of flanking bund	327.7 m
Top width of flanking bund	7.6 m
Maximum width of flanking bund	40.2 m
Maximum discharge capacity	2235.31m^3 sec^{-1} (78940 cusecs)

The morphometric details of the Pick-up dam of Sathanur reservoir has been presented as Table—2.2. The length of the left bank canal is 35.2 km Table—2.3. It consists of 15 distributaries. The total length of the canal and distributaries is about 88 km. The system is currently irrigating about 9,717 hectares of land since 1958 and supplying water to 41 tanks in Chengam sub-division, Thiruvannamalai sub-division and Thirukoilur sub-division in Thiruvannamalai and Villupuram districts, Tamil Nadu.

The farmers in Kallakurichi sub-division in South Arcot district (now Villupuram) experienced severe drought for a long time and the surplus water of Ponnaiyar was draining off to sea without being harnessed for any use. There was a demand for constructing a canal on the right flank of the Pick-up dam. As a result right bank canal was constructed in 1982 for irrigating about 6610 hectares of land.

The length of the right bank canal is 28.64 km. The main canal has been lined along its entire length. There are 22 direct intake sluices and 4 branch canals taking off from the main canal. There are four sub-canals named Moongildhuraipattu, Athiyur, Viriyur and Jambodai. The total length of the branch-canal is about 37.45 km. There are 25 direct distributing gates in the main canal. Around 49 tanks in Chengam and Sankarapuram (formerly Kallakurichi) sub-division are getting benefit from the right bank canal. The discharging capacity of the canal is 6.95m^3 sec^{-1}

Watershed

The catchment of Ponnaiyar between Krishnagiri and Sathanur reservoirs is in a rolling country mixed with hills and plains. The hills are Shervarays, Javadi and Chittori. The catchment of Sathanur reservoir lies between the latitudes 11° 45′ and 12°6 45′ and longitudes 78° 0′ and 78° 50′. The state, district and sub-division wise catchment area details are furnished in Table—2.4. The river basin of Ponnaiyar is approximately 12,986 km^2. The distribution of the basin at salient points on the river is as below:

Catchment at Krishnagiri dam site: 5428.6 km^2

Catchment at Nedungal dam site: 5603.8 km^2

Catchment at Sathanur dam site: 10826 km^2

Catchment atThirukoilur dam site: 2738.6 km^2

Catchment below Thirukoilur dam: 248.3 km^2

Table—2.2 Morphometric details of the Pick-up dam

Location of pick-up dam	7.2 km below dam
Length of pick-up dam	121.9 m
Bye wash	6.1 m
Bed level	El+161.2 m
Crest level	El+166.9 m
MWL front	El+175.6 m
MWL rear	El+174.0 m
Maximum flood discharge	7955.26 m^3 sec-^{1} (2,80,940 cusecs)
Scour vents	
Left side sill	El+162.2 m
Vents	3 nos.
Size of vents	1.5 x 1.8 m
Right side sill	El+165.0 m
Vents	3 nos.
Size of vents	1.5 m X 1.8 m
Head sluices	
Sill	El+165.0 m
No. of. Vents	3 nos.
Size of vents	2.7 m x 1.8m
Full supply discharge	11.327 m^3 sec^{-1} (400 cusecs)

Table—2.3 Details of the canal system for Sathanur reservoir

Total length of main canal	35.2 km
Length of lines portion	35.2 km
Full supply depth	1.5 m
Full supply discharge	11.327 m^3 sec $^{-1}$ (400 cusecs)

Table—2.4 Catchment details of Ponnaiyar basin

State	District	Block	Extent of catchment in m^2					
			At Krishnagiri dam	Total	At Sathanur dam	Total	At confluence	Total
Karnataka			2260.6	2261.3	–	–	–	–
Andhra Pradesh	Chittoor	Chittoor	41.5	41.5	31.5	31.5	–	–
Tamil Nadu	Dharmapuri	Hosur	646.9	–	–	–	–	–
		Krishnagiri	422.7	–	648.3	–	–	–
	Salem	Dharmapuri	–	–	384.6	–	–	–
	North Arcot	Harur	–	1069.7	1373.8	2406.6	7.9	7.9
	South Arcot	Salem	–	–	111.2	111.2	–	–
		Chengam	–	–	209.8	–	433.5	–
		Tiruppathur	–	–	566.4	776.2	–	–
		Tiruvannamalai	–	–	–	–	420.1	853.6
		Kallakurichi	–	–	27.0	–	248.8	–
		Thirukoilur	–	–	–	–	203.7	–
		Villupuram	–	–	–	–	67.3	–
		Cuddalore	–	–	–	16.8	55.8	575.5
Total catchment				3372.5		3352.5		1436.9

Hence the independent catchment of Sathanur reservoir is 5397.5 km^2 (2084 m^2). The total catchment of Sathanur reservoir is 5397.5 km^2. The map depicting the Ponnaiyar river basin is given as Figure—2.2. Major tributaries joining Ponnaiyar river below Krishnagiri dam till Sathanur dam are as follows:

1. Puliampatty river starts from Chitteri hills and joins Ponnaiyar on it's right bank near Echampadi, north of Kambainallur;
2. Vaniaru originates from the Shervarays near Yercaud and joins Ponnaiyar on its right bank near Muthiampatty;
3. Kovil river starts from Chitteri hills and joins Ponnaiyar on its right bank near Periyapatty; and
4. Pambaru starts near Yelagiri hills and joins Ponnaiyar on its left bank near Pavakkal.

Within the water spread the following tributaries join the Ponnaiyar (a) Vediappan kovil odai (b) Ramakkal odai (c) Pulianur aru (d) Mottur aru (e) Anitallam odai and (f) Valayar odai. Other than these, about 150 minor streams and rivulets feed Ponnaiyar. The general aspect of the catchment area is one of the cultivated plains and valleys interspersed with sharply rising hills, which are in some cases merely boulders (Figure—2.3).

Water Spread

The entire water spread of Sathanur reservoir is about 18.2 km^2 (4500 acres) lying within the Ponnaiyar reserve forest. The river stretches along the deep course from the dam to the end of the water spread for about 21 km as illustrated in Figure—2.3. The periphery of the water spread above full reservoir level is surrounded by thick forest.

Command Area

The 35.2 km long left bank canal is designed to irrigate 9713 hectares of land in Thiruvannamalai and Villupuram (previously North Arcot and South Arcot) districts. The 28.6 km long right bank canal is designed to irrigate 8,499 hectares of land in Chengam

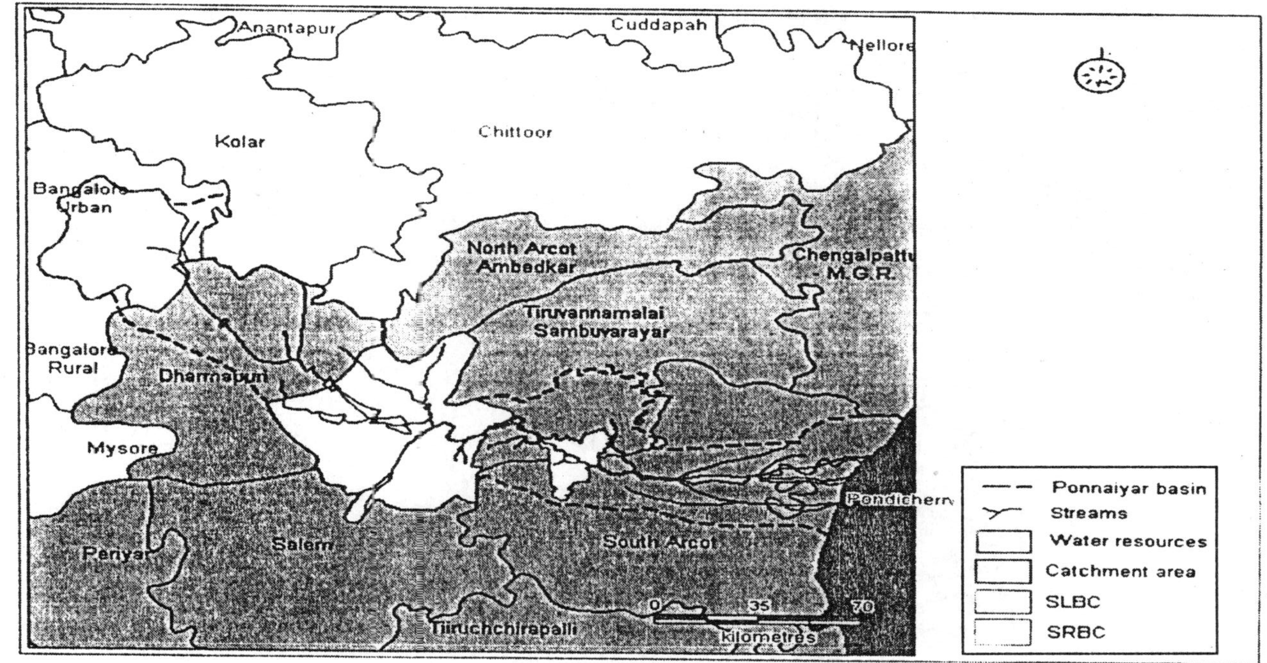

Fig. 2.2: Detail of Ponnaiyar basin

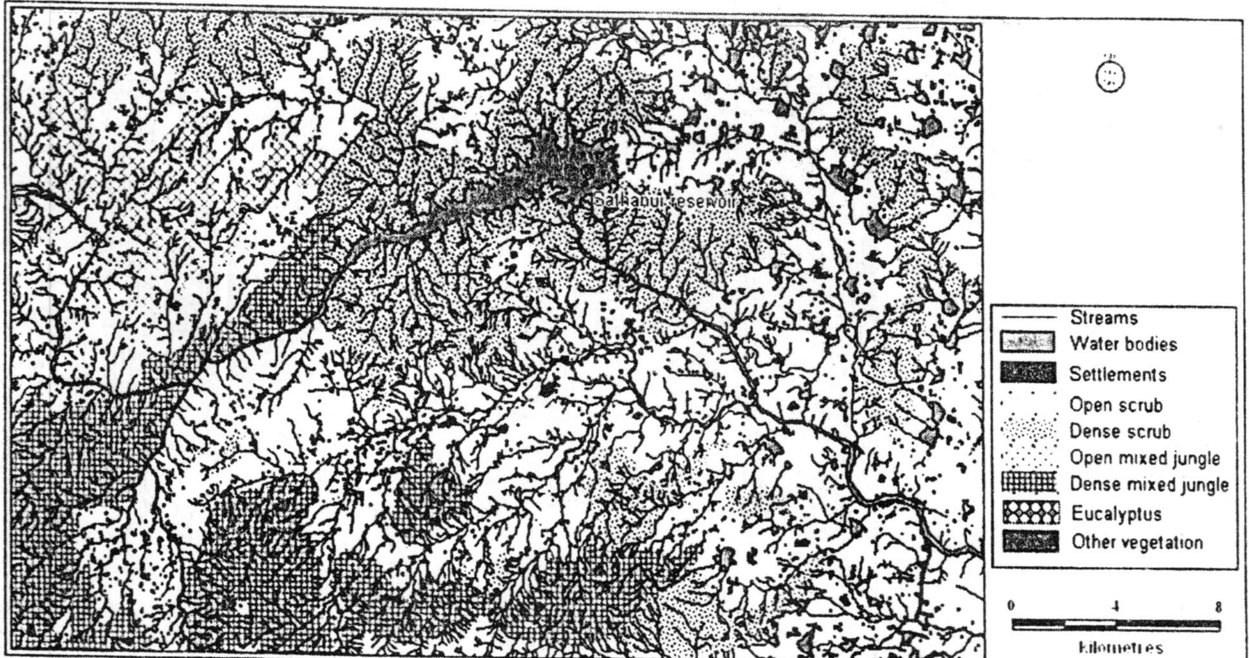

Fig. 2.3: Map showing the Sathanur reservoir water spread and the arterial streams feeding the reservoir

and Kallakurichi sub-divisions. The list of villages and tanks that are benefited, their extent of area under irrigation and canal number is presented in Tables—2.5 to 2.11. The Sathanur Command Area (SCA) is represented in Figure 2.4. The villages of SRBC and SLBC commands complete with their respective canal system and tanks are illustrated as Figures—2.5 and 2.6.

Geology

The geological details of Ponnaiyar basin are illustrated in Figure—2.7. The area is predominantly built up with granite and gneisses rocks of archean period. The granite is of very good quality and extensive out crops and masses of it are commonly found. The chief components of rocks are hornblende and feldspar. Foliation is seldom seen. In the plains of reserve forest, quartz is found. The diamond granite is also found in scattered pockets in the areas of Chitteri hills in Dharmapuri and Krishnagiri sub-divisions. Charnokite rocks of archean period are also seen in some area. At the tail end of the basin, pockets of sand stone, clays pebble of tertiary period, and limestones of cretaceous period are found. Alluvium and sand dunes of quaternary period are also seen.

Climate

Temperature

Command

SCA falls under the tropical belt. The climate in general is hot, April and May being the hottest months of the year. The temperature rises to 34°C during these months. November and December are the coldest months when the ambient temperature falls to 22°C.

SLBC Command

Chengam Sub-division

Table—2.5 Village-wise proposed area under canal irrigation

S. No	Name of the village	Area in ha.	Canal number
1.	Olagalapadi	2.99	Main
2.	Edathanur	120.21	DY1
3.	Allapanoor	82.46	DY2
4.	Agarampallipattu	205.58	DY2
5.	Sadakuppam	147.69	DY4
6.	Unnamalaipalayam	91.57	DY4
7.	Thenmudiyanur	109.98	Main
8.	Radhapuram	37.68	–
9.	Thenkarimbalur	472.74	DY5
10.	Valavachanur	157.75	DY4
11.	Kottaiyur	353.40	DY4
12.	Varkkilapattu	9.38	Main
13.	Serpapattu	45.53	Main
14.	Maluvampattu	110.38	DY7
15.	Vanapuram	183.81	Main
16.	Kunglianatham	96.72	DY5
17.	Perunthuraipattu	122.80	DY5
18.	Edakkal	265.80	DY5
19.	Peraiyampattu	118.43	DY9
	Total	2815.15	

Thiruvannamalai Sub-division

Table—2.6 Village-wise proposed area under canal irrigation

S. No	Name of the Village	Area in ha.	Canal number
1.	Palayanur	166.47	DY 11B
2.	Velayampakkam	254.23	DY 11B
3.	Kallottu	107.21	DY 11B
4.	Kandiankuppam	272.96	DY10
5.	Thachampattu	133.00	DY11A
6.	Allikondapattu	65.93	DY11A
7.	Navapattu	276.59	-
8.	Thalayampallam	315.70	DY11B
9.	Nariapattu	251.93	DY11B
10.	Parayampattu	128.87	DY12
11.	Sakkarathamadai	56.96	DY12
12.	Athipadi	104.28	-
13.	Devanur	38.89	-
14.	Periakallipadi	327.68	DY11a
15.	Pavapattu	229.11	DY13
16.	Kattampoondi	350.34	DY14
17.	Pappambadi	72.14	DY15
18.	Aradapattu	97.08	DY15
19.	Pavithram	269.21	DY15
	Total	3518.17	

Thirukoilur Sub-division

Table—2.7 Village-wise proposed area under canal irrigation

S. No	Name of the village	Area in ha.	Canal number
1.	Melandhal	478.49	DY4
2.	Kangaiyanur	218.34	DY5
3.	Jambai	375.14	DY10
4.	Pallichandal	128.57	DY10
5.	Murukkambadi	186.56	DY11B
6.	Kongamnur	99.32	DY5
7.	Athiyandal	138.10	–
8.	Devaradiyarkuppam	221.39	DY11B
9.	Seilankuppam	92.49	DY10
10.	Sithapattinam	210.94	DY10
11.	Manalurpet	57.72	DY10
	Total	2207.06	
	Chengam Sub-division		2815.2 ha.
	Thiruvannamalai sub-division		3518.8 ha.
	Thirukoilur sub-division		2207.1 ha.
	Total		8541.1 ha.

SRBC Command

Chengam Sub-division

Table—2.8 Village-wise proposed area under canal irrigation

S. No	Name of the village	Area in ha.	Canal number
1.	Thiruvadathanur	240.82	Main
2.	Purthurcheekadi	85.81	Main
3.	Rayandapuram	288.56	Main
4.	Thondamanur	113.55	Main
5.	Elayankani	7.56	–
	Total	737.30	

Sankarapuram Sub-division

Table—2.9 Village-wise proposed area under canal irrigation

S. No	Name of the village	Area in ha.	Canal number
1.	Melsiruvalur	386.30	Main
2.	Vadakeeranur	190.53	B4
3.	Rayasamudram	73.69	Main
4.	Mookanur	47.79	Main
5.	Pakkam	242.12	B2
6.	Olagalapadi	369.41	B1
7.	Kaduvanur	336.78	B2
8.	Mangallam	348.78	B1
9.	Arulambadi	139.26	B1
10.	Vadaponparappi	207.80	B1
11.	Vadamamandur	320.21	B1
12.	Arumparampattu	432.87	B1
13.	Sirpathanallur	150.61	B1
14.	Jambodai	570.71	B1
15.	Thiruvarangam	155.48	B1
16.	Maniyandal	15.33	B2
17.	Sirpanandal	395.58	B1
18.	Arkavadi	22.27	B4
19.	Athadur	8.65	B4
20.	Edathanur	42.27	B2
21.	Kallipadi	28.90	B1
22.	Periyakolliyur	208.85	B1
23.	Chinnkolliyur	29.51	B1
24	Sivapuram	60.44	-
25.	Tholuvanthangal	88.94	B2
26.	Athiyur	325.04	B2
27.	Ariyallur	62.04	B2
28.	Nagalkudi	25.95	B2
29.	Vanapuram	75.47	B2
30.	Odiyanthal	32.55	B2
31.	Sirupanaiyur	0.55	B2
32.	Kadambur	106.30	B2

(Table Contd...)

33.	Agaraharapandalam	24.72	B3
34.	S.Kolathur	207.55	B3
35.	Varagur	116.69	B3
36.	Arur	148.35	B3
37.	Thimmendal	59.09	B3
38.	Kidagudayampattu	89.99	B3
39.	Chellakayakuppam	46.75	B3
40.	Viriyur	198.80	B3
41.	Moongilduraipattu	395.80	B4
42.	Poruvalur	423.09	B4
43.	Erudayampattu	106.53	B4
44.	Por-aspattu	113.71	B4
45.	Vadasiruvalur	59.89	
46.	Ramarajapuram	17.70	
47.	Olagadayampattu	3.76	
48.	Elayanarkuppam	36.73	B1
	Total	7680.79	
	Chengam Sub-division		737.30 ha.
	Sankarapuram		7680.79 ha.
	Total		8418.09 ha.

SLBC command

Table—2.10 Village-wise area proposed under tank irrigation

S. No.	Name of the tanks	Area in ha.
1.	Edathanur	15.96
2.	Allapanoor	18.66
3.	Agarampallipattu	14.81
4.	Sadakuppam	9.16
5.	Kottaiyur	31.86
6.	Thenkarimbalur	84.84
7.	Melandhal	129.65
8.	Serapapattu	31.09
9.	Vanapuram	40.32

(Table Contd...)

10.	Kunglianatham	20.00
11.	Athipadi	10.82
12.	Perayampattu	11.98
13.	Palaiyanur	64.15
14.	Kandiyankuppam	31.65
15.	Maluvamapttu	13.72
16.	Pallichandal	16.42
17.	Jambai	58.19
18.	Sellankuppam	116.42
19.	Manalurpet	54.99
20.	Sithapatinam	107.80
21.	Athiyandal	20.88
22.	Velyampakkam	30.09
23.	Murukkambadi	21.93
24.	Konganamoor	20.32
25.	Kallottu	17.01
26.	Navapattu	28.80
27.	Allimundapattuthangal	6.77
28.	Sakkarathamadai	11.62
29.	Thalayampallam	37.72
30.	Kattampoondi	58.42
31.	Parayampattu	16.29
32.	Pavapattu	53.79
33.	Thachampattu	26.95
34.	Periyakallipadi	62.21
35.	Pavithram	67.60
36.	Aradapattu	27.34
37.	Andhapattu	27.34
38.	Kolakudi	42.56
	Total	1455.61

SRBC Command

Table—2.11 Village wise area proposed under tank irrigation

S. No.	Name of the tanks	Area in ha.
1.	Rayandapuram	59.94
2.	Viyappanur	18.29
3.	Vettuthangal	31.97
4.	Motten	3.26
5.	Moongilthuraipattu	35.35
6.	Porasapattu	12.18
7.	Mananthangal	13.33
8.	Melsiruvallur	4.77
9.	Thenaporai	4.90
10.	Peramakundam	11.92
11.	Melisiruvallur (mid supply)	57.04
12.	Sirpathanallur	17.59
13.	Mangalam	32.44
14.	Thiruvarangam	45.46
15.	Olagalapadi	8.59
16.	Vadakeeranur	448.62
17.	Kallipadi	16.21
18.	Vadamamandur new	19.51
19.	Arumparampattu	21.29
20.	Vadamamandur peria tank	42.20
21.	Arulambadi	12.21
22.	Jambodai	5.92
23.	Maniyandal	15.33
24.	Elaiyanarkuppam	36.73
25.	Sirpanandal	172.23
26.	Pakkam New Tank	43.19
27.	Chinnakolliyur	29.51
28.	Sirupanaiyur big tank	94.61
29.	Periyakolliyur	62.99
30.	Kadambur big tank	106.30

(Table Contd...)

31.	Nagalkudi	26.55
32.	Odiyanthal	32.55
33.	Vanapuram	76.47
34.	Athiyur thangal	31.84
35.	Tholuvanthangal	31.84
36.	Kaduvanur	129.93
37.	Athiyur big tank	60.96
38.	Ariyalur	28.37
39.	Varagur	27.52
40.	S.Kolathur small tank	23.47
41.	S.Kolathur big tank	48.56
42.	Chellakuppam	10.12
43.	Arur	47.35
44.	Vadasiruvalur	59.89
45.	Thimmanndal	17.40
46.	Kidakudayampattu	28.33
47.	Viriyur	36.42
48.	Arasampattu	18.21
49.	Periyakolliyur thangal	9.14
	Total	1824.42

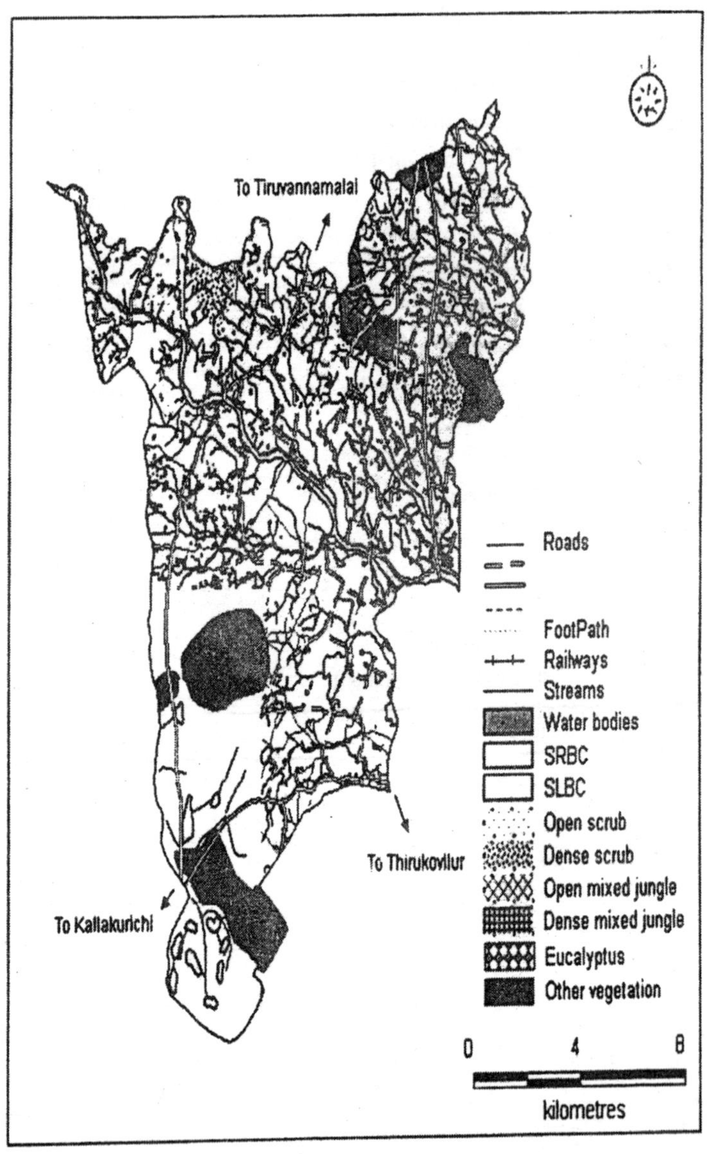

Fig. 2.4: Sathanur Reservoir Project (SRP) Command Area

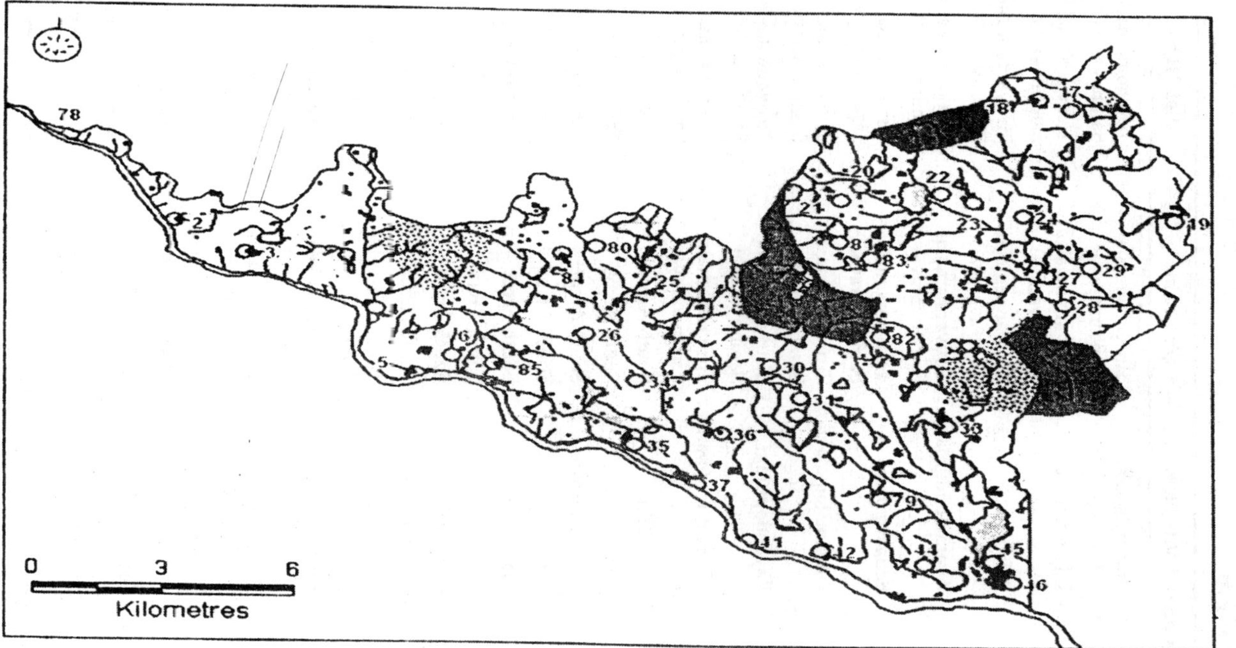

Fig. 2.5: Map representing the current status of SLBC command region with canal system and tanks

Name of the villages as indicated in the Figure 2.5 and Figure—2.6.

Village-Code	Vill_Name	Dist_name	Dist_HQ
1	Putturchekkadi	Tiruvannamalai Sambuvarayar	Tiruvannamalai Sambuvarayar
2	Edattanur	Tiruvannamalai Sambuvarayar	Tiruvannamalai Sambuvarayar
3	Allappanur	Tiruvannamalai Sambuvarayar	Tiruvannamalai Sambuvarayar
4	Agarampallippattu	Tiruvannamalai Sambuvarayar	Tiruvannamalai Sambuvarayar
5	Sadakuppam	Tiruvannamalai Sambuvarayar	Tiruvannamalai Sambuvarayar
6	Unnamalaipalaiyam	Tiruvannamalai Sambuvarayar	Tiruvannamalai Sambuvarayar
7	Rayandapuram	Tiruvannamalai Sambuvarayar	Tiruvannamalai Sambuvarayar
8	Mungilturaippattu	Tiruvannamalai Sambuvarayar	Tiruvannamalai Sambuvarayar
9	Poruvallur	South Arcot	Cuddalore
10	Porasappattu	South Arcot	Cuddalore
11	Irudaiyampattu	South Arcot	Cuddalore
12	Adanur	South Arcot	Cuddalore
13	Mangalam	South Arcot	Cuddalore
14	Arulambadi	South Arcot	Cuddalore
15	Olagalappadi	South Arcot	Cuddalore
16	Vadakiranur	South Arcot	Cuddalore
17	Nadupattu	Tiruvannamalai Sambuvarayar	Tiruvannamalai Sambuvarayar
18	Kolakkudi	Tiruvannamalai Sambuvarayar	Tiruvannamalai Sambuvarayar
19	Pavittiram	Tiruvannamalai Sambuvarayar	Tiruvannamalai Sambuvarayar
20	Pavupattu	Tiruvannamalai Sambuvarayar	Tiruvannamalai Sambuvarayar
21	Parayampattu	Tiruvannamalai Sambuvarayar	Tiruvannamalai Sambuvarayar
22	Kattampundi	Tiruvannamalai Sambuvarayar	Tiruvannamalai Sambuvarayar

(Table Contd...)

23	Sukkampalaiyam	Tiruvannamalai Sambuvarayar	Tiruvannamalai Sambuvarayar
24	Pappambadi	Tiruvannamalai Sambuvarayar	Tiruvannamalai Sambuvarayar
25	Vanapuram	Tiruvannamalai Sambuvarayar	Tiruvannamalai Sambuvarayar
26	Perunduraipattu	Tiruvannamalai Sambuvarayar	Tiruvannamalai Sambuvarayar
27	Periyakallapadi	Tiruvannamalai Sambuvarayar	Tiruvannamalai Sambuvarayar
28	Chinnakallapadi	Tiruvannamalai Sambuvarayar	Tiruvannamalai Sambuvarayar
29	P.K.Pudur	Tiruvannamalai Sambuvarayar	Tiruvannamalai Sambuvarayar
30	Palaiyanur	Tiruvannamalai Sambuvarayar	Tiruvannamalai Sambuvarayar
31	Kallottu	South Arcot	Cuddalore
32	Velaiyampakkam	South Arcot	Cuddalore
33	Murukkambadi	Tiruvannamalai Sambuvarayar	Tiruvannamalai Sambuvarayar
34	Edakkal	Tiruvannamalai Sambuvarayar	Tiruvannamalai Sambuvarayar
35	Melandal	South Arcot	Cuddalore
36	Kandiyankuppam	South Arcot	Cuddalore
37	Kangiyanur	South Arcot	Cuddalore
38	Sripadanallur	South Arcot	Cuddalore
39	Vadamamandur	South Arcot	Cuddalore
40	Arumbarampattu	South Arcot	Cuddalore
41	Pallisandal	South Arcot	Cuddalore
42	Jambai	South Arcot	Cuddalore
43	Tiruvarangam	South Arcot	Cuddalore
44	Sellankuppam	South Arcot	Cuddalore
45	Siddapattanam	South Arcot	Cuddalore
46	Manalurpettai	South Arcot	Cuddalore
47	Elayanarkuppam	South Arcot	Cuddalore
48	Jambadai	South Arcot	Cuddalore
49	Chinna Maniyandal	South Arcot	Cuddalore
50	Eduttanur	South Arcot	Cuddalore
51	Sirpanandal	South Arcot	Cuddalore
52	Chinna Kolliyur	South Arcot	Cuddalore

(Table Contd...)

53	Odiyandal	South Arcot	Cuddalore
54	Kadambur	South Arcot	Cuddalore
55	Vanapuram	South Arcot	Cuddalore
56	Arur	South Arcot	Cuddalore
57	Athiyur	South Arcot	Cuddalore
58	Chellamkuppam	South Arcot	Cuddalore
59	Kaduvanur	South Arcot	Cuddalore
60	Kallipadi	South Arcot	Cuddalore
61	Kidakudayampattu	South Arcot	Cuddalore
62	Mananthangal	South Arcot	Cuddalore
63	Maniyandal	South Arcot	Cuddalore
64	Motten	Tiruvannamalai Sambuvarayar	Tiruvannamalai Sambuvarayar
65	Pakkam	South Arcot	Cuddalore
66	Periyakolliyur	South Arcot	Cuddalore
67	S.Kolather	South Arcot	Cuddalore
68	Sirpathanallur	South Arcot	Cuddalore
69	Sirupanaiyur	South Arcot	Cuddalore
70	Tevaparai	South Arcot	Cuddalore
71	Thimmanendal	South Arcot	Cuddalore
72	Thiruvadathanur	Tiruvannamalai Sambuvarayar	Tiruvannamalai Sambuvarayar
73	Tholuvanthangal	South Arcot	Cuddalore
74	Varakur	South Arcot	Cuddalore
75	Viriyur	South Arcot	Cuddalore
76	Vadaponparappi	South Arcot	Cuddalore
77	Arausarampattu	South Arcot	Cuddalore
78	Olagalapadi	Tiruvannamalai Sambuvarayar	Tiruvannamalai Sambuvarayar
79	Devaradiarkuppam	South Arcot	Cuddalore
80	Matuvampattu	Tiruvannamalai Sambuvarayar	Tiruvannamalai Sambuvarayar
81	Nariyaputtu	Tiruvannamalai Sambuvarayar	Tiruvannamalai Sambuvarayar
82	Navampattu	Tiruvannamalai Sambuvarayar	Tiruvannamalai Sambuvarayar
83	Thalayanpattu	Tiruvannamalai Sambuvarayar	Tiruvannamalai Sambuvarayar
84	Thenkanimpalar	Tiruvannamalai Sambuvarayar	Tiruvannamalai Sambuvarayar
85	Valavachanur	Tiruvannamalai Sambuvarayar	Tiruvannamalai Sambuvarayar

Fig. 2.6: Map representing the current status of SRBC command region with canal system and tanks

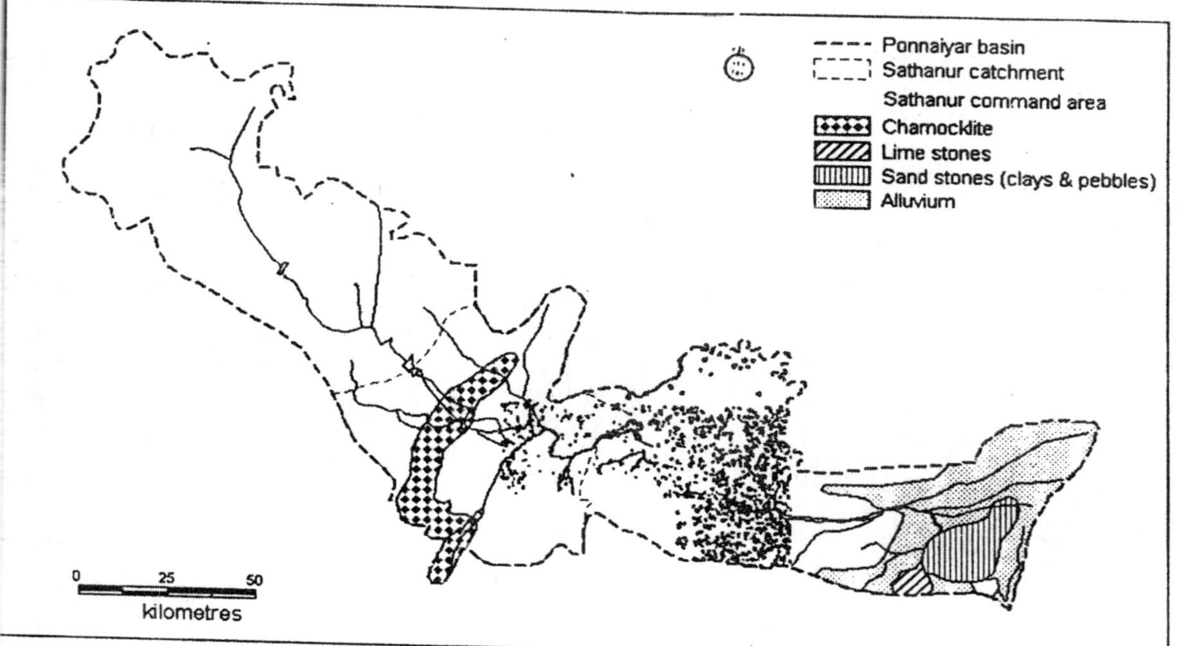

Fig. 2.7: Geological map of Ponnaiyar basin

Catchment

The catchment is interspersed with several hills. The peak of the catchment is in Yercaud. The elevation of Shervarays and Javadi hills are nearly 1000 m, hence the climate is generally cool for nine months of the year in the upper reaches of Thirupathur sub-division. On the plains, during the months of March to May, it is very hot and dry with occasional gusty storms in the afternoons. The climate generally becomes peasant after few showers that usually come in May. Again, during the months of October to February the climate becomes cool and pleasant, supplemented by mist and dew at night. The average maximum temperature recorded is 32°C. The average minimum temperatures recorded in plains and hills are 24°C in plains and 10°C in hills respectively.

Precipitation

Catchment

The average annual rainfall varies from 800 mm to 1000 mm as recorded in the several stations in the catchment (Figure—2.8). The whole of the catchment gets rainfall both from southwest and northeast monsoons. Most of the rainfall received in this tract is from June to December. October and November are the rainiest of the months. A considerable amount of dew falls during December to February. The subsequent months are dry. The details of the annual precipitation recorded in the various stations in the catchment region for various periods has been presented as Figures 2.8-2.14 and Tables—2.12 and 2.13. The average, standard deviation and coefficient of variation values have also been presented. Precipitation values show the minimum coefficient of variation thus, more consistency at Sathanur dam site. The values are maximum in Singarapet and Palacode stations. 48.4 per cent respectively. The trend of annual precipitation in various regions in the catchment is presented in Figures—2.8–2.14. A slight declining trend is observed in the annual precipitation at the Sathanur dam site (Figure—2.8). Declining trend is also observed for the annual precipitation values in Uthangarai, Tirupathur, Royakotta, Palacode, Chengam and Dharmapuri regions for the period of 1960-1982. The values recorded in Krishnagiri and Singarapet depict an upward trend (Figures—2.9 and 2.10 respectively).

Table—2.12 Annual precipitation (in mm) recorded in Sathanur dam

Month/Year	1977	1978	1979	1980	1981	1982	1983	1984	1985	1986	1987
January	–	–	–	–	3.00	–	–	10.20	106.00	90.00	13.00
February	9.50	17.00	42.00	–	–	–	–	117.80	–	4.00	–
March	–	–	–	30	69.00	–	–	61.50	6.00	–	112.00
April	–	10.00	11.00	6.10	17.00	2.00	1.00	83.70	38.50	1.00	–
May	80.50	19.00	78.00	58.50	191.00	87.00	62.50	–	17.00	46.00	79.00
June	60.00	2.00	39.00	9.00	24.10	21.00	66.50	15.00	96.00	47.50	76.00
July	15.50	8.00	45.00	92.50	77.00	29.00	28.50	244.00	113.00	18.00	35.00
August	267.50	8.00	61.00	87.50	98.00	65.00	281.20	9.00	188.00	120.00	110.00
September	205.00	262.00	292.50	32.50	199.20	119.00	395.00	265.50	204.00	140.50	197.50
October	286.60	109.08	52.10	78.50	371.00	37.00	101.00	37.50	106.00	84.00	200.00
November	363.50	226.00	486.00	133.50	104.00	123.00	26.00	200.00	140.00	126.00	86.00
December	3.10	112.00	22.00	18.10	31.00	4.00	257.80	44.50	20.00	32.00	143.00
Total	1291.10	773.80	1128.60	546.00	1184.20	487.00	1219.5	1088.7	1034.5	7.90	1051.00

(Table Contd...)

	1988	1989	1990	1991	1992	1993	1994	1995	1996	1997	1998	1999	2000	2001
January	–	–	59.00	–	42.00	–	–	41.00	–	1.00	–	0	5	0
February	–	–	2.00	3.00	–	–	–	–	1.00	–	–	0	21	0
March	114.00	53.00	57.00	–	–	–	–	2.00	–	1.00	–	0	0	0
April	65.50	-	48.00	34.00	–	–	1.20	11.00	42.00	5.00	2.00	0	4.5	28
May	40.50	75.00	62.00	8.00	42.00	20.00	105.20	128.5	98.40	126.00	41.50	44	12.5	16
June	36.00	55.00	23.50	80.00	56.00	116.00	2.00	87.00	163.00	95.00	17.00	40	0	6
July	26.00	168.00	26.00	15.50	71.00	43.00	35.00	89.00	12.00	15.00	91.00	26	5.5	111.5
August	219.00	13.00	41.00	85.00	47.00	60.00	184.80	127.00	158.00	28.50	200.00	60	141.1	40.1
September	242.00	179.00	212.90	138.50	155.00	100.50	91.00	76.00	324.00	95.50	146.00	50.2	85.3	130
October	134.00	131.00	76.00	320.50	194.00	166.70	225.20	95.00	179.00	217.00	204.00	200	123.8	63
November	39.00	167.50	144.00	325.00	535.00	149.00	267.00	24.00	137.50	226.00	178.00	94	21.7	36
December	74.00	36.00	45.00	3.00	9.00	197.10	07.00	–	399.00	79.00	103.00	58	140.3	33
Total	990.00	877.50	7964.00	1012.50	1151.00	852.30	918.40	680.50	1612.90	889.00	982.50	572.2	560.7	463.6

Table—2.13 Annual precipitation (mm) recorded at Sathanur Command area (SCA)

Year	Annual precipitation (mm)	Year	Annual precipitation (mm)	Year	Annual precipitation (mm)
1962	986.37	1976	808.20	1990	723.50
1963	1103.00	1977	1284.00	1991	1079.40
1964	1308.00	1978	754.10	1992	842.20
1965	1160.00	1979	779.20	1993	844.80
1966	1449.00	1980	414.24	1994	1077.20
1967	922.10	1981	886.21	1995	888.00
1968	610.72	1982	638.22	1996	1754.40
1969	1265.24	1983	986.37	1997	1372.40
1970	1019.06	1984	1199.60	1998	853.80
1971	963.26	1985	1188.80	1999	742.71
1972	1294.84	1986	888.19	2000	522.60
1973	752.40	1987	1042.00	2001	463.6
1974	540.32	1988	798.20		
1975	570.24	1989	845.60		

Average 970.70 mm
Standard deviation 274.80
Coefficient of variation 28.30%

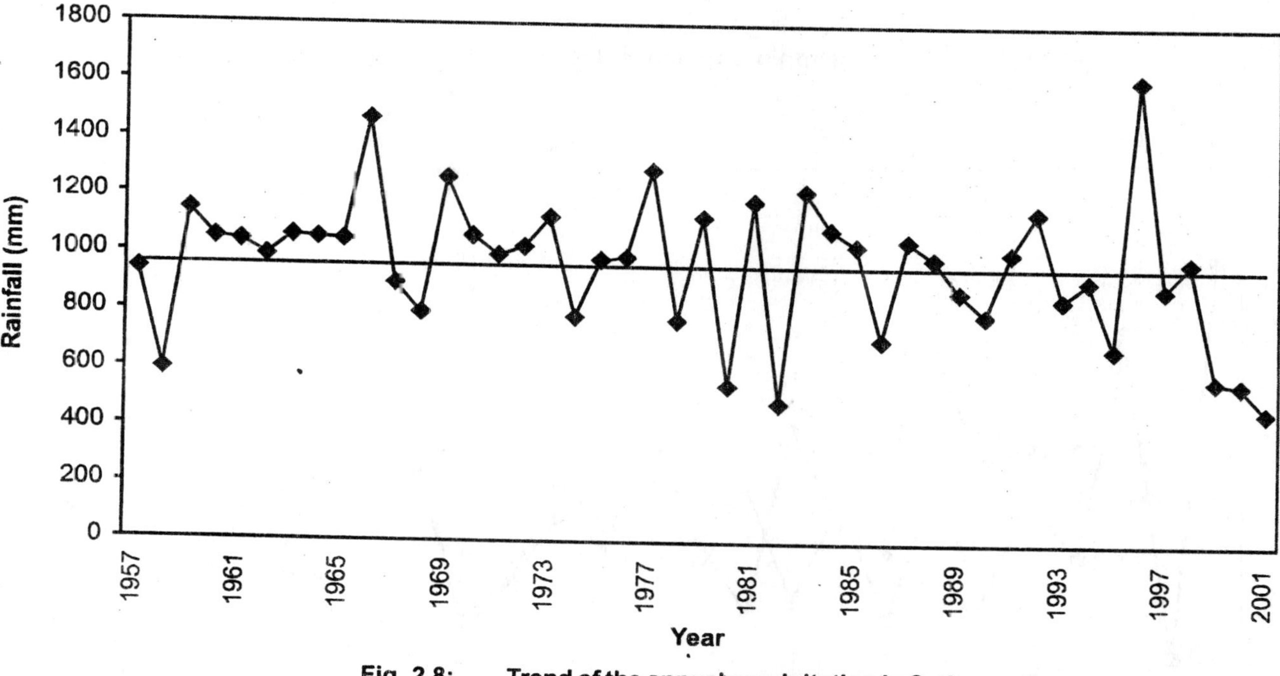

Fig. 2.8: Trend of the annual precipitation in Sathanur dam

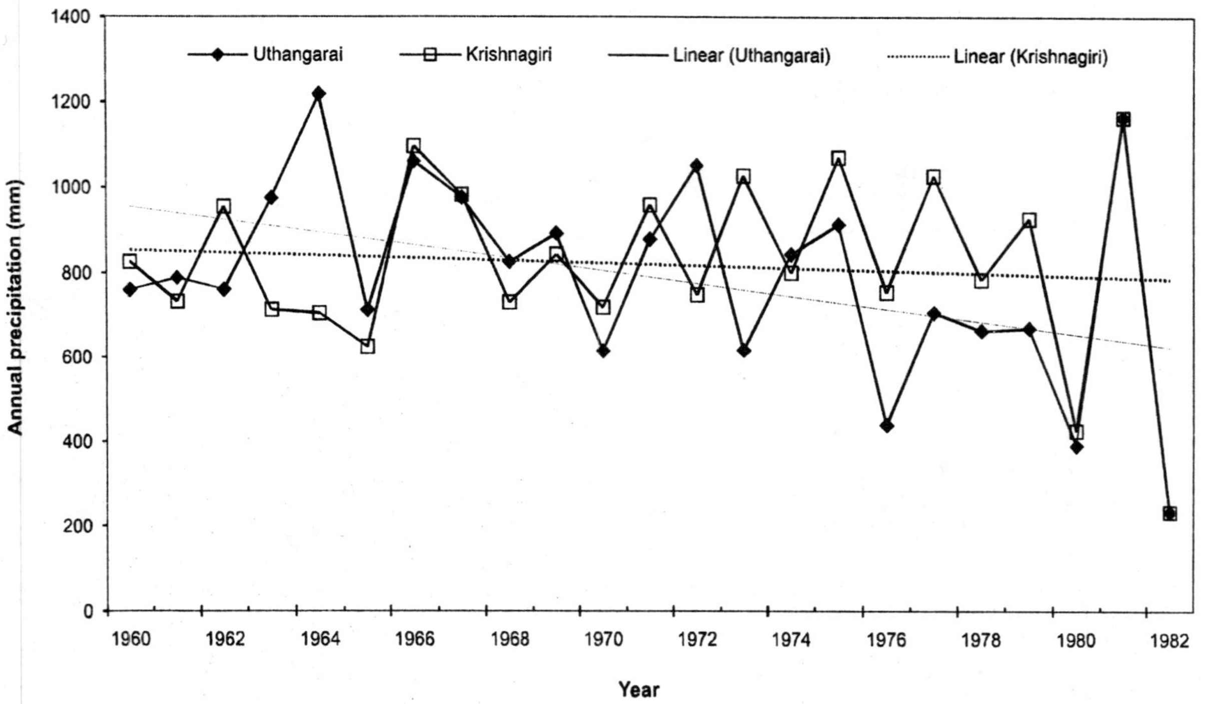

Fig. 2.9: Trend of the annual precipitation in Uthangarai and Krishnagiri

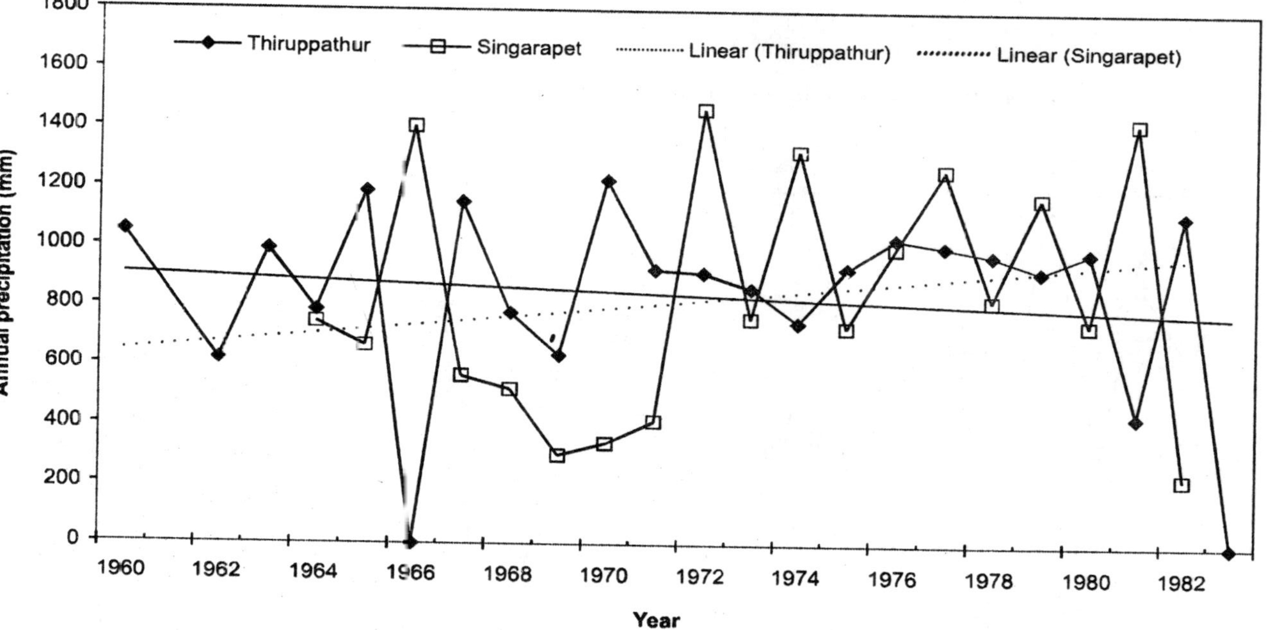

Fig. 2.10: Trend of the annual precipitation in Thirupathur and Singarapet

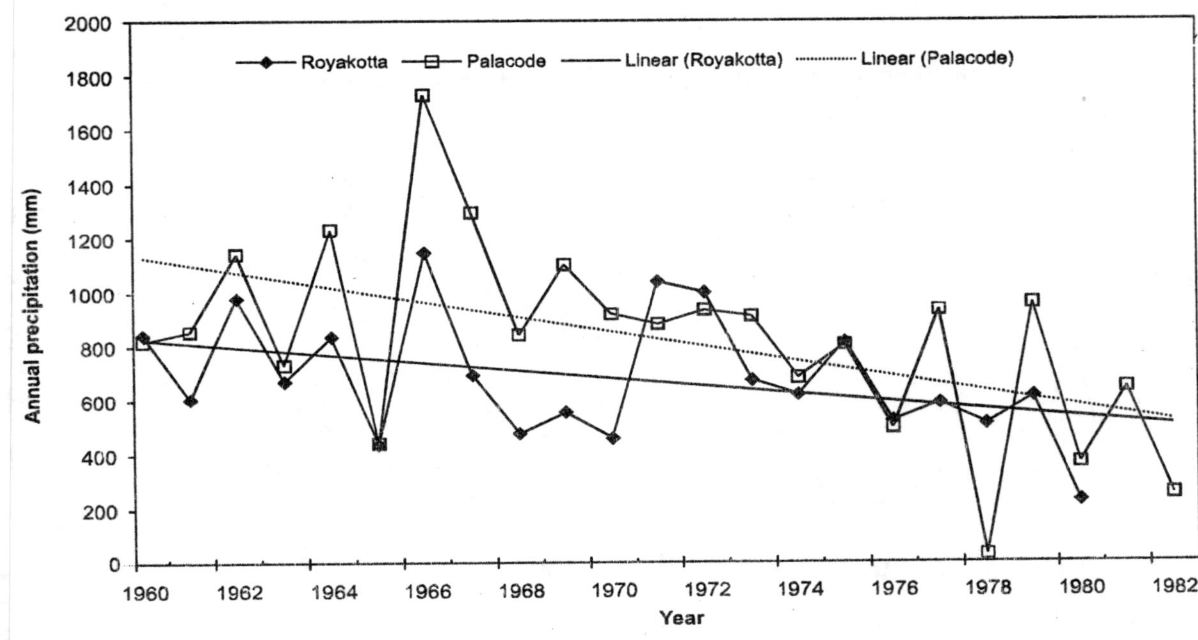

Fig. 2.11: Trend of the annual precipitation in Royakotta and Palacode

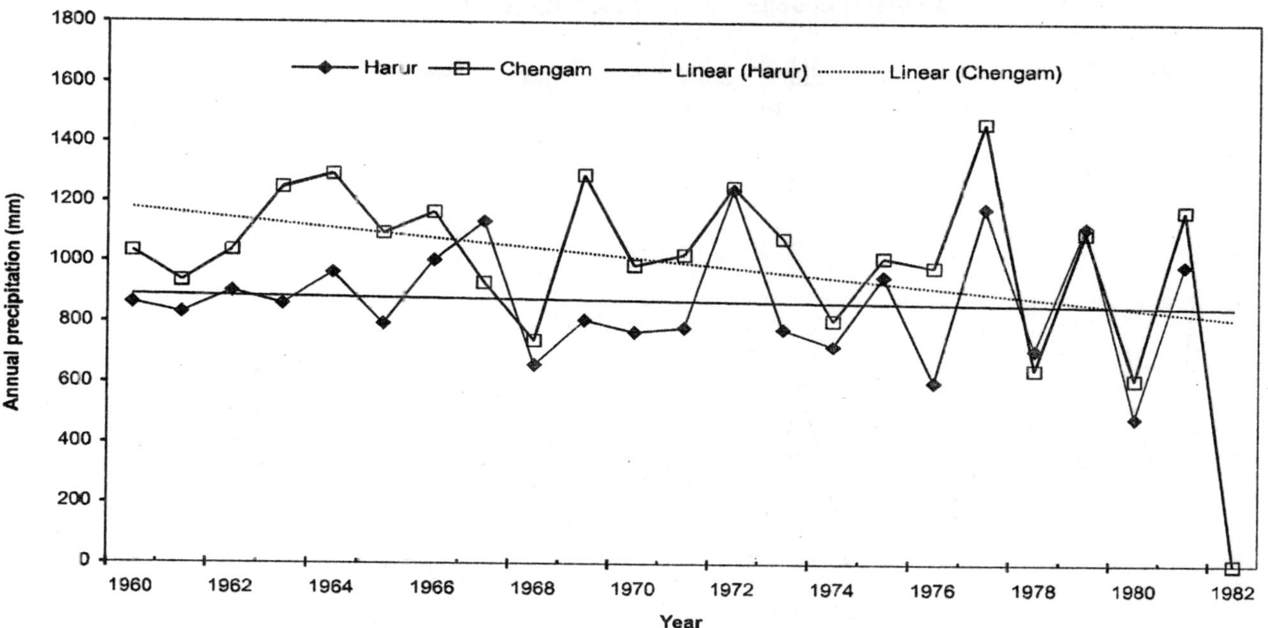

Fig. 2.12: Trend of the annual precipitation in Harur and Chengam

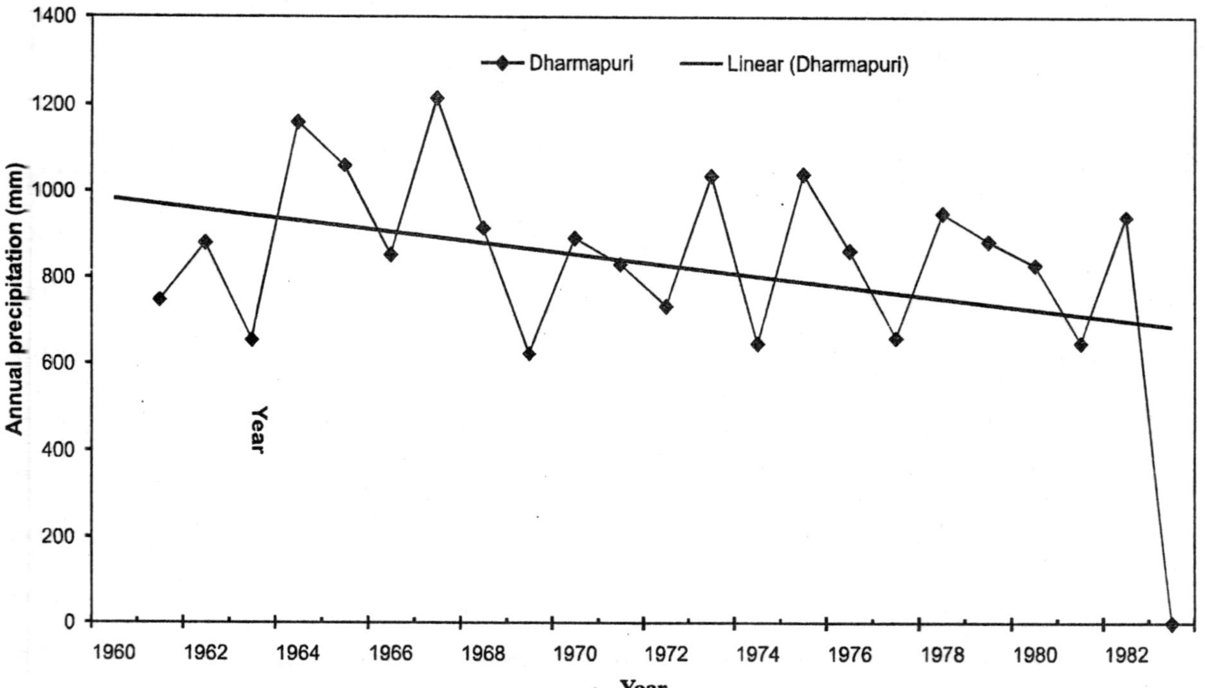

Fig. 2.13: Trend of the annual precipitation in Dharmapuri

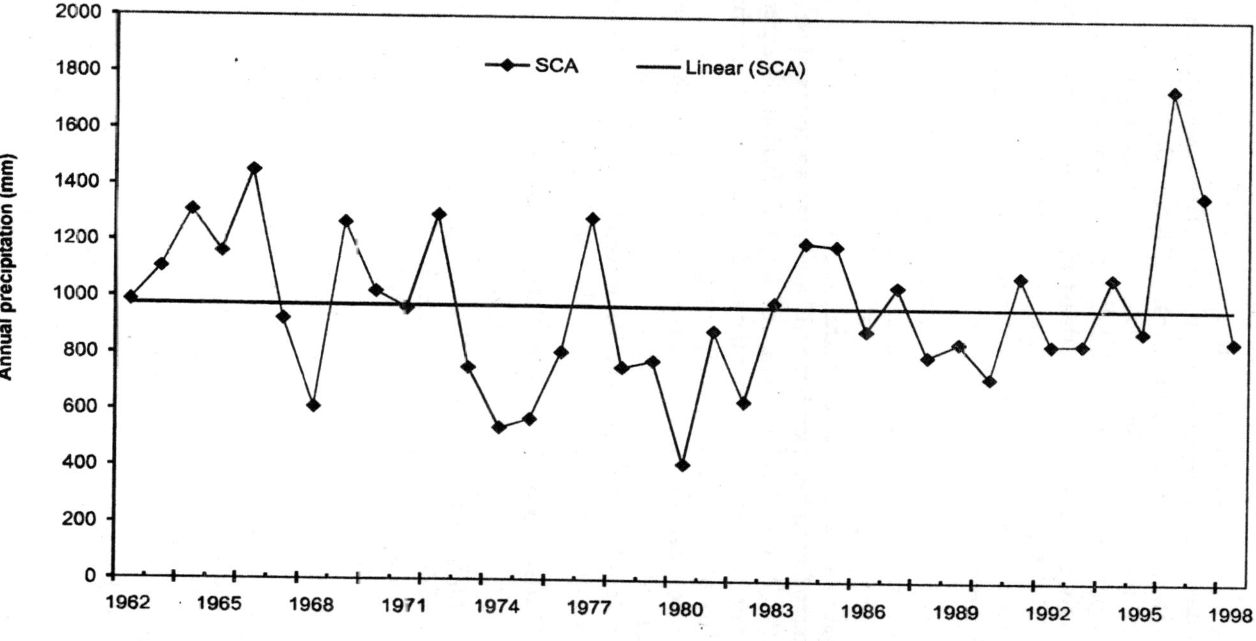

Fig. 2.14: Trend of the annual precipitation in SCA (Thiruvannamalai)

Sathanur Reservoir

The hydrograph of Sathanur reservoir for the period 1977 and 2001 has been presented as Figure—2.15. From the hydrograph it can be inferred that the maximum rainfall is recorded at Sathanur during the months of August till December, which accounts for almost 71 per cent of the total rainfall received in an annual year. Interestingly, Sathanur receives atleast 19.75 mm rainfall almost all months of the annual year. Also, from this hydrograph it is concluded that Sathanur reservoir doesn't benefit from the Southwest monsoon.

The average annual rainfall varies between 464mm to 1613mm as recorded in the Sathanur reservoir (Figure—2.8). The whole of the catchment gets rainfall both from Southwest and Northeast monsoons. Most of the rainfall received in this tract is from June to December. October and November are the rainiest of the months. A considerable amount of dew falls during December to February. The subsequent months are dry. Table—2.12 presents the details of the annual precipitation recorded in Sathanur reservoir for various periods. Precipitation values show the minimum coefficient of variation thus, more consistency at Sathanur dam site. A slight declining trend is observed in the annual precipitation at the Sathanur dam site (Figure—2.8).

Inflow, Outflow and Reservoir Water Level

The inflow and outflow details (in cusecs) of reservoir for a period of 1977-2001 are presented as Tables—2.14 and 2.15 and Figure—2.15 to Figure—2.18. The monthly variation in the water level of the reservoir for a period of twenty-four years has been presented as Table 2.16.

The Sathanur reservoir, as it receives maximum rainfall during Northeast monsoons, there is a fall during the months of November and December (Figure—2.16). The water level of the Sathanur reservoir is least during the months of April-August due to the summer months.

The inflow pattern of Sathanur reservoir (Figure—2.17) indicates that there was a remarkable inflow of water into the reservoir during the months of October, November and December. While the inflow was very poor during the months of January till July, which can be attributed to the approaching summer.

Table—2.14 Inflow (in cusecs) recorded in Sathanur dam

Month/Year	1977	1978	1979	1980	1981	1982	1983	1984	1985	1986	1987
January	925	4621	7713	8139	123	1828	182	5778	5351	4672	2359
February	270	2847	3054	1572	–	393	268	8361	1077	1009	651
March	–	2211	969	555	555	188	117	4258	580	688	645
April	492	1670	346	331	186	36	40	834	275	496	484
May	1366	1685	26	763	2612	996	2757	105	1375	635	754
June	1709	906	1460	1350	515	995	3916	213	1519	977	1302
July	1321	1182	3478	873	5254	383	459	3657	841	513	347
August	11697	446	5904	746	5826	605	5130	802	5791	1211	1136
September	4079	9667	59006	362	99036	3814	27667	2876	11541	20052	8307
October	77601	21496	30459	3753	81807	1057	4416	30565	7160	61857	35250
November	188641	58210	164115	10740	72933	5651	3830	8974	6853	39265	4397
December	40903	19388	47612	1454	13551	621	24123	3655	3610	6971	9940
Total	329004	124329	324142	30638	282398	16567	72905	69778	45973	138346	65572

(Table Contd...)

Month/Year	1988	1989	1990	1991	1992	1993	1994	1995	1996	1997	1998	1999	2000	2001
January	1955	269	542	477	2830	1515	6286	2263	59	18471	10007	2463	1797	2101
February	211	101	378	336	1107	197	667	313	–	836	565	2917	35	0
March	555	459	505	363	499	576	495	347	–	1772	25	171	0	0
April	448	329	329	306	227	391	1096	270	104	2729	970	67	0	6598
May	633	391	427	315	495	564	1243	1488	220	1550	1012	857	0	372
June	402	500	288	1553	1637	1449	645	1234	936	347	195	58	0	116
July	1386	6192	395	289	2080	577	1522	2161	23	138	932	881	35	1307
August	9977	243	499	432	416	874	2119	2168	9626	92	18427	2749	486	902
September	53114	2059	6290	1801	3710	771	538	3504	74878	10754	21690	458	23102	22219
October	18522	9876	747	57345	6730	38670	10542	25043	124811	4502	74345	71984	101884	78972
November	1189	6341	1744	304902	6141	37831	53165	21322	29455	95204	50315	19813	12621	17149
December	592	950	1396	25480	5656	105321	2814	583	201680	72480	57446	11534	6130	1508
Total	88984	27710	13540	393599	31528	188736	81132	60696	441792	249595	235929	113952	146090	131244

Table—2.15 Outflow (in cusecs) recorded in Sathanur dam

Month/Year	1977	1978	1979	1980	1981	1982	1983	1984	1985
January	4524	8543	11850	12007	7525	12234	1447	2670	2650
February	9720	17879	18411	14985	2245	17601	1202	4330	12220
March	3834	20410	17332	20252	2096	19610	1789	8350	4800
April	712	17838	22350	19156	1430	14745	1576	15086	780
May	–	10670	13039	12076	–	7226	302	7793	–
June	–	2100	1110	–	–	2050	–	2880	–
July	–	–	–	–	–	–	–	–	–
August	–	–	–	–	–	–	–	–	–
September	–	–	–	–	31969	–	–	–	–
October	17114	14433	34494	–	70654	–	13894	2600	4000
November	1989983	6698	168739	–	67764	–	13000	3900	6500
December	32255	15311	42974	–	12632	–	9400	19500	–
Total	267142	113882	330299	784761	196315	73776	42610	67109	30950

(Table Contd...)

Month/Year	1988	1989	1990	1991	1992	1993	1994	1995	1996	1997	1998	1999	2000	2001
January	5225	7050	7825	2000	9822	26734	5018	16704	13074	20927	18055	0	16889	21574
February	18800	6675	–	–	26500	5850	16624	12500	6275	15900	21183	0	21300	11707
March	4700	3100	–	–	14100	16200	16080	13375	–	25100	22900	29437	32150	21900
April	–	–	–	–	9900	14790	22570	4700	–	17400	18989	11530	10850	5050
May	–	–	–	–	7610	4340	4800	–	–	7200	1617	0	0	0
June	–	–	–	–	–	–	–	–	–	–	–	0	0	0
July	–	–	–	–	–	–	–	–	–	–	–	0	0	0
August	–	–	–	–	–	–	–	–	–	–	–	0	0	46
September	–	–	–	–	–	–	–	–	8252	–	–	0	0	0
October	149000	–	–	–	–	–	–	–	125044	–	39789	18643	60820	50465
November	–	–	–	265454	–	–	–	12000	26386	90067	15311	19452	14450	13099
December	9600	–	23000	23383	4580	99611	18850	15500	55622	68465	52859	2995	600	10100
Total	87325	16825	10825	290871	72512	177525	83942	74779	234653	245059	190703	82057	157059	133941

Table—2.16 Water level (in feet) recorded in Sathanur dam

Month/Year	1977	1978	1979	1980	1981	1982	1983	1984	1985	1986	1987
January	74.20	116.80	116.85	116.95	69.65	114.45	69.85	89.60	82.70	84.56	97.55
February	57.70	109.65	111.70	110.40	64.40	104.75	66.40	91.85	70.05	79.35	81.20
March	45.15	97.65	102.75	98.60	58.85	91.40	66.20	88.15	62.20	62.45	73.55
April	40.95	84.95	90.20	79.70	53.95	76.75	57.30	73.65	59.85	56.00	68.05
May	43.05	75.00	75.10	62.55	57.55	66.75	60.25	61.10	60.85	52.95	67.90
June	46.10	72.30	74.50	63.40	56.95	63.55	65.25	53.95	62.00	53.80	68.70
July	47.70	72.45	77.40	63.40	64.65	62.80	64.55	59.65	61.95	52.80	67.80
August	68.20	71.75	82.10	63.20	71.30	62.35	70.60	59.60	69.35	54.50	68.40
September	72.75	81.75	116.90	62.55	116.85	66.95	94.10	63.15	81.35	84.95	78.85
October	117.00	81.85	117.00	67.05	116.60	66.70	86.10	91.45	85.60	104.25	91.85
November	117.00	117.00	117.00	78.95	118.95	73.75	75.95	94.45	83.85	114.15	87.65
December	118.75	119.00	119.00	79.65	119.00	73.40	87.80	81.10	86.00	104.95	94.65

(Table Contd...)

Month/Year	1988	1989	1990	1991	1992	1993	1994	1995	1996	1997	1998	1999	2000	2001
January	92.05	78.15	59.80	59.00	115.75	108.05	116.70	98.55	77.95	117.90	115.85	106.51	115.98	109.70
February	71.15	67.15	59.00	57.90	103.95	98.45	106.20	88.00	69.70	111.65	106.85	108.4	106.07	103.19
March	62.15	58.80	98.20	56.60	94.65	85.75	96.70	72.00	67.90	98.50	92.60	95.4	93.05	92.33
April	60.70	57.55	56.95	55.15	84.30	70.50	77.10	60.20	66.90	87.7.	74.80	50.8	78.26	80.79
May	60.30	56.45	55.95	53.45	74.15	61.40	70.30	60.25	66.05	80.85	73.45	83.43	71.48	81.33
June	59.40	55.65	54.30	56.25	75.30	62.40	69.35	63.15	67.20	80.70	73.10	25.06	70.68	81.22
July	60.90	70.00	52.85	54.55	76.75	62.10	70.35	65.75	66.05	79.85	73.40	84.0	69.94	75.59
August	75.75	68.85	51.65	53.30	76.10	62.40	72.30	68.45	79.25	79.15	91.60	86.94	70.09	77.0
September	111.50	70.90	69.85	56.30	79.40	62.90	72.00	73.05	117.00	92.05	104.00	86.63	71.08	88.72
October	93.30	61.60	65.70	115.80	85.45	100.35	83.50	95.40	117.00	114.20	117.00	102.91	111.70	97.63
November	93.40	74.25	67.60	118.70	118.50	118.10	114.50	102.65	117.00	117.00	117.00	116.27	110.49	116.5
December	86.00	74.45	68.70	119.00	119.00	119.00	107.35	92.00	119.00	119.00	119.00	102.72	116.24	86.03

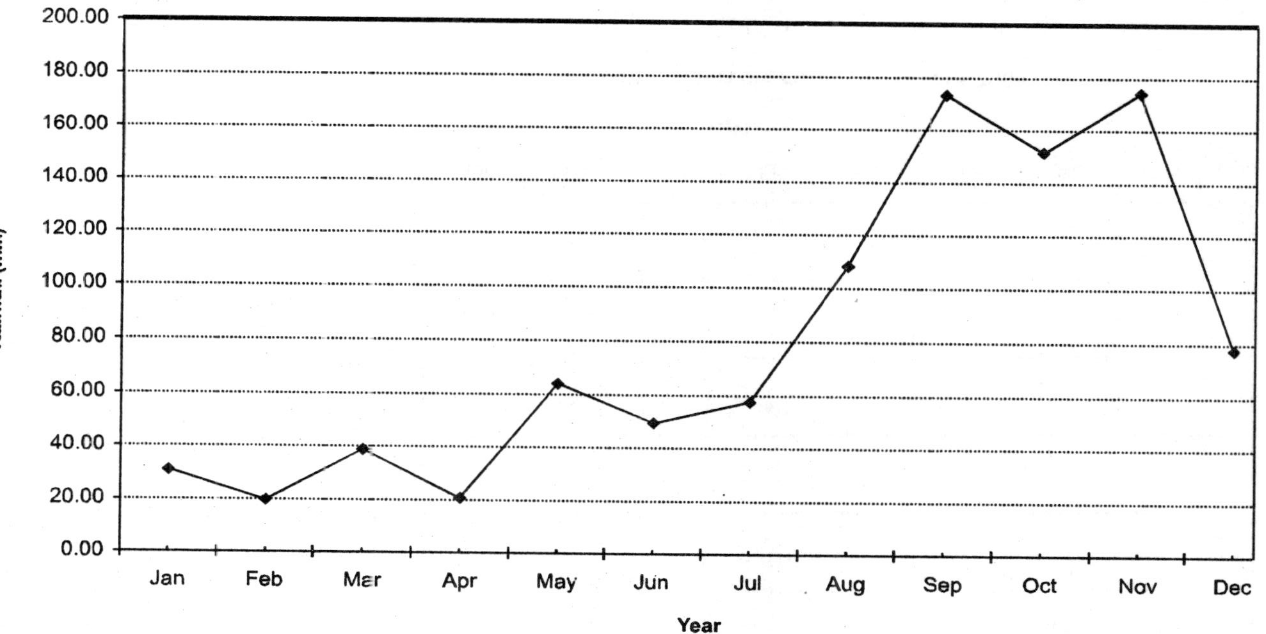

Fig. 2.15: The hydrograph at Sathanur during the period 1977 and 2001

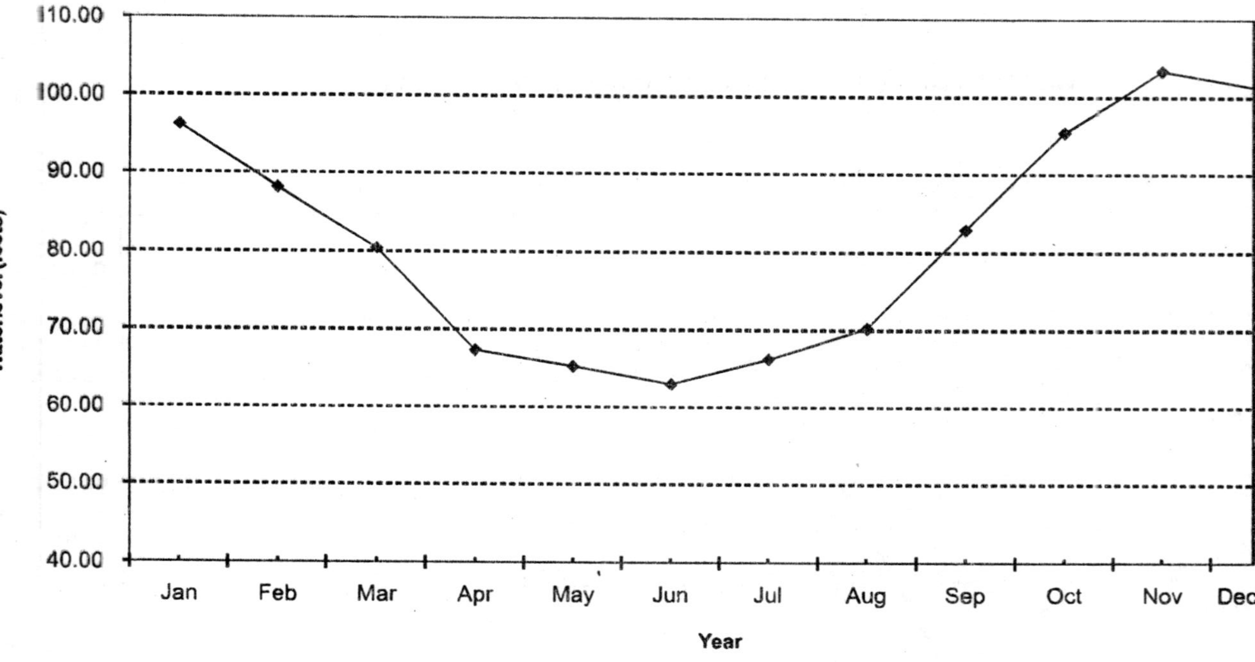

Fig. 2.16: The water level of Sathanur reservoir during 1977 and 2001

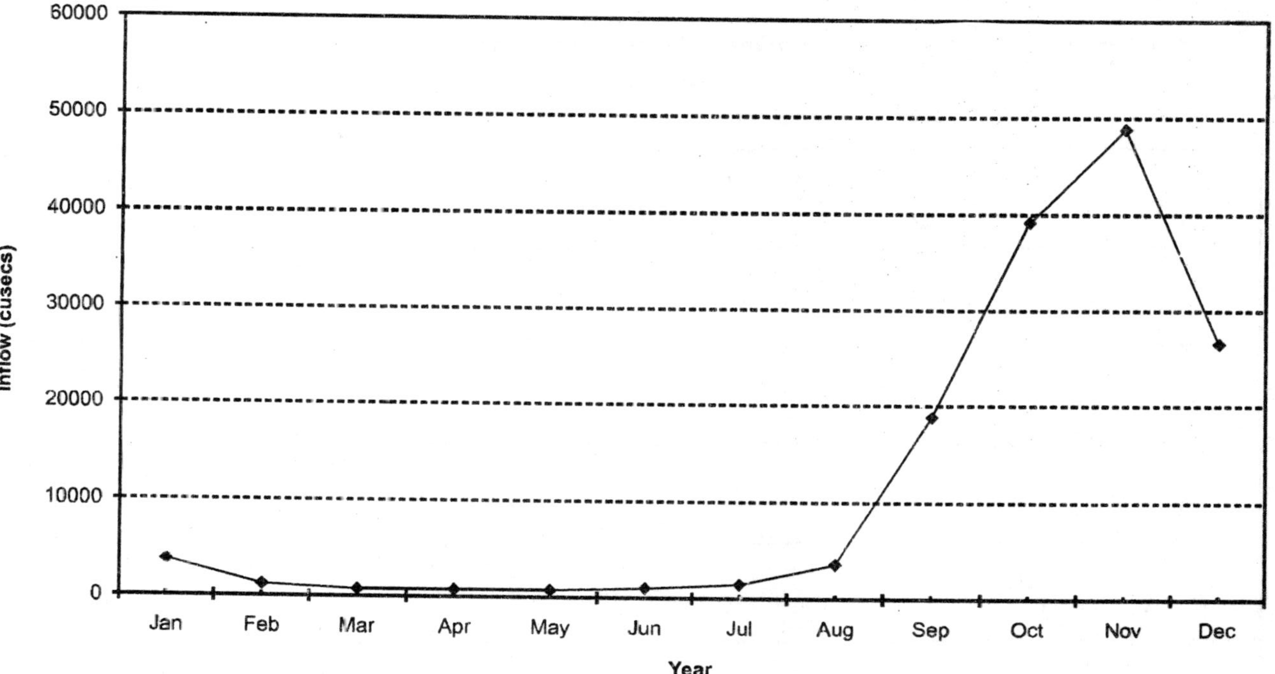

Fig. 2.17: The inflow of water into Sathanur reservoir during 1977 and 2001

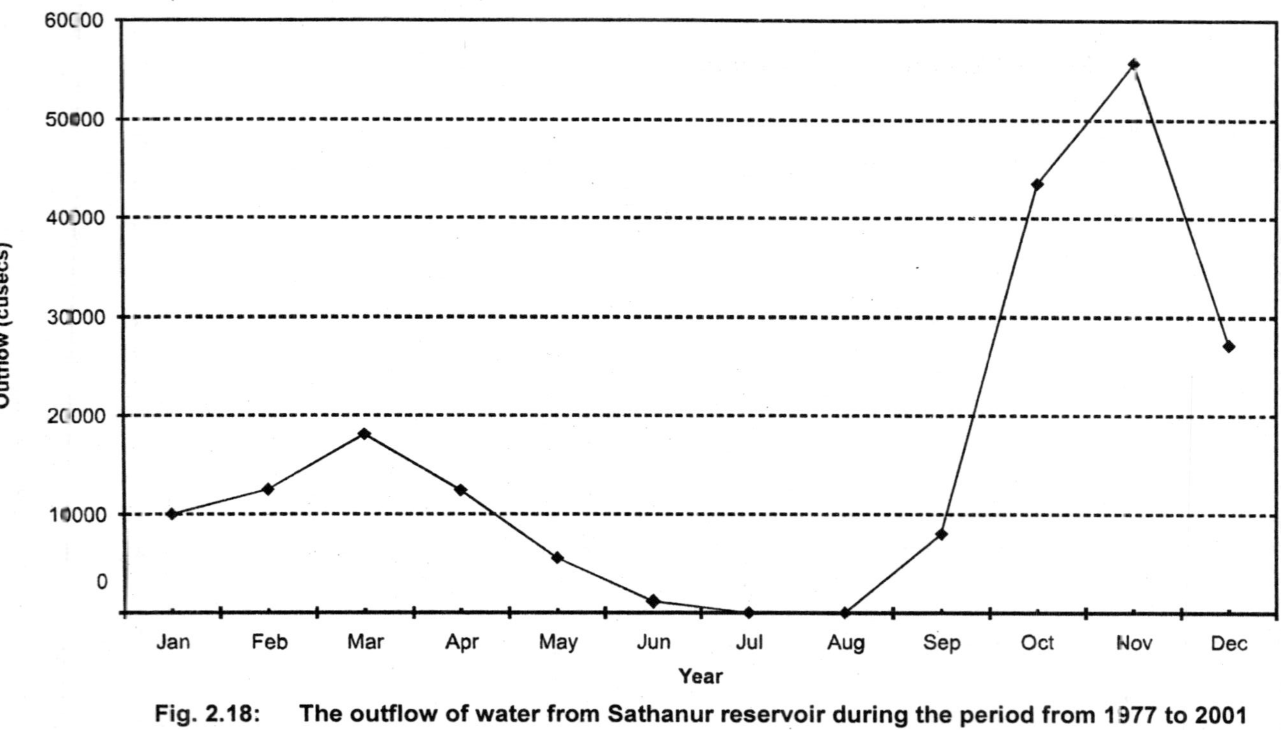

Fig. 2.18: The outflow of water from Sathanur reservoir during the period from 1977 to 2001

The outflow pattern of water from the reservoir (Figure 2.18) suggests that a copious amount of water has been discharged from the Sathanur reservoir during the months of October, November and December. This is also due to the Northeast monsoon, which would have brought a heavy rainfall to the reservoir. It is also noticed that there was no outflow during the summer months from June to August.

Physiographic Relief and Drainage

The irrigation command area of the reservoir has a gently sloping and undulating topography. The areas of SRBC command are at higher elevation compared to most of the regions in SLBC command. The latter has low-lying slopes, thus posing drainage related problems. Owing to the elevated slopes, the SRBC command is well drained.

Section—II Impacts, Upstream

3

Upstream Impacts

INTRODUCTION

The environment of a Water Resources Project (WRP) can be significantly influenced by other WRPs apart from the general environmental factors operating in the upstream. These would influence the quality and quantity of water that is dammed downstream.

We have made an attempt to identify the environmental conditions prevailing upstream of the Sathanur project. The direct and indirect implications of the upstream environment on the project and the eco-restoration measures adopted by the authorities have also been evaluated.

Status

Prior to the construction of the Krishnagiri and Sathanur reservoir projects, there were no storage reservoirs in the Ponnaiyar river basin except for a small barrage across the tributary Markandanadhi. It was built in 1942 to irrigate an area of 400 ha. The main river and its tributaries had only four barrages across them. Currently two of those, i.e. Aliyalam and Nedungal, are situated upstream of Sathanur reservoir and the other two Thirukoilur and Ellischoultry are located downstream, irrigating a vast tract of land under direct (canal) and indirect (tank) system (Figure—3.1).

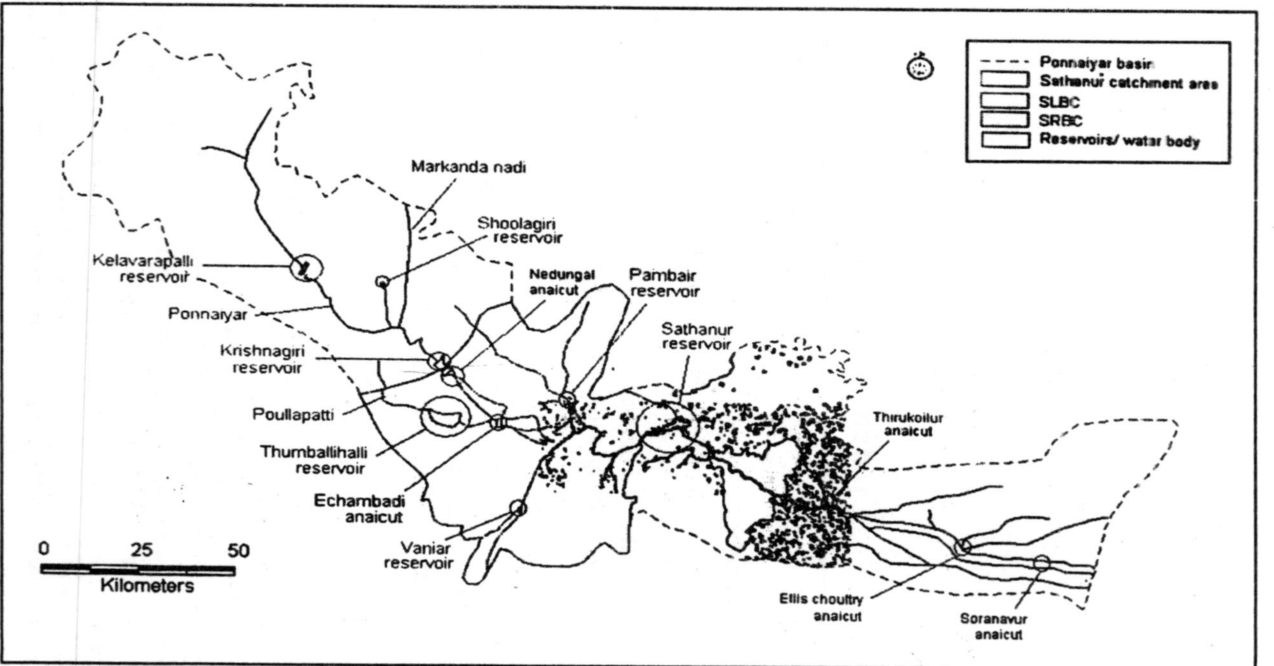

Fig. 3.1: Location of various projects in Ponnaiyar river basin

Krishnagiri project lies 113 km upstream of Sathanur dam near Krishnagiri town, Dharmapuri district. It has 3647 ha of land under direct canal irrigation, covering some thirteen villáges and 920 ha under indirect tank irrigation. The supply is regulated for two crop seasons. Nedungal barrage lies about 16 km below Krishnagiri project.

The Ponnaiyar river basin is approximately 12986 km^2. Total catchment of Sathanur reservoir is 10826 km^2 and the independent catchment is 5397.5 km^2. The basin and catchment details are furnished in Table—2.1 (Chapter 2).

The important projects below Krishnagiri reservoir till Sathanur are Pambar reservoir on the Pambaru river, a main tributary of Ponnaiyar river, about 3 km from Uthanagarai and Vanairu reservoir near Harur constructed on the Vanairu river. Vanaru is one of the major tributaries of Ponnaiyar and serves the irrigation demands of its region.

Paddy (*Oryza sativa*), ragi (*Eleucine coracona*), and millets (*Pennisetum typhoides*) are the major crops cultivated upstream of the Sathanur reservoir. On river bank betel (*Piper betel*) vine are grown. Pulses (*Phaseolus mungo, P.aureus, Cajanus cajan* etc.) millets, and mango (*Mangifera indica*) orchards on the foothill slopes are grown as monsoon crops.

The catchment of Ponnaiyar between Krishnagiri and Sathanur reservoirs is a rolling country mixed with hills and plains. The hills are Shervarays, Javadi, and Chitteri. The catchment upstream of the Sathanur mostly lies in Ponnaiyar reserve forest. The forest consists of dense scrub and open mixed tropical trees like *Acacia amara, Acacia initia, Eucalyptus* hybrid, *Acacia sundra, Azadirachta imitea, Zizyphus xylophyrus* etc. The extreme heat and drought conditions supplemented with insufficient rains in the tract have reduced the reserve forest into a degraded forest, as evident from the thorny trees that abound. The soil is red gravelly loam to loam in the forest region.

IMPACTS

The major portion of the independent catchment, upstream Sathanur reservoir, consists of dry, degraded, deciduous open forest, rocky terrain, uprising hills and agricultural plains. The soil erosion hazards are thus exacerbated by these environmental conditions prevailing upstream. Cattle grazing and illegal felling of the trees in the forest, as confirmed by the authorities, lead to soil loss. The efforts of the authorities to afforest the denuded natural forest fail due to the extreme hot and drought-like conditions. The sediment-laden water flows downstream and the sediments get trapped in the Sathanur reservoir. An increase in the solid contents in the reservoir has been observed over the past years (Figure—9.6, Chapter 9).

Sedimentation studies undertaken for the Krishnagiri and Sathanur reservoir (Chapter 5) reveal that the, specific erosion from the gross catchment of Sathanur reservoir is higher compared to that of Krishnagiri reservoir despite the fact that Krishnagiri project is situated upstream of the Sathanur hence should trap more silt than the Sathanur reservoir.

The reservoirs and their command areas upstream Sathanur project indirectly affect the water quality of the Sathanur reservoir. All the reservoirs upstream Sathanur reservoir has been constructed to meet the irrigation demands of their respective commands. Thus, the agricultural practices upstream and the agricultural runoff draining into the Ponnaiyar river lead to the alterations in the water quality downstream as evident by the increased value of alkalinity, chlorides, hardness, phosphorus, nitrogen (nitrate) and other ions in the Sathanur reservoir (Chapter 9).

Ameliorative Steps Taken by the Authorities

To check the erosion from the catchment, several silt retention dams have been constructed upstream of the reservoir. The forest officials often undertake afforestation programmes but most of them fail to deliver results because of prevailing dry and hot conditions. Only thorny, deciduous and sturdy trees with meager canopy cover survive in these conditions, as explained by the forest officials.

Section—III Impacts on Reservoir and Catchment

4

Inundation, Seismicity and Hydro-electricity

INUNDATION

Introduction

Creation of a new reservoir by damming a river may submerge large tracts of land, which might earlier be inhabited, to a smaller or greater degree. This necessitates shifting of people living in the area-to-be submerged before a new dam is commissioned. The extent and complexity of the relocation and the rehabilitation mode can vary widely from location to location. In some situations, this aspect can become one of the key elements in the popular opposition to a new dam. Likewise, the importance of the area to be inundated may vary from location to location. When such area happens to be of very great ecological significance, the popular opposition can even successfully stop the project, as happened with the Silent valley project, Kerala (CSE, 1982).

Status

The Sathanur Reservoir Project (SRP) is located in the Ponnaiyar reserve forest (Chapter 2). This entailed clearing of some portion of the forest. There was no human displacement, as the affected area had no human habitation. It may be mentioned that the period during which SRP was commissioned (1958 and 1992), much less attention was bestowed on the ecological damage caused by such inundation of reserve forest as is done now-a-days.

Impacts

Unlike several other dam-based projects where the area-to-be-inundated has human habitation and a public-owned property, necessitating complex and often litigious acquiring by the government, SRP is situated in an entirely government-owned area. There is no human habitation. The SRP occupied 1066 ha of forest land and a state-to -state transfer of 0.26 million rupees was made from the SRP authorities to the state's Department of Forests.

The cleared forestland was utilized for the following purpose:

- Major portion of the land was brought under Sathanur saddle surplus course and river course;
- Land was utilized for the development of park in the vicinity;
- An approach road from Sathanur village to the dam was constructed;
- A road to the pick-up dam was laid clearing the thick jungle;
- Land was utilized for the development of the site with all government offices and residential colony coming up at the site.

Compensatory afforestation programme was undertaken in the reserve forest from 1958 onwards. No endangered species of the flora were reported in the forest, which actually is a tropical dry deciduous forest.

For the construction of hydro-electricity project another 3.85 hectares of the forest area was cleared. The following number and species of the trees were cut.

Species	Number
Albizia amara	1480
Cassia auriculata	312
Azadirachta imitea	4
Ficus bengalansis	2
Total	1798

The officials have taken up compensatory afforestation programme involving the planting of (*Azadirachta imitea*) *Tamarindus indica* and *Acacia sp*. in the forest.

SEISMICITY

Introduction

Creation of artificial reservoir leads to the major modification in hydro-geological regime in and around the dam site. This, in turn, may bring about changes in the tectonic stress regime, causing deformation of rocks, and increasing the instability of slopes, eventually leading to seismicity, land slides, collapses and land subsidence.

More than hundred reservoirs from different parts of the world so far have been reported to have induced earthquakes (Guha and Patil, 1990) The phenomenon was first noted in lake Mead (Hoover dam), USA in 1936 (Carder, 1945). In India seismicity associated with the reservoir impoundment had remained only a subject of academic interest till the disastrous Koyna earthquake occurred (Abbasi, 2000). By now seismicity is assessed routinely in most reservoir projects worldwide and has become an integral part of standard Environmental Impact Assessment (EIA) and Environmental Management (EM) procedures.

The present study is an assessment of the potential of Sathanur reservoir in inducing seismicity and the possible risks involved.

Status

Sathanur dam is located in Zone II as per seismic zone map of India. It is not known whether the reservoir mapping prior to the commissioning of the Sathanur Reservoir Project (SRP) was accomplished, as no map is available. It was not verified whether there are any major or minor lineaments at or near the dam site. There have been no instances of tremors or earthquakes in the past as per the information provided by the governmental staff manning the SRP.

The reservoir area is in a flat terrain with a few relict hills jutting out here and there. Though out crops of boulders are seen on the surface, there have not been any instances of landslides in the region and the possibility of such slides occurring in the future are slim due to the absence of thick and steep soil slopes. Occasional release of loose bo lders is not ruled out but it will have no effect on the storage and safety of the reservoir rim and dam structure as per the assurances of the project officials.

A brief note on the pre-project investigation with respect to the seismic status of the region is presented in the following section. Sathanur dam and its neighborhood is located on the archean suite of rocks consisting of granites, gneisses and charnockites. These are all considered to be very hard and strong rocks, which form the basement of the southern peninsula. During the pre project investigation stage in 1943, geologists from the Geological Survey of India examined the dam site. The salient findings are summarized below.

Geology at the Sathanur Dam

The main rock type in the area is charnockite and its variants. The typical charnockite is blue feldspar. It undergoes metamorphosis and assumes granulitic texture in several places. The foliation strike. Is ENE-WEW with a steep dip (80°) towards S by E or SE. Intruding into these charnockites and gneisses are magnetite-quartzite bands, minor pegmatite and quartz veins. The charnockites are jointed. Except for three major sets of joint systems, no structurally weak features such as faults and major shear zones were reported. The minor sheared and weathered joints encountered at the site during the investigation were treated by backfilling with concrete as described in subsequent sections.

Pre-project Investigation in Brief

The survey consisted of drilling forty seven major test pits along the dam alignment besides several other trenches and pits over the area. The over burden varied from 0 to 6 m (0 to 20 feet). In fifty per cent of the test pits, the overburden reported was only 0 to 1.5 m (0 to 5 feet) including the ones on the riverbed where the spillways are located. About eighteen test pits had the depth of weathering ranging from 0-3 m (0 -10 feet) in the right flank. All the shear joints, soil filled joints and weathered pockets in the foundation were treated adequately with concrete. Drilling was done by means of calyx holes and wagong drill holes for consolidation as well as curtain grouting. Water tests were done prior to grouting. The intake of grout was reported to be minimal in most cases thereby indicating that the joints were tight at depths below the foundation.

Between L.S. 198.1 m and L.S. 204.2 m on the left flank, three horizontal joints were reported. Removing the weathered material treated these and backfilling with concrete and subsequent grouting through pipes embedded in the backfill concrete.

Impacts

During the inspection of the drainage gallery and the drains of the right earthen dam by a panel in 1989 seepage was found to be very less, the maximum seepage being 20.4 litres per month at full reservoir level (FRL). Less seepage was reported to be indicative of the effectiveness of the consolidation and curtain grouting and also the good quality of the masonry. No leaching of lime was reported in the rear face of dam housing the drains. The panel suggested that the seepage water from the masonry portion be analyzed periodically for hardness, particularly the lime fraction to confirm that no leaching of lime takes place at any time. The masonry dam was reported to be in a good shape by the experts. No erosion, scour or damage of any kind on the masonry portion due to the spillway energy dissipation arrangements in the main dam and the right flank was reported. It was decided during the panel review that the GSI would be approached for better understanding of the seismic status of the region. The criteria for design of solid gravity dams has undergone drastic changes since 1958, the year when the Sathanur dam was commissioned. The panel therefore had suggested that the design of the dam be rechecked as per the latest relevant Indian Standard Institution (now Bureau of India standards BIS) codes with emphasis on the seismic consideration.

The seismic status of the Sathanur reservoir is unclear, as no detailed studies with modern assessment techniques have been undertaken. The BIS has classified India into five types of zones *vis-a-vis* siesmicity (IS : 1893 – 1984) : I to V, the ascending order representing an increased risk of seismicity. The location of the Sathanur dam in the zone II reflects moderate susceptibility of the area to the seismic hazard.

HYDRO-ELECTRICITY

Introduction

India, being endowed with numerous perennial and seasonal water resources, has immense potential for the hydro-electricity development. It ranks fifth in the world in terms of exploiting non-conventional hydropower (Prasad, 1999). However the growth in the hydro power sector has not been as phenomenal as anticipated due to several constraints, the note-worthy being the apprehensions regarding its' feasibility and utility in the long run. The controversy on the ecological impacts and social justice seem to be never ending (Abbasi et al, 2000). Yet hydropower is, to this date, the only renewable energy source, which has proved to be economically and environmentally viable.

This section briefly evaluates the impact of the hydro-electricity unit of the Sathanur Reservoir Project (SRP) in terms of its potential, economy and environmental impacts.

Status

The 7.5 MW hydro-electricity unit of the Tamil Nadu Electricity Broad (TNEB) is built near the Sathanur dam at an estimated cost of Rs 300 million (Plate 4.1). It was commissioned in September '99. The construction work on the project had started in 1995-96. The salient features of the project are represented in Table 4.1 and an account of cost-benefit analysis is presented in Table 4.2. Sulzer Flovel Private Limited in joint venture with Sulzer of Germany had setup the unit.

Plate 4.1: Hydro-electricity generation unit

Table—4.1 Salient features of the Sathanur hydro-electricity unit

Power house	
Installed capacity	1 x 7.5 MW
Turbine type	Vertical kaplan
Maximum gross head	36.20 m
Design head	27.74 m
Maximum discharge	32.54 cumecs
Full reservoir level	222.20 m
Maximum tail water level	+184.95m
Minimum tail water level	+183.60 m
Keel of centerline of distributor	+183.25 m
Intake	
Number of river sluices to be utilized for power generation	2
Size of river sluice	1.524 x 1.829 m
Velocity inside the sluice barest for peak discharge	7.75 m sec^{-1}
Velocity in pen stock for peak discharge	6.63 m sec^{-1}
Penstock	
Number of penstock pipes upto 'y' junction	2
Length of penstock pipes upto 'y' junction	30 m
Maximum discharge through penstock	32-54 cumecs
Length of steel liner inside the sluice	30 m

Table—4.2 Cost-benefit analysis of the hydro electricity

Total cost of the project	Rs. 3 billion
Annual hydro-electricity generation	15.03 million units
Hydro-electricity generation expenses per unit	Rs. 1.20
Cost per unit	Rs. 1.48
Annual return	Rs. 22.2 millions
Annual expenses	Rs. 18.03 millions
Profit incurred annually	Rs. 4.21 millions

Impacts

During the construction phase of the hydro-electricity unit, the health officials had to work very hard in preventing out break of malaria in the region. This aspect has been discussed in Chapter 7. The details regarding the clearing of the forest, lands for the construction purpose have been presented in Section—4.1.

The 7.5 MW project fulfils a mere 0.2 per cent demand of the state's electricity generation (5000 MW) and 0.8 per cent of the hydro-electricity generation. The generation period is expected to last for four to five months during the time when water is released from the reservoir for irrigation purposes downstream. According to the arrangements, the water from the two of the five river sluices will run the turbine before being let downstream. The pick-up dam being situated 7 km downstream of the Sathanur dam, the warm waters exiting from the turbine shall have ample time to reach ambient temperature. There shall, thus, be no adverse impact on the aquatic flora or fauna due to water temperature. According to the officials, owing to the sound proof arrangement in the powerhouse, the noise made by the turbine won't affect the fauna of the reservoir. The six to seven months period of no generation will be devoted to the maintenance and repair works of the powerhouse. The hydro-electricity has been priced at Rs. 1.48 per unit, which is a lot cheaper than the conventional thermal electricity. The electricity will be initially supplied to meet the demands of Sathanur, Pachal and Thandarampet regions.

5

Sedimentation

INTRODUCTION

A reservoir is a veritable sedimentation tank (Abbasi, 1997a). The sediments brought in by the feeder streams and the run-off from the reservoir catchment get an opportunity to settle down in the quiescent waters of the reservoirs. This results in two impacts of far reaching consequences. The first is that the settled sediments reduce the storage. The second impact is caused when the de-sedimented water from the reservoir is released into the river down stream and in the irrigation canals. The water tends to reacquire its sediment load and erodes the banks of the river or the canal carrying it (Abbasi, 1991).

Sedimentation also affects the chemistry, biology and the thermal characteristics of the reservoir water. High concentration of suspended matters result in low primary production because of restricted light penetration. The availability of dissolved oxygen may be limited due to high sediment oxygen demand either as COD or BOD.

The importance of the sedimentation studies was felt when the reservoirs constructed in Egypt, India, Japan, China, USA, Australia and elsewhere showed a tendency to silt to varying degree depending upon a number of factors. Records show that the rate of silting in some reservoirs is alarming. For example, the

Yamoka reservoir on the Teuryh river in Japan, having an original capacity of 176.6 Mm^3 has silted to as much as 85 per cent of it capacity in just 13 years (Bhatia et al, 1993). The Nizam Sagar reservoir across the river Manjira has lost more than 60 per cent of its live storage in about 50 years of its' operation. The reservoir across the Tungabhadra near Hampi, is supposed to be fast losing its live storage as a result of which the utilization from the reservoir has fallen to 100 TMC from the original 170 TMC, i.e., nearly 35 per cent in about 45 years (Murthy 1997).

The various hydraulic aspects to be considered and determined in connection with an exhaustive study of sediment motion have been established long back and include settling velocity (Rubey, 1933), surface drag on the stream bed, bar resistance, velocity distribution on a sediment bed (Keulegan, 1938), suspension, bed load, interrelation between bed and suspended load, bed load function, tractive force equations, similarity, saltation (Bagnold, 1936), etc. Methods for conducting sedimentation surveys have been reported by Gottschalk (1952).

No two reservoirs behave in a like manner in the context of sedimentation. Very high, normal and very low rates of sedimentation are observed in reservoirs. Some of the natural factors affecting sediment yield of water sheds and reservoir sedimentation are water spread area, topography of the water shed, vegetative cover density, intensity of rainfall in the catchment and the trap efficiency of the reservoir.

Studies conducted in 1976 and 1982 on the Sathanur project had revealed that it had started to lose its' live storage capacity, though not in an alarming way. A brief account of the studies is presented in the following section. An assessment has been made to bring out the relevance of the studies in the current scenario. The eco-restoration measures adopted by the authorities to check siltation and the effectiveness of those measures have also been evaluated.

Status

Study on the sedimentation of the Sathanur reservoir is based on the sedimentation reports acquired from Institute of Hydraulics and Hydrology, Poondi, and other related data collected from

PWD, Sathanur dam. To conduct sedimentation studies in the reservoirs of the Tamil Nadu, the Tamil Nadu Government had formed a Water Shed Management Board in the year 1975. Detailed sedimentation studies of the Sathanur reservoir were conducted done in the years 1976 and 1982 with the following objectives:

1. Determining the current capacity of the reservoir and thus the loss in storage capacity;
2. Determining the rate of sedimentation and the current trap efficiency of the reservoir;
3. Finding out the distribution pattern of sediment and to prepare current depth-area-capacity relationship;
4. Determining the probable useful life of reservoir;
5. Determining the characteristics of sediment deposit;
6. Comparing the results obtained with other reservoirs.

IMPACTS

Procedures Adopted for Sedimentation Studies in Brief

It may be pertinent to briefly recapitulate the various methodologies followed for the sedimentation survey 1972 and 1986.

As a first step, for conducting sedimentation studies of the Sathanur reservoir, an accurate water spread map prepared at the time of construction, was collected. A reconnaitry survey of the whole water spread was made so as to mark a paper location of silt ranges in sufficient numbers in the available water spread map in which cross sections were taken to assess the current capacity of reservoir and thereby the volume of sediment deposited. The ends of the ranges were monumented in the field above the full reservoir level so that range monuments could be traced even after a considerable period of time. For locating the position of range pillars in the water spread map, triangulation survey of water spread connecting all the range pillars was conducted. By conducting check level survey from a permanent bench mark, the reduced levels of the range pillars were fixed. The cross section of the range lines were found out by echo sounder and ground survey by levelling depending upon the water level in the reservoir. From the cross sections, the current bed level was worked out, plotted in the map, and contours were drawn from which the current capacity was calculated by different methods.

Thus, the sedimentation survey of Sathanur reservoir, conducted in 1976 and 1982 involved the following steps:

1. Reconnaissance and layout of ranges
2. Preparation, casting and erection of range monuments
3. Check level survey
4. Plane table survey
5. Range survey – hydrographic and ground survey
6. Collection of sediment samples.

Reconnaissance and Layout of Ranges

A reconnaitry of the whole water spread covering a length of about 21 km was initially made. In the available water spread map, 28 ranges were marked on the main river course, to the extent possible parallel to each other and perpendicular to the river. The spacing between the ranges were also marked across the mouths of the important and considerable streams joining the main river, with it the water spread.

Preparation, Casting and Erection of Monuments

To identify the ranges in the field, the ends of the ranges were monumented with a permanent type of monument. As the water spread is bounded by thick and tall forest growths, the range monuments were of R.C.C 1:2:4 pillars with base concrete of 1:4:8 and considerably high above the ground. Totally 80 range pillars were pre-cast at the dam site.

Check Level Survey

Check level survey was carried out from the Public Works Department benchmark along the peripheries of both the flanks of reservoir watershed, covering a total distance of about 40 km and benchmarks were established on the range pillar bases, which were very useful for taking topographic survey along the ranges at later stage.

Plane Table Survey

In the absence of completion plan showing the correct dam alignment, plane table survey was found necessary for transferring the triangulation station points from the field to the map for the further mapping purposes. Hence plane table survey was carried out along the centre line of dam alignment starting from the saddle to the end of masonry dam and also the field triangulation station points.

Triangulation Survey

Triangulation survey is employed to accurately determine the relative positions of a system of widely separated points on the surface of the earth and also their absolute position. To determine and fix up accurately the location of range pillars in the map and to get exactly the horizontal distance between the range pillars, Triangulation survey was carried out connecting all the range pillars from a base line. Altogether 184 triangles were formed in the system of survey for fixing the accurate position of the range pillars.

Range Survey

Survey along the ranges was conducted to get the cross sections between range pillars. This consists of hydrographic survey and ground survey. Hydrographic survey, in combination with ground survey, is necessary if the survey is conducted during low water level. For the ranges where there is no water, ground survey alone suffices to get the cross sections between the range pillars.

Hydrographic Survey by Echo Sounder

Due to low water level in the reservoir during December 1976, i.e. between EI 687. 45 feet and 688.70 feet as to against full reservoir level, EL 729.00 feet, it was necessary to conduct both hydrographic and ground survey. Out of 28 ranges along the main river, 16 ranges were covered by hydrographic survey and the remaining ranges by ground survey.

Collection of Sediment Samples for Analysis

Altogether 64 sediment samples were collected along the ranges from various representative places. Normally along a range 3 samples were taken, one at centre of river course and two other at left and right side of river course. At the tail end of the river course, one representative sample along ranges was collected. These sediment samples were collected using "Grab type" sediment sampler where there was water and by digging pits where there was no water. The collected samples were sent to Soil Mechanics and Research Division, Madras, to assess the mechanical properties and agricultural properties.

Computation of Reservoir Capacity and Sediment Volume

After the preparation of contour map, map pillar location map and grid maps in different years, the capacity of the reservoir below full reservoir level and thereby the volume of deposited sediment was proposed to be worked out using the following different methods:

1. Contour area interval method
2. Modified prismoidal method
3. Grid method

Contour Area Interval Method

In this method, from the contour map already prepared to a scale of 1 cm = 100 m, the successive areas enclosed by the contours, starting form the contour, were calculated by counting the full and partial squares and converting to the scale of map. From this, the capacity between the successive unit elevations were worked out arithmatically by taking the average contour areas and multiplying by unit height i.e., contour interval. From the successive cumulative volume starting from the lowest elevation, the elevation capacity relationship was established up to the full reservoir level. The difference between the original and present capacity was the total volume of sediment deposited for intervening period. The capacities at different elevations arrived by this method are shown in Table—5.1.

Modified Prismodial Method

In this method, the volume below the lowest contour was worked out by end area method. Using the following formula, for each succeeding higher contours, the volumes were worked out as:

$$Vx = \left[\frac{2H}{6}\right] (A+4B+c)-Vy$$

where

Vx – volume between contour B&C

H – contour interval

B – area of mid surface

A – area of bottom surface

C – area of top surface

Vy – volume between contours A and B previously determined

The cumulative volumes at different area elevations till full reservoir level, worked out by this method during 1976 survey and 1982 survey, are presented in Tables—5.1 and 5.2.

Grid Method

In this method, the accuracy of the result depends on the size of the grid. As far as possible, smaller the size of grids, greater the accuracy. This method is widely followed for the following advantages:

1. By adopting this grid system, the coordinates of range pillars can be worked out which will be very much useful for plotting the range pillar location more accurately and quickly, avoiding plotting error by usual methods.
2. Only by this method details of location of silt and scour can be marked in the map while by other methods, the net effect only can be obtained.

3. For calculating the capacity by range method using general formula and by constant factor method, it was essential that the ranges in the field should be parallel, a condition which is very difficult to obtain normally.

Due to the above mentioned advantages over other methods, the capacity arrived by grid method was considered to be accurate and adopted for further analysis. However to justify the effectiveness of grid method, capacity worked out by the contour area interval method and prismoidal formula methods had been compared. The Tables—5.1 and 5.2 give the comparison of the results obtained by various methods, during the year 1976 and 1982 surveys.

Table—5.1 Comparison of area-capacity of Sathanur Reservoir by various methods, 1976.

S. No.	Reservoir water level in meters	Area in Mm²	Capacity in Mm³		
			Contour area interval Method	Modified prismoidal formula method	Grid Method
1.	186	0.0000	0.00000	0.00000	0.0031
2.	189	0.0531	0.04261	0.04261	0.0379
3.	192	0.3648	0.66946	0.66946	0.0379
4.	195	0.6993	2.26561	2.25421	1.1727
5.	198	1.4448	5.48176	5.27626	3.8470
6.	201	2.5599	11.48881	11.29261	8.7399
7.	204	3.7755	20.99191	20.73616	17.8276
8.	207	5.4113	36.77211	34.36581	28.5980
9.	210	7.4044	53.99566	53.56126	47.1110
10.	213	9.5386	79.41016	78.93337	72.4059
11.	216	12.8917	113.05561	112.01176	104.3729
12.	219	16.6797	157.41271	156.71847	150.9227
13.	222.2	20.4517	216.82295	215.40937	211.6510

Table—5.2 Comparison of area-capacity of Sathanur Reservoir by various methods, 1982

S. No.	Reservoir water level in meters	Area in Mm^2	Capacity in Mm^3		
			Contour area interval method	Modified prismoidal formula method	Grid Method
1.	186	0.0000	0.00000	0.00000	0.0031
2.	189	0.0004	0.00600	0.00060	0.0000
3.	192	0.3839	0.69585	0.66585	0.0000
4.	195	0.7778	2.43240	2.41345	0.1059
5.	198	1.7180	6.18210	5.90895	1.1860
6.	201	2.7209	12.84045	12.78410	4.7690
7.	204	3.9718	22.87750	22.48235	11.4262
8.	207	5.4582	37.02450	36.85040	23.0080
9.	210	7.6159	56.63565	55.90285	40.8612
10.	213	8.9702	81.51480	81.74240	66.2228
11.	216	12.6610	113.91160	112.06055	101.0096
12.	219	16.3475	157.47435	157.70410	146.0765
13.	222.2	20.4517	216.33947	217.55475	207.30192

Sediment Volume

The comparison of the results obtained by various methods (presented as Tables—5.1 and 5.2) illustrated that the grid method possessed certain advantages over the other methods. Thus, the capacity worked out by this method had been considered for further calculations and to work out the sediment volume. The original capacity was calculated from the contour map for the year 1957 by contour area interval method (as worked out for 1st capacity survey and the capacities arrived at by the grid method were used for the years 1976 and 1982 for further calculations.

Original capacity of the reservoir for the year 1957: 234.828 Mm^3

Capacity of reservoir for the year 1976 (1st capacity survey): 211.651 Mm^3

Capacity of reservoir for the year 1982 (2nd capacity survey): 207.302 Mm³

Loss in storage capacity

1976-23.177 Mm³

1982-27.526 Mm³

Percentage of loss in storage capacity i.e., percentage of silt deposition:

1976-9.87%

1982-11.72%

Percentage of annual silting rate for

1976-0.519%

1982-0.4688%

Total original capacity-234.8280 Mm³

Dead storage capacity-0.1232 Mm³

Live storage capacity-234.70 6038 Mm³

Loss in live storage-9.82%

Depth-Area-Capacity Relationship

Depth-wise capacities and areas for the years 1957, 1976 and 1992 are furnished in Tables—5.3, 5.4 and 5.5 and represented as Figures—5.1, 5.2 and 5.3. The capacity of the reservoir is reduced from 234.825 Mm^3 to 211.651 Mm^3 in 1976 and still to 207.302 Mm^3 in 1982. The total loss in storage capacity is 11.72 per cent and annual average loss in storage capacity is 0.4688 per cent.

Table—5.3 Area and capacity at different elevations for the years 1957, 1976 and 1982

S. No.	Contour level	Area in Mm^2			Capacity in Mm^3		
		1957	1976	1982	1957	1976	1982
1.	186	0.1060	0.0000	0.0000	0.1382	0.0031	0.0000
2.	189	0.3175	0.0531	0.0004	0.7684	0.0379	0.0000

(Table Contd...)

1	2	3	4	5	6	7	8
3.	192	0.5785	0.3648	0.3839	0.1123	0.1659	0.0000
4.	195	1.3231	0.6993	0.7778	4.9838	1.1727	0.1059
5.	198	2.0003	1.4448	1.7180	9.9795	3.8470	1.1860
6.	201	3.0516	2.5599	2.7209	17.5811	8.7399	4.7690
7.	204	4.0740	3.7755	3.9718	28.2915	17.8276	11.4262
8.	207	5.9268	5.4113	5.4582	43.3340	28.5980	23.0080
9.	210	7.2876	7.4044	7.6159	63.1865	47.1110	40.8612
10.	213	10.2333	9.5386	8.9702	89.5265	72.4059	66.2228
11.	216	13.9254	12.8917	12.6610	125.3320	104.3729	101.0096
12.	219	17.5372	16.6797	16.3475	173.1563	150.9227	146.0765
13.	222.2	21.5079	20.4517	20.4517	234.8279	211.6510	207.30192

Table—5.4 Depth-wise capacity at different elevations

S. No.	Contour level in meters	Depth in meters	Capacity 1957 in Mm^3	Capacity 1976 in Mm^3	Capacity 1982 in Mm^3
1.	183.5	0	0	0	0
2.	186	2.51	1.38×10^5	3.14×10^3	0
3.	189	5.51	7.68×10^5	3.79×10^4	0
4.	192	8.51	2.11×10^6	1.66×10^5	0
5.	195	11.51	4.98×10^6	1.17×10^6	1.05×10^5
6.	198	14.51	9.98×10^6	3.85×10^6	1.18×10^6
7.	201	17.51	1.76×10^7	8.74×10^6	4.76×10^6
8.	204	20.51	2.93×10^7	1.70×10^7	1.14×10^7
9.	207	23.51	4.33×10^7	2.86×10^7	2.30×10^7
10.	210	26.51	6.32×10^7	4.71×10^7	4.09×10^7
11.	213	29.51	8.95×10^7	7.24×10^7	6.62×10^7
12.	216	32.51	1.25×10^8	1.04×10^8	1.01×10^8
13.	219	35.51	1.73×10^8	1.51×10^8	1.46×10^8
14.	222.2	38.71	2.35×10^8	2.12×10^8	2.07×10^8

Table—5.5 Sediment deposit depth-wise distribution, 1976

S. No.	Contour level in meters	Depth in meters	Sediment deposited volume x 10^6 m^3 1976	Sediment deposited volume x 10^6 m^3 1982
1.	183	0.00	0.0000	0.0000
2.	186	2.51	0.1351	0.1382
3.	189	5.51	0.7305	0.7684
4.	192	8.51	1.9464	2.1123
5.	195	11.51	3.8111	4.8779
6.	198	14.51	6.1325	8.7935
7.	201	17.51	8.8412	12.8121
8.	204	20.51	11.2639	11.8653
9.	207	23.51	14.7360	20.3260
10.	210	26.51	16.0695	22.3193
11.	213	29.51	17.1206	23.3037
12.	216	32.51	20.9599	24.3242
13.	219	35.51	22.2336	27.0798
14.	222.2	38.71	23.1769	27.5260

Table—5.3 shows that there is a gradual increase in reduction of capacity till El + 201.00 and considerable variation up to full reservoir level. The dead storage of 0.1232 Mm^3 has been fully lost while the loss in live storage is 27.4028 Mm^3 i.e. 11.67 per cent.

Classification of Reservoir Types

Based on the analysis of 30 reservoirs in the United States of America, Boreland and Millor have classified the reservoirs into four standard types. This classification is based on the depth to capacity relationship and is presented as follows:

m	Reservoir type	Standard classification
1–1.5	Gorge	IV
1.5–2.5	Hill	III
2.5–3.5	Flood plain Foot Hill	II
3.5–4.5	Lake	I

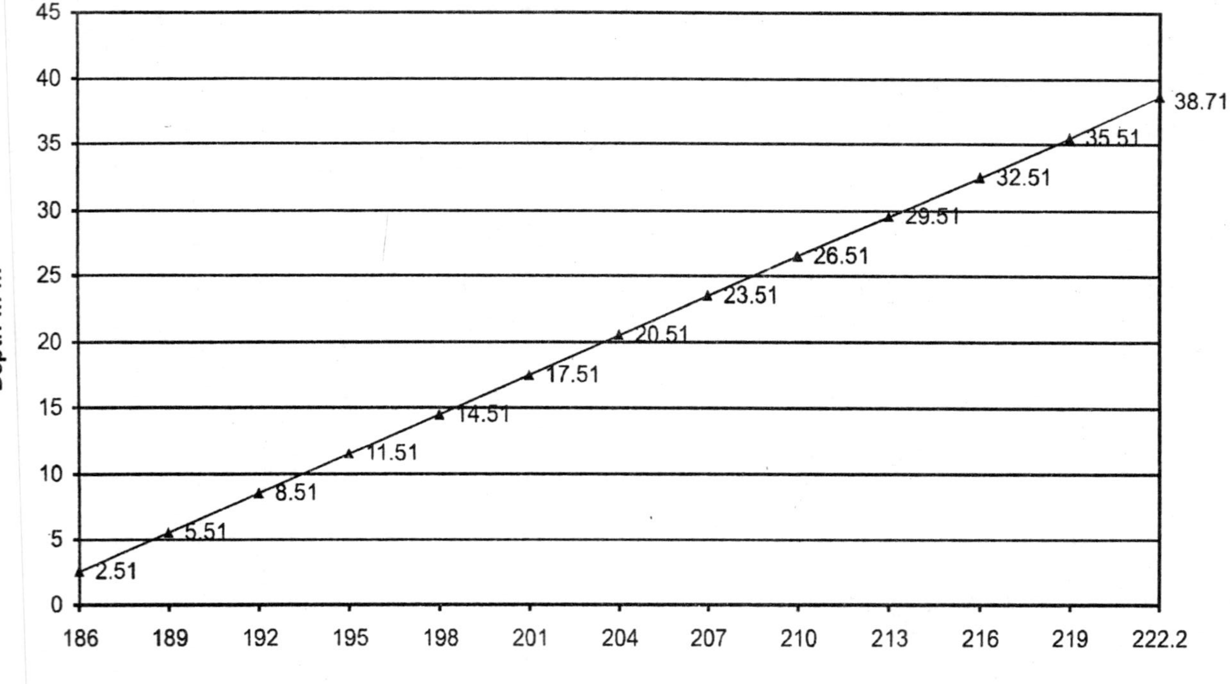

Fig. 5.1: Sathanur reservoir contour level vs depth

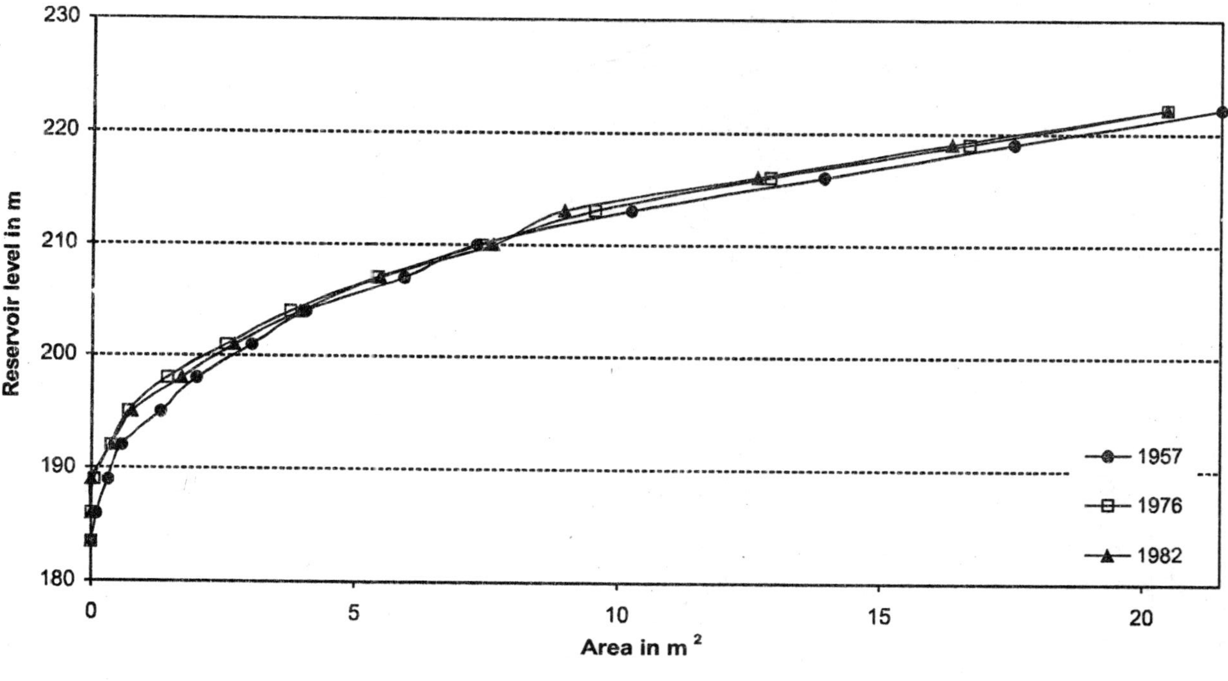

Fig. 5.2: Reservoir contour level vs area

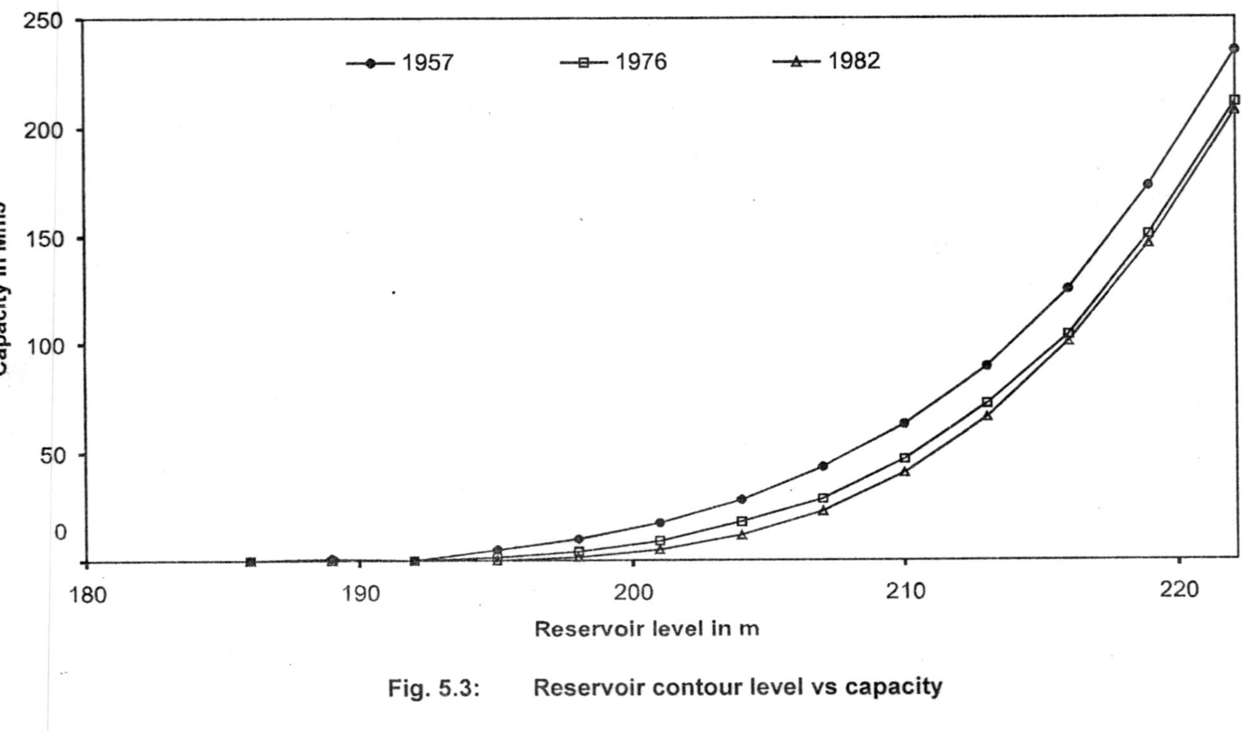

Fig. 5.3: Reservoir contour level vs capacity

Where, m is the reciprocal of the slope of the line obtained by plotting reservoir depth as ordinate and reservoir capacity as abscess in log-sheet (Figure—5.5).

Applying this classification criteria to Sathanur reservoir in the year 1957, two slopes are formed, 'm' value being 2.13 and 3.05 respectively indicating that the reservoir was in transition from the hill type to the flood plain – foot hill type; vis a vis standard classification, III to II.

For the year 1976, the values of 'm' being 3.86 and 4.27 for lower and upper elevations respectively indicate that the Sathanur reservoir has become the lake type (standard classification I). For the year 1982 also the value of 'm' was 4.76, which indicates that the reservoir has finally attained classification I (Figure 5.5).

Sediment Distribution

The Figure—5.4 depicts that at 25 per cent of depth there is 9 per cent sediment deposition (11% for 1976) and at 50 per cent of depth there is 80 per cent deposition (73% in 1976). This indicates that more of the incoming sediments are deposited at higher depths of reservoir. The data is furnished as Table—5.6.

Table—5.6 Percentage of distribution of sediment vs depth (%) above the river bed

S.No.	Contour level in meters	Depth in meters (Cumulative)	Percentage depth (Cumulative)	Sediment quantity (Cumulative)	Percentage of sediment (Cumulative)
1.	183	0.00	0.00	0.0000	0.00
2.	186	2.51	6.48	0.1382	0.50
3.	189	5.51	14.23	0.7684	2.79
4.	192	8.51	21.98	2.1123	7.67
5.	195	11.51	29.73	4.8779	17.72
6.	198	14.51	37.48	8.7935	31.95
7.	201	17.51	45.25	12.8121	46.55
8.	204	20.51	52.98	11.8653	43.11
9.	207	23.51	60.73	20.3260	73.84
10.	210	26.51	68.48	22.3193	81.08
11.	213	29.51	76.23	23.3037	84.66
12.	216	32.51	83.90	24.3242	88.37
13.	219	35.51	91.73	27.0798	98.38
14.	222.2	38.71	100.00	27.5260	100.00

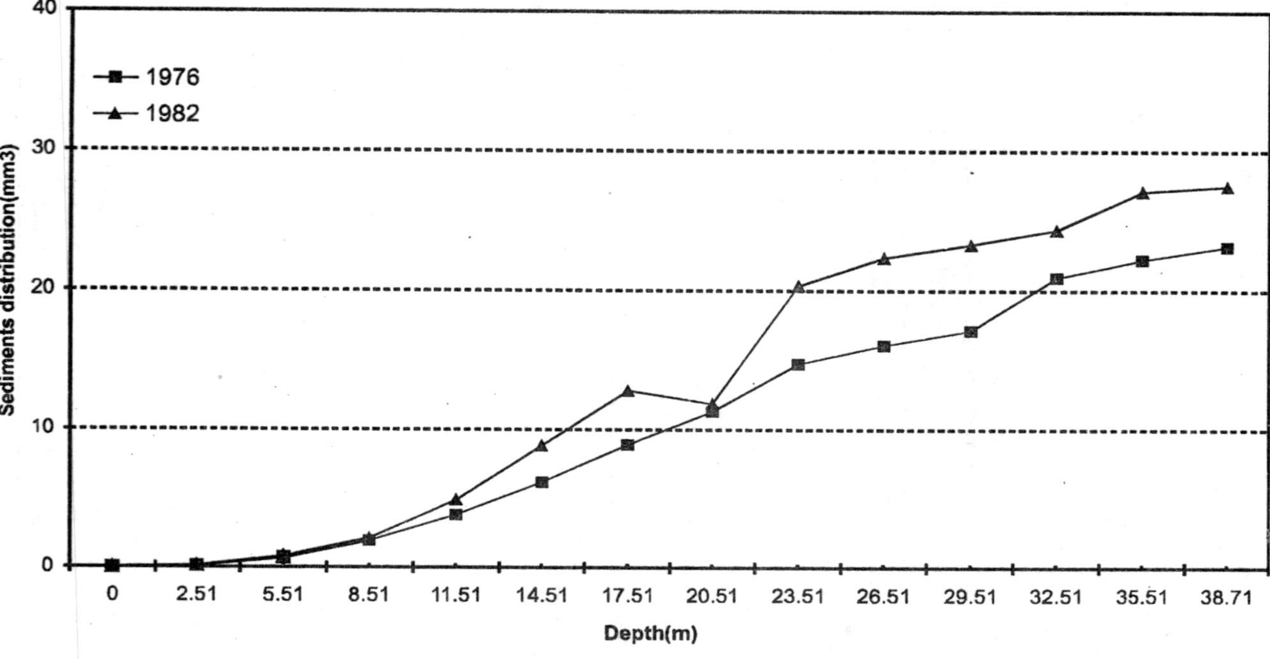

Fig. 5.4: Depth Vs sediments distribution (%)

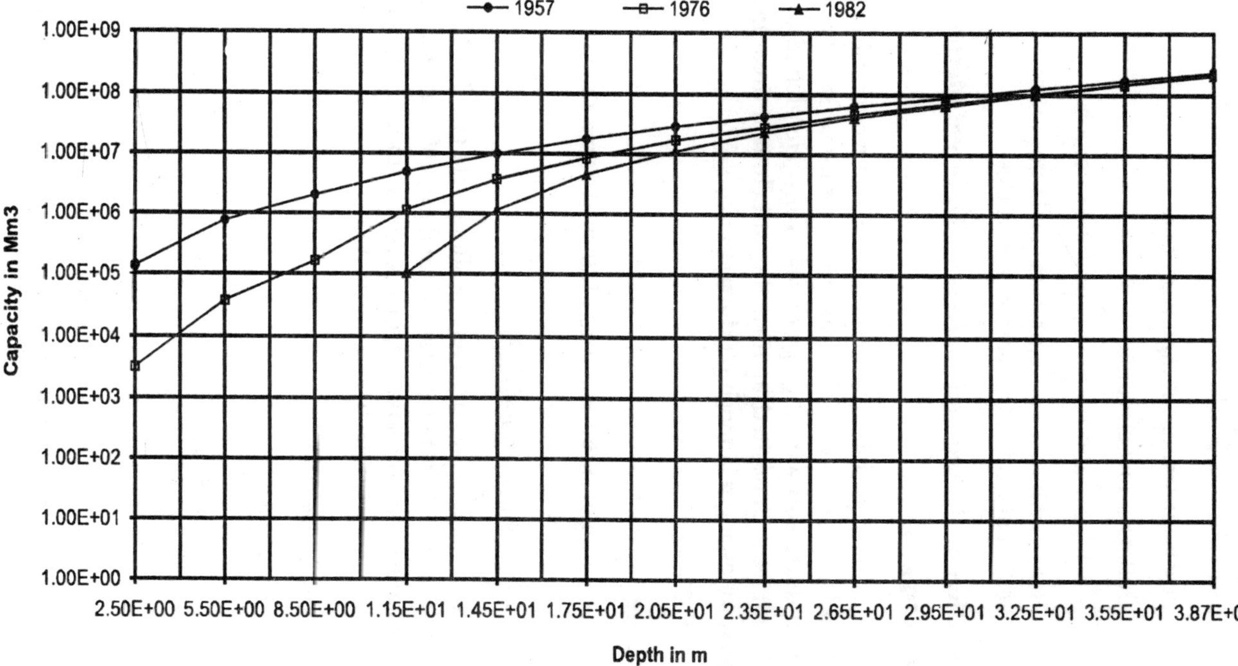

Fig. 5.5: Classification of Sathanur reservoir

The factors favouring deposition in the higher depths of the reservoir were considered to be:

1. Reservoir storage at high elevation
2. Outlets at low elevation
3. Much vegetation at the head of the reservoir
4. Heavy sediment concentration and large portion of the sediment load of the sand size particles

Computation of Sedimentation Rate by Different Methods

Khosla's Method

Khosla analyzed the data from various reservoirs in India and abroad and observed that the annual rate of sediment deposition decreased with the age of reservoirs. He plotted the annual sediment deposited against the cathchment area and suggested following relationship (Mutreja 1985).

$Qs\text{-}0.323/A^{0.28}$

Where

Qs - annual silting rate from 100 km^2 of water shed areas

A - catchment area (km^2)

Based on his findings an average silting rate of 0.036 Mm^3 100 km^{-2} at catchment area was adopted for design of reservoirs in India. Sathanur reservoir data, when assessed with this method, shows less extent of silting: 0.02 Mm^3 100 km^{-2}.

Hachiro Kira's Method

Kira's Formula is:

$$Ch_i = 0.214 \left[\frac{C^{0.473}}{I} \right]$$

Where

C_{hi} - sediment rate in per cent

C - original capacity in Mm^3

I - average annual inflow in Mm^3

When applied to Sathanur reservoir, it yields the sedimentation rate

$$\text{Chi} = 0.214 \left[\frac{234.828^{0.473}}{577.431}\right] = 0.3275\%$$

Trap Efficiency Estimation

Trap efficiency was estimated by different methods and they are presented as follows:

(a) GottaChak's Curves

According to this method, the trap efficiency of Sathanur reservoir which was 89 per cent in 1957 is reducing by 1 per cent every 12 years.

(b) Brune's Curve

Brune has related the trap efficiency of storage type reservoir to the rate of capacity to annual inflow after analyzing data from 44 reservoirs. According to this curve, trap efficiency of Sathanur reservoir which was 95 per cent in 1957 was reduced to 94.3 per cent in 1976 and by the same extent in 1982.

Life of Reservoir as Computed by Various Methods

Hachiro Kira's Method

The equation using capacity–inflow ratio developed by Hachiro Kira is given as

$$\text{Ys} = 467 \left[\frac{(\text{C})^{0.473}}{\text{I}}\right]$$

Where,

Ys - average number of years during which the silt fills up the reservoir

C - original capacity of the reservoir

I - average inflow into the reservoir

According to this method the life of the Sathanur reservoir is about 305 years.

Taylor Method

The equation given by Taylor to find out the capacity of reservoir after 'n' years of life:

Vn - VR^n

Where,

Vn - capacity after n years

V - original capacity of the reservoir

R - ratio of the reservoir capacity at the end of one year to that of the previous years which is assumed to be constant

Accordingly, capacity after 100 years is worked out to be 146.6055 Mm^3 which is 62.43 per cent of its original capacity.

Similarly, capacity after 200 years has been computed to be 91.527 Mm^3 which is 38.98 per cent of its original capacity. Normally, the useful life of a reservoir is taken till its capacity is reduced to about 30 per cent of it original capacity. The useful life of the Sathanur reservoir, before its capacity recluses to 30 per cent, has been computed to be 255.56 years.

Varshney's Method (Varshney, 1986)

Varshney has analysed the data of Indian reservoirs and given the relationship among capacity inflow ratio, trap efficiency and storage loss in a tabular form as presented in Table/15.7.

For a capacity-inflow ratio of 35.56 per cent nearer value of storage loss from the Table—5.7 is 0.5 per cent, which is very close to the loss of capacity actually worked out 0.4680 per cent. According to this method the life of the Sathanur reservoir is 200 years i.e. 100/0.5. But considering the reservoir will not serve its purpose if the capacity is reduced to 30 per cent of its original capacity, the life of Sathanur reservoir is 140 years 70/0.5.

Trap Efficiency Ratio for Different Reservoir Capacities

In this method, for different capacities, the capacity-inflow ratio and hence by the use of Brune's curve, trap efficiencies are determined. Using the average trap efficiency, annual sediment load trapped is calculated; dividing the value interval (reduction in volume by the annual sediment load trapped) will give the number of years required to fill up this volume interval. Likewise all the volume interval years are summed up to get the total life of reservoir. The average annual inflow of silt of Sathanur reservoir is taken as 1.1633 Mm^3 and average annual sediment deposition is 1.1010 Mm^3.

Table—5.7 Relationship between capacity inflow ratio, trap efficiency and storage loss

S. No.	Capacity inflow in %	Trap efficiency in %	Annual loss of reservoir capacity in %
1.	2	45.0	–
2.	10	87.0	–
3.	20	93.0	3.0
4.	50	97.0	0.5
5.	100	98.0	0.14
6.	200	98.3	0.035
7.	500	98.3	–

The average annual inflow is taken as 577.431 Mm^3 and useful life of reservoir is worked out on the assumption that the reservoir will not serve its purpose when the capacity is reduced to 30 per cent of its original 1976 capacity,

Average annual inflow of silt 1.289 Mm^3

Average annual inflow (19 years) 596.422 Mm^3

According to this the life of the Sathanur reservoir is 153 years as depicted in Table—5.8 and represented in Figure—5.6.

As per the 1982 sedimentation survey, the annual loss of reservoir capacity is 0.4988 per cent. The trap efficiency of a reservoir decreases with age, as the reservoir capacity is reduced. Thus, the percentage loss of storage capacity will also be reduced with age (Table—5.9). Even then the life of the Sathanur reservoir has been calculated to be 140 years keeping the fact in view that he reservoir will not serve its useful purpose after 70 per cent loss in its original capacity.

The 1976 survey had established the life by the similar method to be 135 years.

Characteristics of Sediment Samples Analysed

The results of the analysis of sediment samples have indicated that the sediments up to 11.06 km from dam are loamier type followed by silt-loamy in some places. Beyond 11.06 km, up to the tail end the sediments are sandy. These facts are also in correlation with the results obtained by mechanical analysis. The samples were found to display the following agricultural properties:

Electrical conductivity	-	normal
pH value	-	normal
Nitrogen	-	50 per cent samples depicted low values and another 50 per cent high.
Phosphorous	-	high
Potash	-	medium

Table 5.8 Computing life of Sathanur reservoir by various methods

S. No.	Capacity in Mm³	Capacity inflow ratio	Trap efficiency in %	Average trap efficiency in %	Annual Sediment trapped in Mm³	Reduction in volume in Mm³	Years to fill
1.	234.828	0.40667	95	–	–	–	–
2.	211.651	0.367	94.29	94.625	1.100	23.177	21.07
3.	175.000	0.363	93.93	94.110	1.095	36.651	33.47
4.	150.000	0.260	93.21	93.570	1.088	25.000	22.98
5.	125.000	0.216	92.41	92.675	1.078	25.000	23.19
6.	100.000	0.173	91.43	91.785	1.068	25.000	23.41
7.	70.450	0.122	87.14	89.285	1.039	29.550	28.44
						Total	152.56

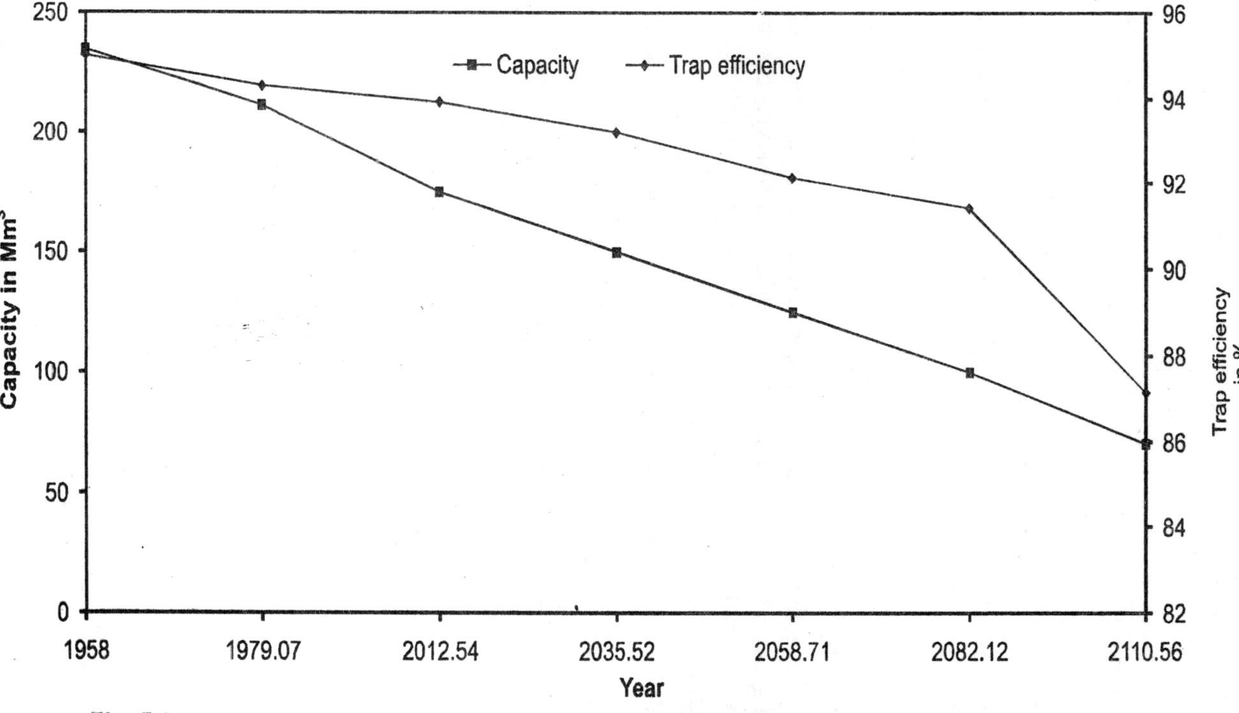

Fig. 5.6: The life of Sathanur reservoir estimated by Trap efficiency method

Table—5.9 Loss of annual capacity in reservoir

Range of annual loss of Capacity	Number of reservoirs within range	Range of data length
Less than 1%	6	10-56
1 to 5%	19	10-81
More than 5%	19	6-73

Comparison of Sathanur Reservoir Sedimentation to that of Krishnagiri Reservoir

Krishnagiri dam was the first dam constructed across Ponnaiyar river in the Ponnaiyar basin and is situated 112 km (70 miles) upstream of Sathanur reservoir project.

From the Table 5.10 it can be concluded that the independent and gross catchment of areas of the Krishnagiri reservoir are the same i.e. 5428 km^2 whereas Sathanur reservoir has an independent catchment of 5397.55 km^2 and gross catchment of 10826.0 km^2.

Table—5.10 Comparison of specific erosion of Sathanur catchment with that of Krishnagiri catchment

Name of reservoir	Catchment area	Original capacity in Mm^3	Present capacity Mm^3	Loss in capacity in 19 years Mm^3	Percentage loss in capacity	Percentage average annual rate of siltation	Specific erosion in mm m^{-2}
Krishnagiri Gross & Independent	5428.6 km^2	66.0	54.26	11.74	17.79	0.94	2.16
Sathanur							
1. Independent	5397.55 km^2	234.828	211.651	23.177	9.87	0.519	3.23
2. Gross	10826 km^2						

The loss in the capacity over a period of 19 years in the Krishnagiri reservoir is 11.74 Mm^3 while that of Sathanur reservoir for the same period is 23.177 Mm^3. Since the silt produced from the gross catchment of Sathanur reservoir is trapped in both the reservoirs so the volume of silt produced from the catchment of

Sathanur is actually the sum total of the volume of silt deposited in both the reservoir. Therefore the specific erosion from the gross catchment of Sathanur reservoir is calculated as 3.23 mm m^{-2} and that from the Krishnagiri reservoir is 2.16 mm m^{-2}. Thus it can be inferred that the catchment of Sathanur gets more silt than that of Krishnagiri catchment.

It is generally believed that most of the silt carried by the river gets trapped in the upper reaches of the reservoir, but in the case of Sathanur reservoir, which is downstream of the Krishnagiri reservoir, it gets more silt than the Krishnagiri reservoir. This phenomenon stems from the land use pattern of the Sathanur catchment which calls for an extensive soil conservation measures in the area.

Salient Findings of the Studies

1. The current capacity of Sathanur reservoir as on 1982 survey has been reduced from 234.82 Mm^3 during the year 1957 to 207.302 Mm^3. The loss in total capacity over 25 years is 27.526 Mm^3 i.e. 11.72 per cent. The average annual loss in capacity is 0.4688 per cent Average annual silting rate is .217 mm/100 sq. km. which is less than the silting rate adapted for design purposes in India i.e. 0.036 mm/100 sq. km.
2. Classification of Sathanur reservoir with standard classification of reservoir indicates it to be lake type (Type-I) while based on sediment distribution pattern to be flood plains foot hill type (Type II).
3. Sediment deposition is more in the higher depths than in the lower depths. At 50 per cent depth 12 per cent of deposition and at 75 per cent depth 80 per cent of deposition are seen.
4. Capacity-water shed ratio has decreased from 43.5x1000 $m^3 m^{-2}$ to 39.2x1000 $m^3 km^{-2}$ over 19 yrs. in 1977 and still reduced to 38.080x1000 $m^3 m^{-2}$ in 1982 over 25 years.
5. Trap efficiency estimation by several methods does not show good correlation. According to Brune the trap efficiency which was 95 per cent during 1957 has reduced to 94.3 per cent during 1976 and was constant in 1982.

6. The usual life of reservoir also estimated by different methods does not depict show good correlation. By trap efficiency method, the reservoir life is worked out to be 153 years.
7. The rate of silting has been the same from the time of completion of reservoir.
8. Comparison of specific erosion of Sathanur reservoir with that of Krishnagiri reservoir reveals that reservoir gets more annual silt because of its location and specific land use pattern.

Present Scenario

Central Water Commission (CWC) Guidelines

After considerable discussions and deliberations, the water planners in India reached a consensus that the reservoirs do not have a single well-defined life (CWC, 1991). They show a gradual degradation of performance without any sudden/non functional stage. Thus, sedimentation, hence consequent reduction in the capacity of the reservoir, is a gradual process which occurs in the following phases:

Phase I	The reservoir show no adverse effects and is able to deliver the full planned benefits;
Phase II	The reservoir delivers progressively smaller benefits, but its' continued operation for the reduced benefits is economically beneficial;
Phase III	The sedimentation causes difficulties in operation such as jamming and passage of flow in canals or turbines;
Phase IV	The phase III difficulties become so serious that the operation becomes impossible;
Phase V	The benefits are reduced to such an extent that it is no longer beneficial to operate the reservoirs.

Indian Planning for Reservoir Sedimentation

The earlier assumption by the planners and engineers of the water reservoir projects that sediment would settle within in the dead storage, provided at the time of construction, was not supported by the experiences of the performances of the dams in India and elsewhere. The results indicating considerable differences from the earlier computations started flowing in by 1965. After 1965, the Central Water Commission (CWC) started insisting that the sediment inflow rates be based on the basis of reservoir survey data. It also brought out the need for distributing the sediments , uniformly throughout the reservoir. However no clear guidelines were given.

At around 1974, it was decided that the 50 years sedimentation status should be used in the simulation of sedimentation pattern. In 1982, CWC made mandatory to the state governments to adopt the above said procedures for sedimentation surveys of the major, and medium reservoir projects, in its report of the working group on the guidelines for the preparation of detailed project report of major and medium irrigation projects. For very serious cases of siltation, re-distribution and re-estimation of trapping efficiency in 10 years block was indicated.

In 1987, CWC made all the regulations a national practice (CWC, 1987) where the general guidelines and the concept of multiple life related terms for an efficient management of water resources projects was spelt out.

Management Issues

Siltation studies of the 44, Indian reservoirs undertaken by the CWC indicate that 43 per cent have serious problems another 43 per cent have significant, if not very serious, problems. Only 14 per cent of the reservoirs do not have any significant siltation problem. These studies illuminated the following significant points:

1. For a majority of reservoirs the rate of sedimentation has been more than the assumed at the time of planning;
2. No large reservoir in India has completed its feasible service time;

3. For many Himalayan streams, which carry very heavy loads of sediments, planning of the project with a feasible service time of 75 or 100 years becomes difficult. For hydro-electric projects in particular, it is possible to repay the development costs in few years, and a project can be planned effectively for a shorted period. In Pakistan for example, the Tarbela project has perhaps been planned to lose most, of its capacity in about 50 years;

4. A large number of hydro-electric and even irrigation projects are planned as pondages where the capacity inflow ratio or detention period can be of the order of a few days to a month. For many such projects, most of the capacity is against crest gates. There is a belief amongst planning engineers that for such structures, where the gates would be kept open during the high flow-high sediment inflow period, no sedimentation would occurs above the crest of the gates. Although there is enough empirical evidence to indicate that sedimentation does occur above the crest level, simple methods to indicate the new regime of the lives upstream of the dam, and the ultimate pondage available for regulation in spite of sedimentation, are not available. For the Gauriganga Hydro-electric project in the Himalayas, India, for which the capacity is of the order of 5 million m^3 with a sediment inflow of 3.2 million m^3 a general stabilization of the million m^3 was seen after 10 years.

Sathanur Reservoir

The dead storage space provided in the Sathanur reservoir was a meager 0.1232 Mm^3 i.e., merely .05 per cent of the reservoir capacity considering the fact that at the time of design of reservoir taking 100 years as the life of the reservoir dead storage space equivalent to the estimated silt deposition for the entire period is provided. From the project report it is understood that there was a proposal for providing five numbers of high level sluices at El + 208.788 m in addition to the sluices, at present functioning at El + 165.928 m in which case the dead storage capacity would have been higher. These high level sluices were not a part of final construction for some unknown reasons.

Trap efficiency of the reservoir is defined as the ratio of sediment retained in the reservoir to the sediment brought in by the river i.e. the percentage of incoming sediment trapped (Ponce, 1990). Trap efficiency is a function of the ratio of storage capacity to inflow. To a lesser degree, it is also a function of sediment characteristics such as texture, density currents, shape of reservoir, method of reservoir operations etc. As the reservoir is filled with sediments its' storage capacity and hence the capacity-inflow ratio decreases with time. Sathanur reservoir confirms to the pattern as the capacity-inflow ratio and hence the trap efficiency marks a gradual decrease over a period of 25 years. The capacity inflow ratio of the Sathanur reservoir is also very less (0.35) which points towards the fact that it is a seasonal storage reservoir. Since capacity to inflow ratio, indirectly provides an index of the residence time of the sediment laden water in the reservoir (Reddy, 1998), small capacity-inflow ratio ensures that much of the sediment will be discharged over the spillways.

The average silting rate of Sathanur reservoir is 0.022 Mm^3 100 km^{-2} as worked out by Khosla's discussed earlier. Based on Khosla's analyses of the sedimentation data of various reservoirs in India and abroad and his subsequent findings, an average silting rate of 0.036 Mm^3 100km^{-2} at catchment area was adopted for the design of reservoirs in India. Compared to it Sathanur reservoir depicts less value of siltation rate

Kira's method computes the siltation rate of Sathanur reservoir as 0.3275 per cent, which is 31.02 per cent less than the observed value of 0.4688 per cent. According to the CWC guidelines, the silting rate of Sathanur reservoir at 0.46 per cent can be bracketed as 'significant but not serious'. Splash erosion and overland velocities are greater on steeper slopes than on flat slopes (Reddy, 1998). Sathanur catchment is surrounded by sleep hillocks with meager, thorny vegetation cover provides ample scope for erosion. The vegetation comprises of dry, deciduous trees because of the extreme hot climate. Prolonged drought periods because of scarcity of rains, have the degraded reserved forest of the Sathanur catchment to a considerable extent. Coupled with this the poor cropping practices upstream, cattle grazing in the absence of enough pasture land cutting of trees provide an agreeable environment for specific erosion which is observed to be 3.23 mm m^{-2}.

The rate of sedimentation observed for the period of 1976-82 was comparatively less. One of the reasons for there may be attributed to the declining trend in the rainfall received in the catchment as inferred from Figures 2.8 to 2.14, Chapter 2.

Total Solids

Reservoir total solids are on a constant increase for the past years as indicated in Figure 9.6 (Chapter 9). Thus the reservoir is still getting silted. A comparison with some other reservoirs (Table—5.11 and 5.12) points out that sedimentation rate of Sathanur reservoir is comparable to that of several others. This points to the generally distressful condition of Indian reservoirs vis a vis siltation.

Eco-restoration Measures Adopted to Bring Down Sedimentation

Since not much can be done to lower the trap efficiency of the reservoir, the only course to decrease the rate of reservoir sedimentation is to adopt effective soil conservation measures. It has been established that the major portion of the silt received in the reservoir is due to the practices upstream of the catchment.

Soil Conservation Measures taken for the Catchment

Afforestation

Afforestation programmes in the form of planting various economical tree species, with large canopy cover such as *Tamarindus indica, Azadirachta imitea* have been taken up by the forest officials. Cutting and felling of the trees is prohibited and the cases are booked under prevailing forest laws.

Silt Detention Dams

Silt detestation dams in the form of check dams have been built upstream of the Sathanur reservoir to check the flow of sediments into the river (Plate—5.1).

Table—5.11 Comparison of reservoirs based on sedimentation surveys

S. No.	Reservoir	Catchment (Km²)	Live storage capacity (Mm³)	Year of survey	Period of record (Years)	Annual storage loss (%)
1.	Mangalam	48.55	24.67	1985	29	0.30
2.	Peruvannamuzhi	108.8	113.45	1986	13	1.66
3.	Malampuzha	47.63	226.96	1985	22	0.16
4.	Peechi	107.00	109.00	1985	26	0.97
5.	Neyyar	140.00	101.15	1997	38	0.71
6.	Sathanur	5397.5	207.3	1982	24	0.50

The fact that rate of siltation of the Sathanur reservoir is comparable or higher than many other Indian reservoirs, points towards the need for the intensification of the control measures stated above. The studies also reveal that even as these measures have controlled siltation to some extent. There is an obvious need to implement these measures much more effectively in future because the rate of siltation of the Sathanur reservoir is still substantial.

Table—5.12 Comparison of reservoirs based on sedimentation survey

S.No	Name	Year of construction	Catchment area (km²)	Gross capacity (Mm³)	Sedimentation rate (Mm³/year)	50% capacity lost (year)	Life of reservoir (years)
1.	Sriramasagar	1970	91750	3172	0.62	1998	56
2.	Nizamsagar	1930	21694	841	0.64	1960	61
3.	Matatila	1956	20720	1133	0.44	2018	124
4.	Hirakud	1956	83395	8105	0.66	2030	147
5.	Girna	1965	4729	609	0.80	2045	161
6.	Tungabhadra	1953	28179	3760	1.01	2019	132
7.	Panchet hill	1956	10966	1497	1.05	2021	130
8.	Bhakra	1958	56980	9800	0.60	2101	287
9.	Maithon	1955	6294	1369	1.43	2031	152
10.	Lower bhavani	1953	4200	931	0.44	2205	504
11.	Mayurakshi	1954	1860	608	1.63	2054	201
12.	Gandhisagar	1960	23025	7740	0.96	2135	350
13.	Koyna dam	1961	776	2988	1.52	3228	2533
14.	Sathanur dam	1958	5397.5	234.8	1.6	2082	153

Plate 5.1: Check dam to control soil erosion

6

Recreation

INTRODUCTION

The artificial lakes and hydraulic structures created for the water resources projects are potential recreational sites. The Bhakhra Nangal project (Punjab, India), Hoover dam (USA), Aswan dam (Egypt) and scores of other dams attract millions of tourist's worldwide. The Vrindavan Gardens (downstream of Krishnaraj Sagar reservoir, Karnataka, India). Jaikwadi Gardens (Jaikwadi project, India) and Ukai Tourist Resort (Ukai dam, Gujarat, India) represent well-known dam-based recreational sites in India. Avifauna and wildlife has been reported to increase in several cases. The phenomenon was witnessed after the completion of artificial lakes by Ramganga dam (U.P, India) Rihand dam (U.P, India) and Matatila dam, U.P, India (Goel and Maitra, 1992).

Whereas Water Resources Projects (WRPs) have the potential of boosting the tourism industry, thereby contributing significantly to the regional or national economy, their recreational and aesthetic resources are sensitive to human disturbances. Without a harmonic balance between the volume and type of tourist activity on one hand and the sensitivity and carrying capacity of the resources on the other, the recreational side of the project may prove to be environmentally harmful and, in the long-run, economically self-defeating.

Among several reports on the impact of tourism on WRP, (Abbasi, 2000) we may mention the report of Tiegland (1999) who has reported on the manifested impacts of tourism due to hydroelectric project and a major road development project when the environmental quality of the area began to deteriorate due to the access to the remote area tourism development generated. A well-planned eco tourism project combines the essentials of conservation of natural sites with economic and recreational benefits (Lindberg, 1991).

An assessment of the tourism potential of the Sathanur Reservoir Project (SRP) is presented in this chapter. Its contribution to the regional economy is also evaluated.

STATUS

Sathanur Dam—a Tourist Spot

Sathanur dam and reservoir are located in a sparsely inhabited natural terrain providing an attractive site for tourists to visit. The dam is flanked on either side by wooded hills presenting a picture of serenity and tranquility. Due to these reasons Sathanur has become one of the favourite getaway spot for the people of the nearby bustling towns and cities.

Not as easily visible to the tourists as the trees (predominantly *Azadirachta imitea, Accacia nilotica, Albizia amara* etc), but extremely important, is the wildlife sheltered by the Sathanur catchment: deer, boar, tigers, rabbits, snakes, lizards, etc.

Main Tourist Attractions

1. *Sathanur dam:* The 40 m high Sathanur dam presents an imposing example of engineering skills, both formidable and admirable. The spillways over which the water cascades down present a hypnotic sight (Plate—6.1).

2. *Ponnaiyar reserve forest:* The hills surrounding the dam, resplendent with the greenery of the flora of Ponnaiyar reserve forest, give a majestic as well as a serene look to the place. Two temples situated on the hills beckon the faithful ones. The reserve forest provides an opportunity to the natural scientists to study bio-diversity.

Plate 6.1: Sathanur dam with water gushing down the spillways

Plate 6.2: Dam surroundings developed into a beautiful garden

3. *Scenery and park:* Plate—6.2 illustrates the lush green grass, grown and maintained by the dam authorities, with beautifully crafted statues, stairs and swings placed in the back drop of reservoir.

4. *Crocodile farm:* A crocodile farm is maintained inside the dam by the Tamil Nadu Forest Department and has mostly the Indian mugger (*Crocodylus palustris*) species, the fresh water crocodiles. The crocodiles were brought to the reservoir from the Crocodile Rehabilitation Centre, Hogenekal, Tamil Nadu. Eggs are taken from the reservoir and incubated in the farm. Every alternate day, the crocodiles are given a feed of fresh meat, procured from the nearby villages. The farm is a major tourist attraction at the dam site as tourists throng to watch these majestic formidable reptilian creatures tamed by man and put in captivity (Plate—6.3).

5. *Swimming pool:* A swimming pool (Plate—6.4) has been constructed inside the dam premises for the swimming and diving enthusiasts. One needs to hardly point it out that the pool is very popular especially during the summer.

6. *Ornamental fishes exhibition:* Approximately thirteen species of ornamental fishes procured from Chennai have been kept for exhibition inside the dam premises. The species are sucker cat, colour gold, blue gold, black m.t, scat, angel, blue moop, owratas, red castle, flying saucer, belloon, skel, and light houser (Plate—6.5).

7. *Mini-zoo:* There is a mini zoo at the entrance of the Sathanur dam, which displays monkeys, deer and other animals in captivity.

IMPACTS

Management and Revenue Generation

The Public Works Department (PWD), Government of Tamil Nadu, in association with Tamil Nadu Forest Department, solely manages the tourism part of the SRP and Tamil Nadu Fisheries Development Corporation. Lawns are by and large kept well

Plate 6.3: Crocodile farm

Plate 6.4: Swimming pool

Plate 6.5: Ornamental fishes exhibition

maintained and clean. Water in the swimming pool is changed once every fifteen days. The solid waste generated by the tourists is incinerated. The drinking water facilities are adequate. Figure 6.1 gives the details of the revenue generated (based on the entrance to the dam over the past 10 years. The revenue fluctuates within 0.2 to 0.3 million rupees. The revenue is charged from the tourists in the form of entrance fee to the dam premises (Rs.0.50 per person), and to the fish exhibition and crocodile farm. The annual revenue generated from crocodile farm is around Rs.70-80 thousand which is enough for the upkeep and maintenance of the crocodiles as reported by the TNFDC officials. The major chunk of the total revenue generated goes towards the maintenance of the dam premises as a whole. The Sathanur dam has not been taken up by the Tamil Nadu Tourism Development Corporation and is completely under the control of PWD, Sathanur dam.

Public Opinion

Fifty tourists, including a few foreigners, were interviewed by us to assess the tourists' views on the present status of the SRP. Suggestions were sought towards the betterment of the Sathanur dam as a tourist spot.

Seventy five per cent of the people visiting the Sathanur dam are generally the rural folks from in and around the place; mostly from Thiruvannamalai, Thirupathur, Villupuram, Cuddalore, Gingee, Pondicherry etc. Educational trips are also arranged for school and college students to the dam site. Since there is no other recreation source in the nearby areas, the dam proves to be a major tourist destination. Approximately 0.6 million people visit the dam site every year (Figure—6.1). A very slight declining trend in the influx of the tourists is observed from the Figure.

The general public opinions were:

- Though Sathanur dam is attractive for sightseeing certain infrastructure facilities are lacking especially adequate transportation, eateries and hotels. There is no facility at all for transportation from the entrance gate to the dam site. The uphill journey proves tiresome to several tourists.

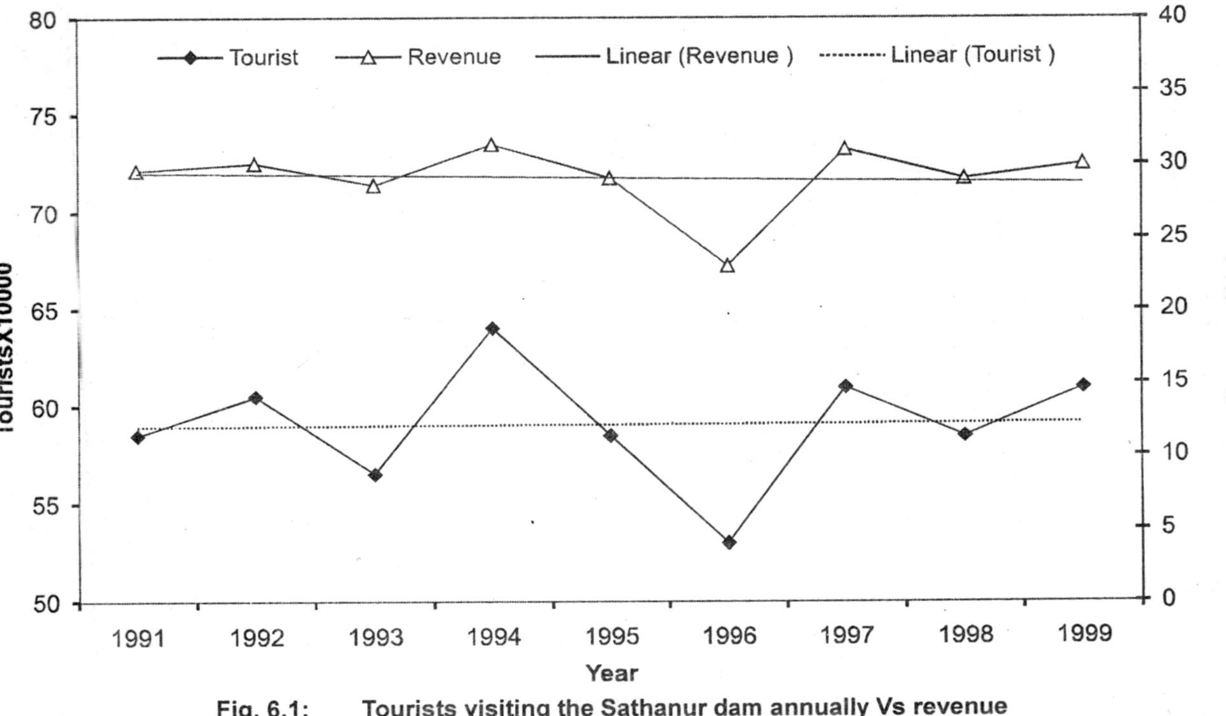

Fig. 6.1: **Tourists visiting the Sathanur dam annually Vs revenue**

- There is no provision of recreational boating in the reservoir.
- There is a small hillock in the middle of the reservoir. It can be developed into an attractive visiting spot for the tourists boating in the reservoir.
- Water sports and other fun games should be introduced for the children.
- More species of animals and birds should be displayed in the mini-zoo.
- Park and lawns need more attention in terms of cleanliness and infrastructure.
- As of now, in the absence of boating and other activities, which require time, a typical tourist visit to Sathanur gets over in a matter of a few hours.

Officials' Comments

Based on the comments of the general public, follow up visits were made to the concerned authorities and public demands were presented to them. According to them:

- Based on the inflow of the tourists and the carrying capacity of the dam site the transportation facilities have been developed;
- Steps are being taken by the management to introduce more of the attractions and to develop and better the Sathanur dam further;
- The officials have barred boating in the reservoir because the water level of the reservoir peaks at 119 ft. during monsoon and storage months and it is never allowed to go down below 46ft. So it is risky boating in this deep reservoir. The risk from crocodiles in the reservoir is also always there.

7

Health

INTRODUCTION

The assessment of the impact of major Water Resources Projects (WRPs) on the human health has become one of the essential components of the overall Environmental Impact Assessment (EIA) procedure. This has been prompted by several reports implicating some of the major WRPs in the outbreak of epidemics or in aggravating the risk of water borne diseases in other forms. Malaria, filiariasis and schistosomiasis are known to have proliferated following water resources development in several areas (Abbasi, 2000). Creation of some 104 small dams in Ghana, led to the prevalence of *Schistosoma haematobium* infection in the local population tripling from 17 per cent to 51 per cent in 38 surveyed areas (WHO, 1993). Incidence of schistosomiasis was also reported around the Aswan dam in Egypt (Shalaby and White, 1988). In Nagarjuna Sagar dam area, A.P. India, fluorsis was found to be endemic and the reason was attributed to high percentage of fluoride in the reservoir water (Varshney, 1986). The dam has also been implicated in the 'knock-knee' disease, called *genu velgum* in the Telugu language (CSE, 1982 Abbasi, 2000), which was believed to be caused by molybdenum. The metal is believed to have reached the humans in large quantities via water and food that became enriched with molybdenum after water logging and resulting anaerobic conditions had caused luxury release of the metal from the soil to overlying water and crops.

In this chapter we have assessed the possible role of Sathanur reservoir, under the current environmental and geo-climatic settings, in exacerbating the hazards of malaria and elephantiasis. These two diseases are known to be caused by the mosquitoes in the region. The study also discusses the preventive measures taken by the authorities in reducing the incidence of water-borne diseases in the reservoir area.

STATUS

The information on the incidence of malaria in the population residing by the dam site and in the village nearest to it, over the period 1994-1999, is presented in Tables—7.1 and 7.2 respectively. Data and other related information were collected from the office of the Directorate of Public Health Services, Thiruvannamalai. The malarial cases involving long term residents, originating solely due to the environmental conditions prevailing at the site, have been tabulated as 'local' and those involving migrants either labourers who come to the area for a short period for work or the fishermen employed with Tamil Nadu Fisheries Development Corporation (TNFDC) have been tabulated as 'imported'.

The population as per the latest census records of the health department has been put as 'indicative population'. There would be fluctuations within a range of ±5 per cent due to births, deaths and some movement of people to or from the village during the six years, for which the data is presented.

Annual Parasite Index

In order to bring the entire information to a common frame of reference, the data on population and malarial cases has been converted to Annual Parasite Index (API). The index represents the number of malarial cases per thousand of population

$$\text{API} = \frac{\text{Number of malaria cases}}{\text{Population}} \times 1000$$

The annual parasite indices for the Sathanur dam and the Sathanur village are plotted along with the annual precipitation, for a period of six years, as Figures—7.1 and 7.2.

IMPACTS

At the Dam Site

The Annual Parasite Index (API) values and the annual rainfall of the command region, as functions of time, are depicted in Figure—7.1. It may be seen that even though the average rainfall has fluctuated over the years, the API hasn't followed that pattern; rather it has steadily declined. The API was 120 in 1994. It marginally fell in 1995, more sharply in 1996 and even more sharply there after.

If no other factor is operative, incidence of malaria should naturally be a direct function of the extent of area under lentic habitats. It is in lentic habitats that mosquitoes get an opportunity to breed and go through their larval and pupal stages. Commissioning of large water resources projects are known to cause increase in the incidence of water- borne diseases primarily by this mechanism. Such projects suddenly make available new lentic habitats leading to proliferation of disease vectors such as mosquitoes, houseflies and snails.

In Sathanur too, one would have expected a continuous increase in the incidence of malaria and other water-borne diseases. As detailed earlier (Chapter 2), the reservoir is encompassed by the hills housing the Ponnaiyar reserve forest. Also a water-spread area of 18.2 km^2 is available at full reservoir level. This area shrinks during the summer months but is never totally dry. Hence extra habitat for the breeding and the early growth stages of mosquitoes is made available by the reservoir all through each year. The forest also offers a potential site for breeding and proliferation of mosquitoes especially during the monsoon months. The marshy and swampy grounds littered with human faeces and cattle dung transform into mosquitoes harbouring sites. Low lying areas of the forest form soakage pits, ditches and other semi-permanent pools of water, where mosquitoes can thrive and breed easily. Yet, the API indicates that there is no consistent trend of increase in the incidence if malaria in the region.

Table—7.1 reveals that 22 per cent of the malaria instances in 1994, 28 per cent in 1995 and 42 per cent in 1996 were 'imported'. The parasites were feared to be brought in the blood of the migrant labourers or fishermen working on the site. These parasites, on finding the ideal combination of vectors and hosts, led to the transmission of malaria in the region. As discerned from Figure—7.1, the API peaked during the years 1994 and 1995. Incidentally these were the years when the construction work on the hydro-electricity generating unit began with the aid of hired labourers from the nearby regions. These migrant labourers are believed to spread the malaria in the region in such high proportions during these periods. This is similar to what had happened at the site of Aswan dam (Bhargave et al, 1992). There was a northward migration of vectors from Sudan towards dam site in this particular case and it is believed that the Sudanese fishermen brought the disease with them in the Aswan dam area. There have been no incidences of filiariasis at the dam site as per the health officials.

At the Sathanur Village

Sathanur village (population 6775), which is 11 km from the dam site, has comparatively low rate of malarial incidence as is evident from Table—7.2. All the incidences of malaria in the population during 1994 and 1995 were imported. API values, depicted as Figure—7.2, also recorded low rates when compared to the API values of dam region (Figure—7.1). There is no consistency between the patterns of annual precipitation and the API values during the study period (Figure—7.1).

The reason for low incidences of malaria in the village population can be attributed to the absence of the environmental conditions conducive for the growth and proliferation of disease vectors as encountered in the reservoir. According to the villagers, there is in general scarcity of water, in spite of the village being located in the proximity of the dam. The village doesn't get water for irrigation purposes from the dam. Thus the location and the geo-environmental conditions of the village offer little scope for the breeding and thriving of mosquitoes.

Table—7.1 Malaria incidence in Sathanur dam

Name of dam	Indicative Population	1994	1995	1996	1997	1998	1999
		32*	98#	46*			
Sathanur	1198	144	138	107	52	19	7
		112#	40*	61#			

* Imported
\# Local

Table—7.2 Malaria incidence in Sathanur village

Name of village	Indicative Population	1994	1995	1996	1997	1998	1999
Sathanur	6775	4*	1*	-	8	2	2

* Imported
\# Local

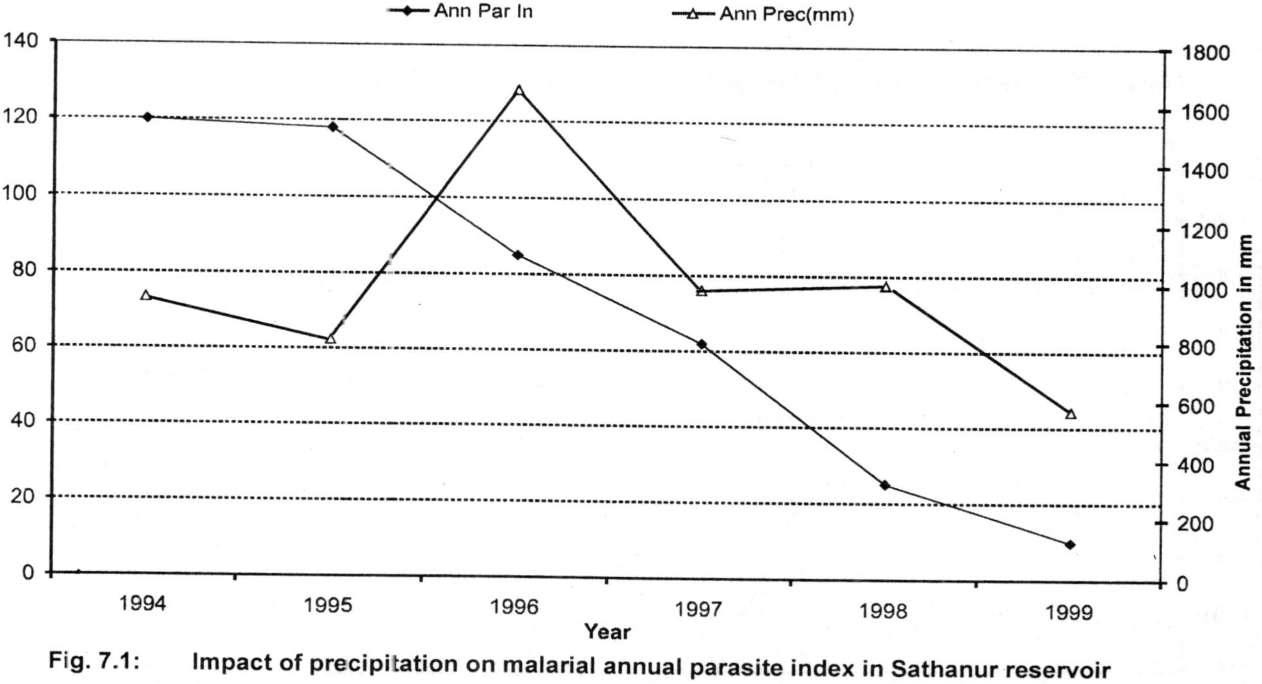

Fig. 7.1: Impact of precipitation on malarial annual parasite index in Sathanur reservoir

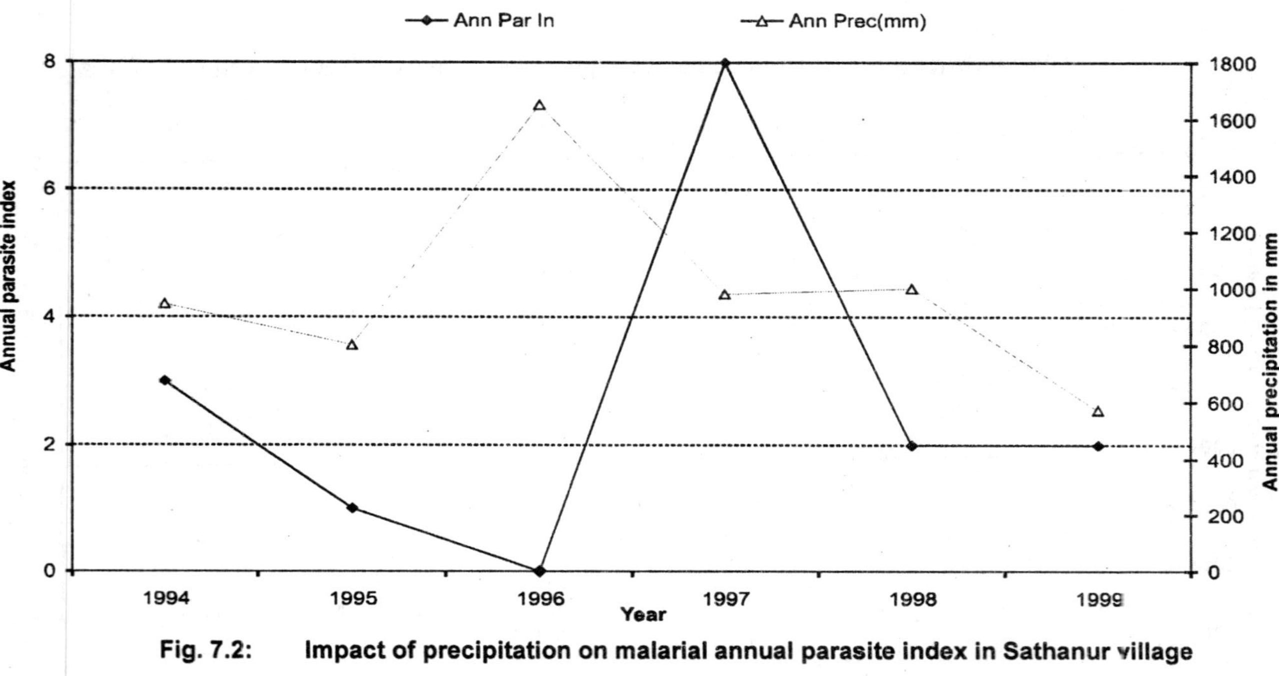

Fig. 7.2: Impact of precipitation on malarial annual parasite index in Sathanur village

The 'import' of malaria during the years 1994 and 1995 can be attributed to the temporary residents who work as labourers in the nearby places. With the introduction of better transportation and communication facilities, following the development of the region as a tourist spot, more and more of the villagers started to earn their livelihood working as hired labourers or fishermen. The nature of their work, the unhygienic surroundings and their socio-economic status, makes them easily susceptible to malarial attacks.

Measures Adopted by the Health Officials in Combating Water Borne Diseases

As discussed in the earlier section there is a steady decline in the API values over the years (Figure—7.1). Even though unusually high rainfall was recorded in the year 1996, there was no unusual increase in API in that particular year. The decline in the API for the period 1994-95 was marginal 4.16 per cent and in 1996 it reduced sharply to 25.69 per cent. After 1996 again there was a significant reduction of 63.89 per cent and the value continued falling till 1999 when it finally reached 95 per cent of it's 1994 value. This remarkable phenomenon of the reduction of API over the years is largely due to the mitigating measures adopted by the health officials in checking the disease from becoming an epidemic. A primary health centre is located at the dam site offering health services to all the nearby villages.

Once the nature of the malaria and its intensiveness was established, the health officials vigilantly started monitoring the whole situation. The health officials became even more vigilant after the construction work of hydro-electricity generation unit started in 1994-95 and there was an influx of migrant labourers at the dam site. The API recorded in that particular year reached exceptionally high i.e. 120. The steps undertaken by them in controlling the spread of the malaria are briefly described as follows:

- Periodic blood smear tests are performed on the work force employed at the Sathanur dam site.
- The hired labourers from the nearby regions are screened for any unusual symptoms before their entering the work point.

- Anti-malarial drugs (mostly chloroquines) are distributed free of cost to the residents at the dam site.
- DDT, BHC and other chemicals are sprayed regularly indoors and outdoors at the dam area to check the menace of mosquitoes.
- The mosquito larvae in the ponds and drains are killed using various larvicide formulations.
- Borrow pits, soakage pits and other semi-permanent water pools in the forest, hills and in the vicinity of the reservoir are periodically monitored and closed.

8

Aquaculture

INTRODUCTION

Aquaculture is the aquatic counterpart of agriculture (Beveridge, et al., 1994). Production of food by this means has increased significantly over the past two decades and currently accounts for some 15 million tones or 17 per cent of the world fisheries (capture and culture) production (FAO, 1993). Reservoir fisheries in south-east Asia are considered oldest in the world based on the fact that the irrigation schemes in south-east Asia has an older tradition, going back farther in history, than in any other region of the world. Aquaculture in reservoirs has been identified as one of the indirect benefits of water resources projects otherwise aimed primarily at generating irrigation facilities and hydropower (Abbasi, 1991, 2000).

A study on the icthyofauna of the Sathanur reservoir has been presented in this chapter. The study assesses the annual commercial production of the fishes over a period of twenty years. An attempt has also been made to establish the impacts of various hydrological factors operating in the reservoir on the annual fish catch.

STATUS

Pisciculture in Sathanur Reservoir

The piscifauna of the Sathanur reservoir, during the period of impoundment, consisted of a very few species of *Cyprinus.carpio* (carps) and catfishes. In the year 1959, the Tamil Nadu Fisheries Department took charge of reservoir fishery. Seeds of fast growing species of fishes such as *Catla catla* (catla), *Labeo rohita (rohu)*, *Cirrhinus mrigala* (mrigal) etc., were stocked in the reservoir and the Indian fisheries act was enforced in the reservoir thus barring fishing other than organized by the government in the reservoir. Due to all these steps, the commercial catch, which was 700 kgs in 1962-63, increased to 214 tones in 1976-77 as reported by the Tamil Nadu Fisheries Development Corporation Limited (TNFDC), functioning at the dam site.

The Corporation came into existence in 1977 preceded by the establishment of a fish seed production centre and a fish farm at the dam in 1959 for the purpose of the production of fish seeds for stocking other similar reservoirs in the state. Consequently the fishery rights of the Sathanur reservoir were transferred to the Corporation and the fish farm was also placed under it.

Functioning of TNFDC

The mandate of Tamil Nadu Fisheries Development Corporation Limited (TNFDC) pertaining to the pisciculture in Sathanur reservoir constitutes the following:

- Exploitation of the fishery of the reservoir
- Marketing the fishes at fixed prices through the retail outlets established in the nearby towns
- Stocking the reservoir with seeds of major *C.carpio* (carps)
- Production of major *C.carpio* (carps) seeds in the fish farm and supplying the surplus seeds to stock other waters in the state

Paid employees and private fishermen are engaged in the exploitation of the fisheries from the reservoir. Each fishing unit consists of two fishermen, one coracle and 20-gill nets. The units operate under the close supervision of the Deputy Manager. While the employee units are paid regular monthly wages by the corporation, the private fishermen units engaged for fishing are paid one third of their daily catches as wages.

The nets used by TNFDC, called 'Rangoon nets', are basically gill nets of mesh size 15 cm without bottom rope and sinkers. The stretched length and depth of each net are 30m and 6m respectively. The fishermen are not permitted to use nets of smaller mesh size to avoid capture of fingerlings and adolescent animals. Operation of nets in the upper reaches of the reservoir during the breeding season is prohibited:

Marketing

The fish catches from the Sathanur reservoir are sold to the public through various retail outlets at the rates fixed by the TNFDC. Currently there are fourteen such outlets functioning at the following places:

Sathanur dam

(During government holidays)

Thiruvannamalai–I

Thiruvannamalai–II

Thiruvannamalai–III

Polur

Chengam

Authoor

Kallakurichi

Vandavasi

Cheyyar

Thirupatthur

Chetpet

Arakkaulur

Sankarapuram

The data pertaining to the annual fish catch of various species and the annual revenue incurred for a period of twenty years was collected from the office of TNFDC. The details of the fish catch for the period 1978-2001 are presented in Tables—8.1, 8.2, 8.3 and 8.4. The catch is represented both in terms of numbers and zoomass. Correlations between the catch of the various species of fish and also between the catch and other hydrological parameters are worked out and the results are given as Tables—8.5 and 8.6 respectively. The statistical analyses of the fish catch are given in Table—8.7.

Table—8.8 presents the average zoomass per fish and the fish catch per hectare of the reservoir for the period of twenty-four years. The details regarding the fish seed production and the total fish catch along with the revenue generated are furnished in Tables—8.9, 8.10, 8.11, and 8.12.

The species wise annual catch and the cumulative catch of the ichthyofauna of the Sathanur reservoir as a function of time, for a period of twenty years, is illustrated in Figures—8.1 to 8.6. The Figures also depict the influence of annual precipitation, if any, on the fish catch.

Figures—8.7 to 8.10 illustrate the correlation between the mean water level of the reservoir and the fish catch. The relationships between the total inflow into the reservoir and the fish catch for various species are illustrated in Figures—8.11 to 8.15. The variations in the zoomass per fish for the piscifauna of the reservoir over the period of twenty-four years are depicted in Figures—8.16 and 8.17.

Figure—8.18 illustrates the percentage wise distribution of the ichthyofauna for a period of twenty years. Figure—8.19 presents the trends of the fish seed production and the revenue earned over the years. The total annual catch of the fishes and the revenue earned there from as a function of time are represented as Figure—8.20.

Table—8.1 Details of commercial exploitation of ichthyofauna in Sathanur reservoir

Year	Catla catla		L.rohita		C.mrigal		C.carpio		L.fimbriatus	
	Number	Yield (Kg)	Number	Yield (Kg)	Number	Yield (Kg)	Number	Yield (Kg)	Number	Yield (Kg)
1978	13202	98015.8	13014	39385.9	6838	8727.7	0	0	12636	15527.5
1979	7403	72267.7	4911	17581.5	9781	16618.9	5	21	2607	3456
1980	5244	50877.4	5087	21665.1	21088	44183.8	3	14.7	4765	6851.7
1981	21911	94191.9	3411	14215.8	38902	78937.2	23	15.5	8803	12619.6
1982	22376	108074.8	3352	11602.9	6336	13304.4	0	0	5180	6472.5
1983	4966	42377	75595	29006.2	19163	39669	0	0	5535	6523.1
1984	934	7246.2	3795	7031.7	8537	13454.7	81	175	2037	1487.3
1985	1741	6453.3	4449	7096.6	12503	18648	404	446.2	3356	2903.6
1986	1531	9144.7	4157	7948.2	5741	9216.4	352	531.6	4096	4468.8
1987	4002	19763.7	3732	8753.5	4104	8017.6	162	593.4	3728	3993.3
1988	8316	28758.2	7790	13598.3	13204	19570.8	112	151.9	2258	2629.6
1989	8870	33966.8	19301	31304.5	24561	34807	133	14.8	3565	3488.2
1990	18128	52374.2	19313	28219.3	10826	15119.3	97	99.4	723	1358.2
1991	10203	36666.1	16309	23355.9	1305	12186.8	43	67.1	1494	1903.4
1992	11190	34386.3	16024	24261.5	12421	14481.1	10	26.9	1471	1709.1
1993	19703	69558.2	11101	19081.8	26861	34794.3	32	82.7	39	1163.1
1994	5988	28650.15	19859	29594.8	10441	18336.8	18	35.7	287	311.3
1995	4039	11713.4	16169	25351.05	7829	12922.48	39	43.5	89	135.8
1996	9999	23215.1	12716	21196.6	6539	10036.6	2	5	77	114.5
1997	15147	46958.7	23733	35599.6	8422	14232.7	4	6	321	436.2
1998	18158	52659.1	27486	43977.1	10571	17971.1	2	3.2	302	453.7
1999	14349	51244.3	14051	27311.6	9713	19838	–	–	0	0
2000	11438	44407.7	30098	45993.21	5386	239246	–	–	0	0
2001	7564	26878	27128	44651	20593	33922.5	–	–	0	0

(Table Contd...)

Year	L.calabasu		C.cirrhosa		Wallago attu		Cat-fish		Others	Total yield
	Number	Yield (Kg)	Number	Yield (Kg)	Number	Yield (Kg)	Number	Yield (Kg)	Yields (Kg)	In tones
1978	3742	3260.3	5573	9839.6	3483	4913.2	1167	2294.4	0	182
1979	5381	5084.3	4890	5966.7	1778	4710.3	1478	2049.8	2	128
1980	9754	10421.4	6628	10394.8	2324	6652.5	5075	6884.5	123	158
1981	11385	11232.3	3351	4993.7	1859	6275	9279	13652.3	1076.5	237.14
1982	4236	3604.8	1299	1426.2	1188	4376.8	4454	7478	223.1	156.81
1983	23526	15289	2141	2501.3	1630	5337.8	15875	13498.7	7545.3	156.75
1984	24527	13472.4	295	171.6	751	2430.8	7520	5389.2	4255.4	53.14
1985	17585	10450.8	200	180.2	634	1763.8	11247	7834.2	1953	52.56
1986	13711	8807.9	68	142.6	1367	3235	8755	7305.1	3045	53.83
1987	36051	23420.4	61	50.5	901	3035.2	21660	15156.1	20205.7	102.99
1988	19943	12931	0	0	3232	6656.65	31803	25034.6	16326.55	126.02
1989	33294	19638.9	0	0	2952	6078	39998	22888.8	17240.5	169.64
1990	12390	7201.6	0	0	3276	6193	20155	15708.6	19143.5	145.32
1991	3855	2188.4	0	0	3267	6409.7	11884	8998.9	13916.2	105.69
1992	11208	6845.1	21	24.1	2917	6845.1	5957	4630.3	12480.6	105.72
1993	19647	12650.95	41	53.7	1447	3058.9	10398	7362.8	17749.4	165.05
1994	27320	17639.15	29	26.8	2122	4365.15	15465	11015.9	30556.3	140.48
1995	20446	13496.4	0	0	851	1931.1	4491	3103.5	11367.2	80.06
1996	22121	14603.5	0	0	733	1314.1	4832	3591	68502.5	142.64
1997	39592	19795.8	0	0	1307	2744.5	4094	2866	11183.4	133.82
1998	37810	22685.89	0	0	750	1875.36	4201	3024.7	7647.2	151.23
1999	15824	16683.8	–	–	939	3054.1	5084	4091.2	7987.6	130.21
2000	20506	19619.7	–	–	1247	3008.5	5976	4308.3	18423.1	375.00
2001	17870	13893	–	–	1421	2586.5	8116	5205	17304.5	127.13

Table—8.2 Details of commercial exploitation of ichthyofauna in Sathanur reservoir, 1999-2000

Months	Catla catla		L. rohita		C. mrigal		L. calabasu		Wallago attu	
	No.	Wt (kg)	No.	Wt (kg)	No.	Wt (kg)	No.	Wt (kg)	No.	Wt (kg)
Apl '99	804	2053	454	994	635	1183.5	1227	1226	65	211
May '99	1141	3286.7	715	1708.5	699	1453.5	1160	1261.5	81	292
Jun '99	846	2767.8	571	1484.1	807	1858.5	563	620.9	39	104.1
Jul '99	1017	3172.0	315	1028.3	1047	2276.4	499	522.5	62	155.7
Aug '99	1268	3539.7	383	1211.8	781	1914.2	743	732.4	84	212.3
Sept '99	1780	4958.7	682	2239.4	505	1396.8	483	499.1	44	144.5
Oct '99	1186	7096.4	949	2996.7	606	1628.1	796	943.1	177	641.5
Nov '99	917	4818.0	1181	2149.0	721	1219.5	872	9975	68	216.5
Dec '99	1117	3179.5	1392	2106.0	854	1327.5	2265	2425.5	70	224.0
Jan '00	1443	3769.5	2358	3340.5	1004	1805.0	2982	3170.5	91	209.5
Feb'00	1726	5432.5	2780	4105.0	1162	2068.0	2152	2232.5	74	320.0
Mar '00	1104	7170.5	2271	3948.3	892	1707.0	2030	2052.3	84	323.0
Total	14349	51244.3	14051	27311.6	9713	19838.0	15824	16683.8	939	3054.1

(Table Contd...)

Months	Cat fish		Total value (Rs.)
	No.	Wt (kg)	
Apl '99	530	439.5	169646.00
May '99	436	424.5	230242.60
Jun '99	127	118.5	193478.20
Jul '99	93	79.4	212737.40
Aug '99	84	66.5	218429.20
Sept '99	41	34.3	258681.90
Oct '99	53	51.5	363972.40
Nov '99	135	139.5	256231.75
Dec '99	543	484.5	259648.75
Jan '00	1374	978.0	353775.75
Feb'00	874	668.5	403262.50
Mar '00	794	6.6.5	456088.65
Total	5084	4091.2	3376195.10

Table—8.3 Details of commercial exploitation of ichthyofauna in Sathanur reservoir, 2000-2001

Months	Catla catla		L. rohita		C. mrigal		L. calabasu		Wallago attu	
	No	Wt (kg)	No	Wt (kg)	No	Wt (kg)	No	Wt (kg)	No	Wt (kg)
Apl '00	2534	9262.8	1627	3411.5	659	1474.3	1313	1290.0	67	220.5
May '00	1853	7563.6	1535	3221.9	565	1199.5	1354	1447.7	147	388.5
Jun '00	995	4178.0	1122	2300.5	452	1061.0	2105	2250.5	199	458.0
Jul '00	939	3535.0	521	1156.5	401	1088.5	2587	2700.0	116	230.5
Aug '00	825	3372.5	447	995.0	289	765.5	1491	1597.0	121	308.5
Sept '00	1104	4701.0	625	1284.5	272	742.0	1327	1277.0	131	327.0
Oct '00	617	2889.5	736	1336.5	223	454.5	1167	1192.0	119	326.5
Nov '00	447	1834.3	2334	3027.0	747	991.0	1465	1374.0	61	161.5
Dec '00	373	1628.0	3823	5255.8	1923	2508.5	1196	1070.5	52	125.5
Jan '01	582	1854.3	4425	6200.5	2705	3615.3	1472	1380.0	103	164.5
Feb'01	563	1789.0	5632	7790.5	2830	3987.0	1931	1607.5	63	136.5
Mar '01	595	1993.5	7271	10213.0	4320	6037.5	3098	2433.5	68	161.0
Total	11433	44409.7	30098	45993.2	15386	239246	20506	19619.7	1247	3008.5

(Table Contd...)

Months	Cat fish		Total value (Rs)
	No.	Wt (kg)	
Apl '00	216	156.0	441476.90
May '00	424	280.8	413492.90
Jun '00	696	523.0	298888.50
Jul '00	370	553.0	246193.00
Aug '00	400	353.0	207753.75
Sept '00	330	268.5	244499.50
Oct '00	267	201.5	173556.25
Nov '00	227	153.0	221757.90
Dec '00	474	304.5	310975.90
Jan '01	600	367.5	393120.15
Feb'01	718	478.0	454906.25
Mar '01	1254	775.5	628977.00
Total	5976	4308.3	4635578.00

Table—8.4 Details of commercial exploitation of ichthyofauna in Sathanur reservoir, 2001-2002

Months	C.catla		L. rohita		C.mrigal		L.calabasu		Wallago attu	
	No.	Wt (kg)	No.	Wt (kg)	No.	Wt (kg)	No.	Wt (kg)	No.	Wt (kg)
Apl '01	495	2339.5	4653	6579.0	2611	3640.0	2074	759.5	71	141.5
May '01	424	1752.5	2925	5547.0	2287	3355.0	.1891	378.0	90	154.0
Jun '01	433	1621.5	3135	4986.0	2716	4416.0	1436	1111.5	170	252.0
Jul '01	739	2947.5	1819	3358.0	2709	5075.0	1366	994.5	181	271.5
Aug '01	532	2170.5	1385	2681.0	1686	3037.0	1199	964.0	166	350.5
Sept '01	643	2435.0	1505	2592.5	1062	1877.0	1231	979.5	0	0
Oct '01	711	2705.0	1168	2177.0	771	1330.5	1022	785.5	150	303.5
Nov '01	600	2557.5	1545	249.5	950	1472.5	817	635.0	108	192.5
Dec '01	502	1925.5	2303	4200.0	1637	2711.0	1181	880.5	121	229.5
Jan '02	458	1805.0	2452	4306.0	1931	3215.5	1622	1285.0	122	230.5
Feb'02	527	2042.5	2014	3771.5	1253	2112.0	1460	1157.0	128	226.5
Mar '02	789	2576.5	2224	4253.5	980	1681.0	2570	1963.0	114	235.5
Total	7564	26878	27128	44651	20593	33922.5	17870	8116	1421	2586.5

(Table Contd...)

Months	Cat fish	
	No.	Wt (kg)
Apl '01	582	355.0
May '01	780	503.5
Jun '01	916	560.5
Jul '01	864	523.5
Aug '01	396	339.5
Sept '01	469	304.5
Oct '01	469	308.5
Nov '01	322	192.0
Dec '01	798	229.5
Jan '02	1014	230.5
Feb'02	688	226.0
Mar '02	788	235.5
Total	8116	2586.5

Table—8.5 Coefficients of correlation between the fish catch and hydrological parameters

	Precipitation	Mean water level	Annual inflow
Catla catla	-0.3	0.3	0.1
L.rohita	-0.1	0.4	0.1
C. mrigal	0.1	-0.2	0
C.carpio	0	-0.3	-0.2
L.fimbriatus	-0.2	0	-0.1
L.calabasu	0.2	0.1	-0.1
C.cirrhosa	-0.2	-0.2	0
Wallago attu	-0.2	-0.3	-0.2
Catfish	0	-0.6	-0.3
Others	0.3	0.3	0.4
Total Catch	0.2	0.2	0

Table—8.6 Coefficients of correlations for the ichthyofauna catch in Sathanur reservoir

	Catla catla	*L. rohita*	*C. mrigal*	*C. carpio*	*L.fimbriatus*	*L.calabasu*	*C.cirrhosa*	*Wallago attu*	*Catfish*	*Others*
Catla catla	1.0	0.3	0.4	-0.5	0.6	-0.3	0.6	0.4	-0.1	-0.2
L.rohita			-0.3	-0.6	0	0.2	0.1	0.1	-0.2	0.3
C. mrigal				-0.3	0.4	0.1	0.4	0.4	0.3	-0.2
C.carpio					-0.1	0.2	-0.3	-0.3	0.2	-0.2
L.fimbriatus						-0.4	0.8	0.4	0	-0.4
L.calabasu							-0.4	-0.4	0.2	0.4
C.cirrhosa								0.4	-0.2	-0.3
Walla goattu									0.5	-0.3
Catfish										0.1
Others										1

Table —8.7 Average, standard deviation and coefficient of variation of the fish fauna in Sathanur reservoir

	Catla catla	*L.rohita*	*C.mrigal*	*C. carpio*	*L.fimbriatus*	*L.calabasu*	*C.cirrhosa*
Average	44158.03	21896.56	21677.9	111.1	3714.6	12129.5	1703.4
Standard deviation	29958.50	10740.9	16758.9	180.9	4061.0	6223.8	3262.3
Coefficient of variation	67.8	49.1	77.3	162.8	109.3	51.3	191.5

	Wallago attu	*Cat fish*	*Others*	*Water level*	*Inflow*	*Rainfall*
Average	4295.3	9036.5	12597.5	81.1	142.08	951.7
Standard deviation	1888.4	6539.5	15327.7	10.3	129.8	251.5
Coefficient of variation	43.9	72.4	121.7	12.68	91.4	26.4

Table —8.8 Fish catch per hectare of Sathanur reservoir

Year	Catch ha^{-1} in kg										
	Catla catla	*L.rohita*	*C.mrigal*	*C. carpio*	*L.fimbriatus*	*L.calabasu*	*C.cirrhosa*	*Wallago attu*	Catfish	Others	Total
1978	54.5	21.9	4.8	0	8.6	1.8	5.5	2.7	1.3	0	101.1
1979	40.1	9.8	9.2	0.01	1.9	2.8	3.3	2.6	1.1	0	71.1
1980	28.3	12	24.5	0	3.8	5.8	5.7	3.7	3.8	0.1	87.8
1981	52.3	7.9	43.8	0	7	6.3	2.7	3.5	7.6	0.6	131.7
1982	60	6.4	7.4	0	3.6	2	0.8	2.4	4.2	0.1	87.1
1983	23.5	16.1	22	0	3.6	8.5	1.4	2.9	7.5	4.2	87.1
1984	4	3.9	7.5	0.1	0.8	7.5	0.1	1.4	2.9	2.4	29.5
1985	3.6	3.9	10.4	0.2	1.6	5.8	0.1	0.9	4.4	1.1	29.2
1986	5	4.4	5.1	0.3	2.5	4.9	0.1	1.8	4.1	1.7	29.9
1987	10.9	4.9	4.5	0.3	2.2	13	0.02	1.7	8.4	11.2	57.2
1988	15.9	7.6	10.9	0.1	1.5	7.2	0	3.7	13.9	9.1	70.0
1989	18.9	17.4	19.3	0	1.9	10.9	0	3.4	12.7	9.6	94.2
1990	20.1	15.7	8.4	0.1	0.8	4	0	3.4	8.7	10.6	80.7
1991	20.4	12.8	6.7	0	1.1	1.2	0	3.6	4.9	7.7	58.7
1992	19.1	13.5	8	0.01	0.9	3.8	0.01	3.8	2.6	6.9	58.7
1993	38.6	10.6	19.3	0.04	0.6	7	0.02	1.7	4.1	9.9	91.7
1994	15.9	16.4	10.2	0.01	0.2	9.8	0.01	2.4	6.1	16.9	78
1995	6.5	14.1	7.2	0.02	0.1	7.5	0	1.1	1.7	6.3	44.5
1996	12.9	11.8	5.8	0	0.1	8.1	0	0.7	1.9	38.1	79.2
1997	26.1	19.8	7.9	0	0.2	10.9	0	1.5	1.4	6.2	74.3
1998	29.3	24.4	9.9	0	0.3	12.6	0	1	1.3	4.2	84
1999	28.13	14.99	10.89	–	–	9.16	–	1.67	2.24	–	–
2000	24.38	25.25	131.36	–	–	10.77	–	1.65	2.36	–	–
2001	14.75	24.51	18.62	–	–	76.28	–	14.20	28.58	–	–

IMPACTS

There are some fifteen species of fish in the Sathanur reservoir. The dominant ones are *Catla catla* (catla) *Labeo rohita* (rohu), *Cirrhinus mrigala* (mrigal), *Cyprinus carpio* (carp), *Labeo fimbriatus. Labeo calabasu, Cyprinus cirrhosa, Wallago attu, Heteropneustes* sp.and *Clarius batrichus.* Except for some species of *Cyprinus carpio* (carp) and the cat fishes all the other species are exotic, brought to the reservoir in 1977-78. Some species like *Argentious carpio* (silver catla), *Murral* sp., *Tilapia mosambica* and few other uneconomical catfishes are also present in the reservoir in low numbers and they have been categorised as 'others'. The category also includes some unidentified fish species.

A detailed study of the catch of the piscifauna of the reservoir from the Figures—8.1 to 8.6 points out towards the fact that the annual catch per species of the fish has undergone wide fluctuations over the years.

The yield of *Catla catla* (catla) after assaying an initial decline during the years 1978-79, attained a peak in 1982 before coming down drastically in 1984 by 93 per cent of its 1982 yield. The yield was steady thereafter till 1986 and 1987 onwards there was a gradual increase in the yield till 1990 which again became low during 1991-92, picked up in 1993 before falling yet again till 1995. From 1995 onwards there has been a marginal increase in the catch. Overall there has been a declining trend in the catla catch (Figure—8.1).

Table—8.9 Details of total catch, fish seed production and revenue generated for the past twenty-four years

Year	Total fish catch		Fish seed production	
	Quantity (Tonnes)	Value (million rupees)	Quantity (Tonnes)	Value (million rupees)
1977-78	182.00	1.0	152.40	.08
1978-79	128.00	0.9	155.15	.08
1979-80	158.00	1.0	45.35	.01
1980-81	237.14	1.5	104.56	.02
1981-82	156.81	1.2	178.10	0.2
1982-83	156.75	1.2	272.00	0.1
1983-84	53.14	0.4	77.70	.06
1984-85	52.56	0.4	255.35	0.1
1985-86	53.83	0.5	384.14	0.6
1986-87	102.99	0.9	402.60	0.7
1987-88	126.02	1.1	573.40	0.6
1988-89	169.64	1.5	600.30	0.8
1989-90	145.32	1.5	500.52	0.9
1990-91	105.69	1.2	347.05	0.3
1991-92	105.72	1.3	258.00	0.5
1992-93	165.05	2.3	209.54	0.5

(Table Contd...)

1	2	3	4	5
1993-94	140.48	2.2	312.00	0.7
1994-95	80.06	1.5	233.05	0.5
1995-96	142.64	2.3	407.50	0.9
1996-97	133.82	3.0	402.00	0.9
1997-98	113.76	2.8	176.75	1
1998-99*	–	–	–	–
1999-2000	130.21	3.3	204.2	1.1
2000-2001	375.00	4.6	181.5	0.7
2001-2002	127.13	-#	118.75	0.5

Data for the year 1998-1999 and the revenue data for the year 2001-2002 has not been provided as these records were available at the source.

Table—8.10 Details of Fish seed production of Sathanur Reservoir, 1999–2000

Months	C.catla	L.rohita	C.mrigal	C.carpio	L.fimbriatus	L.calabasu	C.cirrhosa	Wallago attu	Cat-fish
April	6.50	–	–	–	–	–	–	–	12825.00
May	77.50	–	–	–	0.01	–	–	–	127556.00
June	26.75	6.00	–	–	2.11	–	–	–	180006.00
July	25.25	9.50	–	–	0.81	0.28	–	–	161764.00
August	–	10.50	1.00	–	0.82	–	–	–	59562.00
September	–	–	4.00	2.00	0.58	1.26	–	0.01	79996.00
October	–	–	–	4.00	0.80	0.86	1.34	0.85	139862.00
November	–	–	–	5.00	0.40	0.24	0.75	1.27	99006.00
December	–	–	–	4.00	–	–	–	4.44	158324.00
January	–	–	–	1.00	–	–	–	1.53	54957.00
February	–	–	–	1.00	–	–	–	0.58	37313.00
March	–	–	–	1.00	–	–	–	0.27	1864815.00
Total	136.00	26.00	5.00	18.00	5.53	2.64	2.09	8.95	1129819.15

Table—8.11 Details of Fish seed production of Sathanur Reservoir, 2000–2001

Months	C.catla	L.rohita	C.mrigal	C.carpio	L.fimbriatus	L.calabasu	C.cirrhosa	Wallago attu	Cat-fish
April	1.0	–	–	–	–	–	–	–	13500.00
May	26.5	–	–	–	0.05	–	–	–	67926.00
June	47.5	21.0	8.0	–	0.01	–	–	–	129454.00
July	11.0	17.5	1.0	1.0	0.71	0.08	–	–	131472.00
August	3.0	11.5	3.0	–	0.15	1.79	0.07	0.09	132604.00
September	–	1.0	–	2.0	1.29	0.83	0.56	0.15	76904.00
October	–	–	–	5.5	0.08	0.01	0.90	0.98	42542.00
November	–	–	–	3.0	0.52	0.01	0.01	0.41	30777.90
December	–	–	–	2.5	–	–	–	0.77	31039.00
January	–	–	–	1.0	–	–	–	0.81	5061.00
February	–	–	–	–	–	–	–	0.13	3348.00
March	–	–	–	–	–	–	–	0.09	–
Total	89.0	51.0	13.0	15.0	3.81	2.72	1.54	3.43	789496.90

Table—8.12 Details of Fish seed production of Sathanur Reservoir, 2001–2002

Months	C.catla	L.rohita	C.mrigal	C.carpio	L.fimbriatus	L.calabasu	C.cirrhosa	Wallago attu	Cat-fish
April	–	–	–	–	–	–	–	–	–
May	16.0	9.0	–	–	25.20	0.20	–	–	–
June	20.0	8.5	1.5	–	30.0	0.08	–	–	–
July	9.75	10.50	9.0	–	29.25	0.26	–	–	–
August	–	8.5	6.0	–	14.50	0.05	–	–	–
September	–	–	–	4.50	4.50	–	–	–	–
October	–	–	–	4.25	4.25	–	–	–	–
November	–	–	–	1.40	1.40	–	–	–	–
December	–	–	–	6.35	6.35	–	–	–	–
January	–	–	–	1.50	1.50	–	–	–	–
February	–	–	–	2.00	2.00	–	–	–	–
March	–	–	–	–	–	–	–	–	–
Total	45.75	36.5	16.5	20.00	118.75	0.50	–	–	–

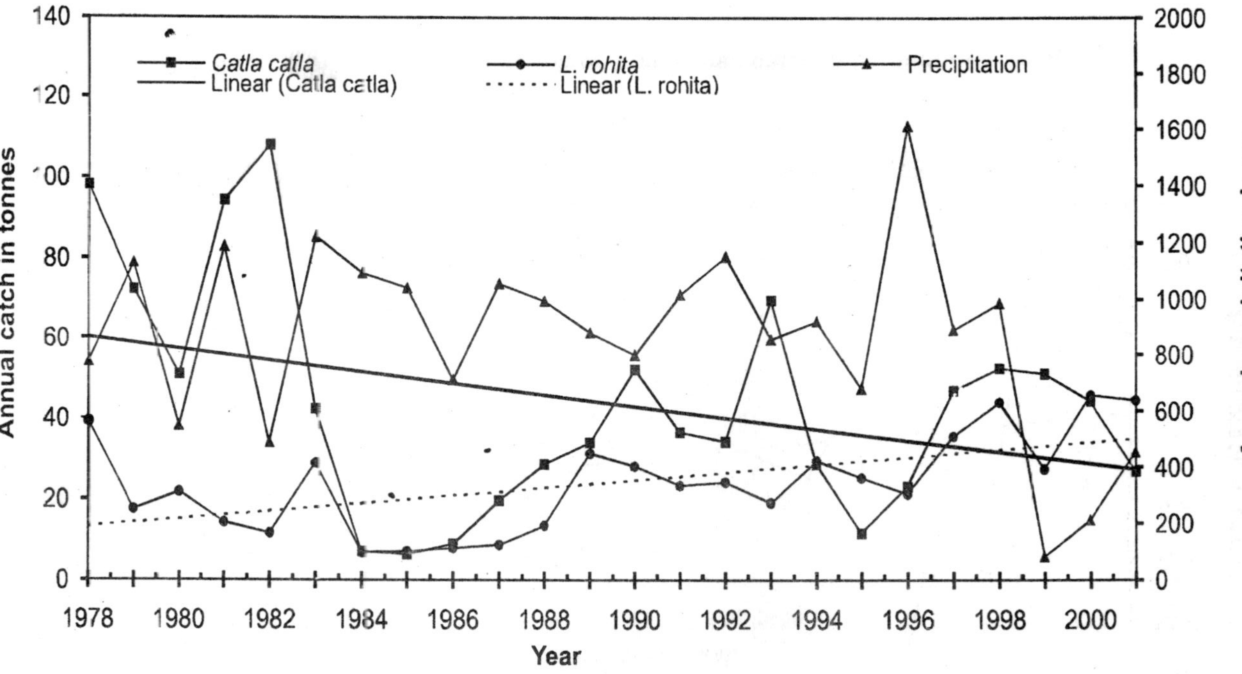

Fig. 8.1: Impact of precipitation on annual fish catch in Sathanur reservoir

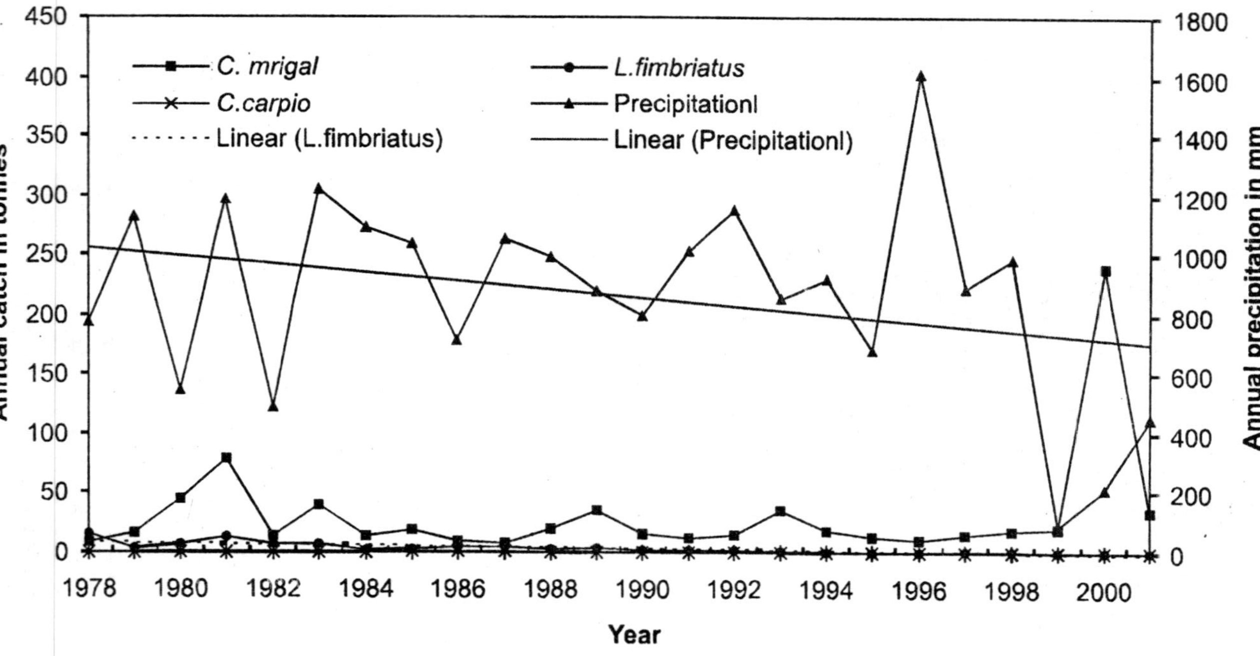

Fig. 8.2: Impact of precipitation on annual fish catch in Sathanur reservoir

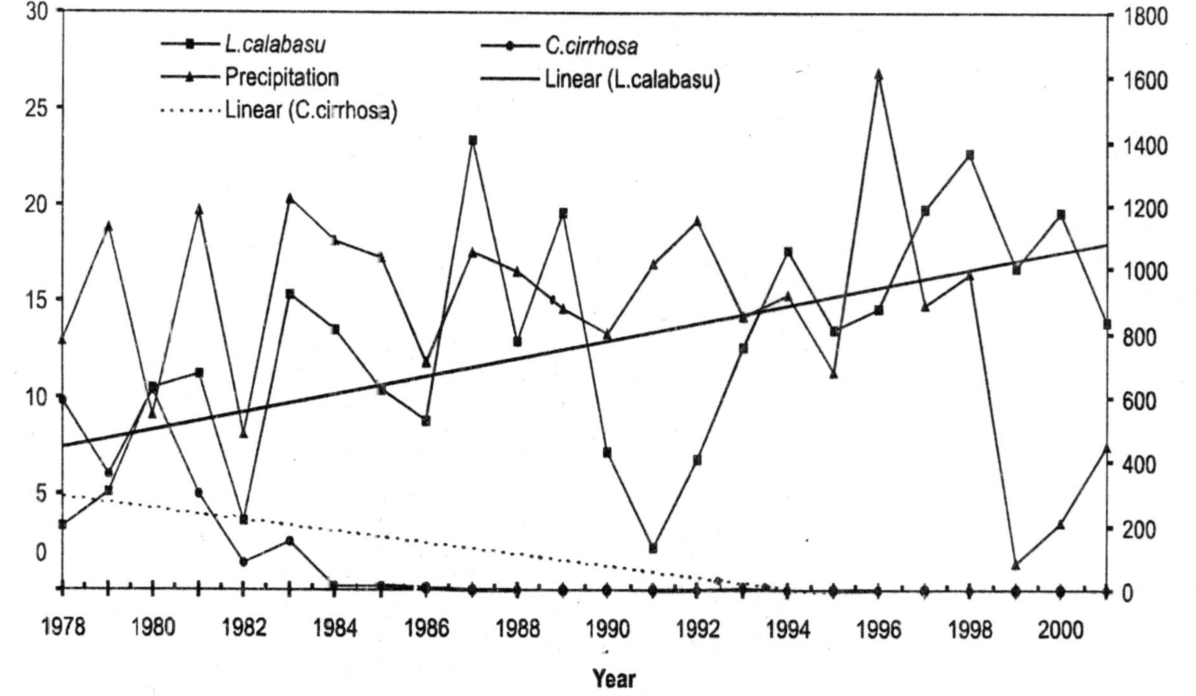

Fig. 8.3: Impact of precipitation on annual fish catch in Sathanur reservoir

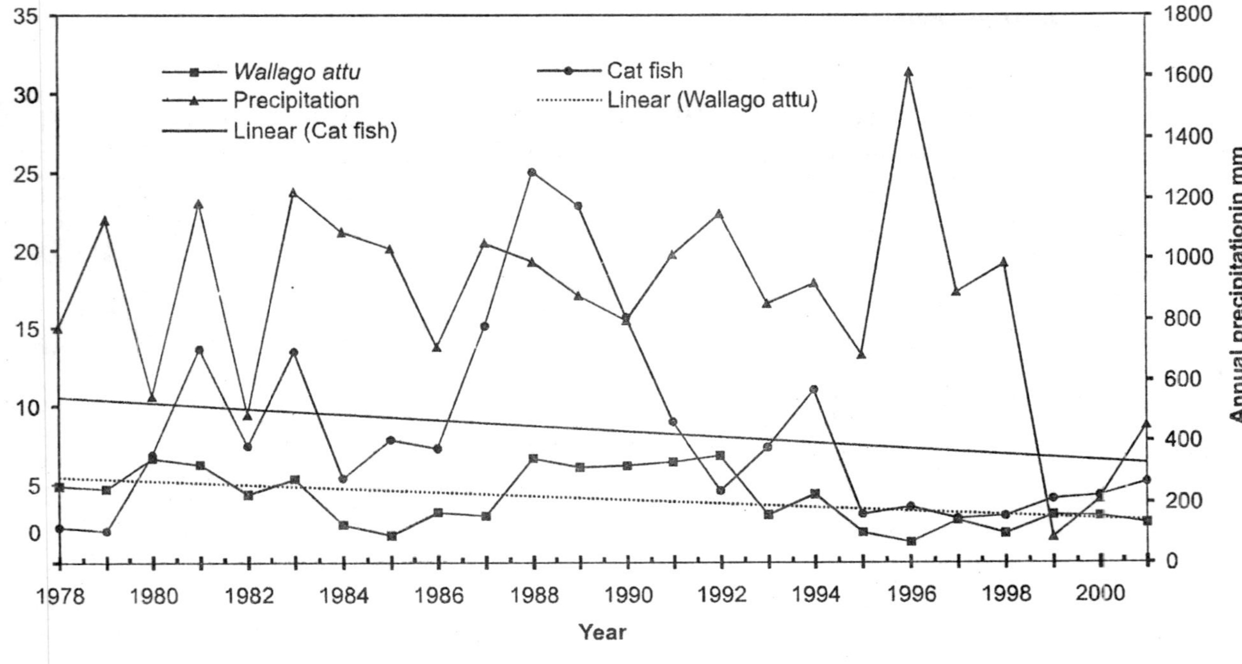

Fig. 8.4: Impact of precipitation on annual fish catch in Sathanur reservoir

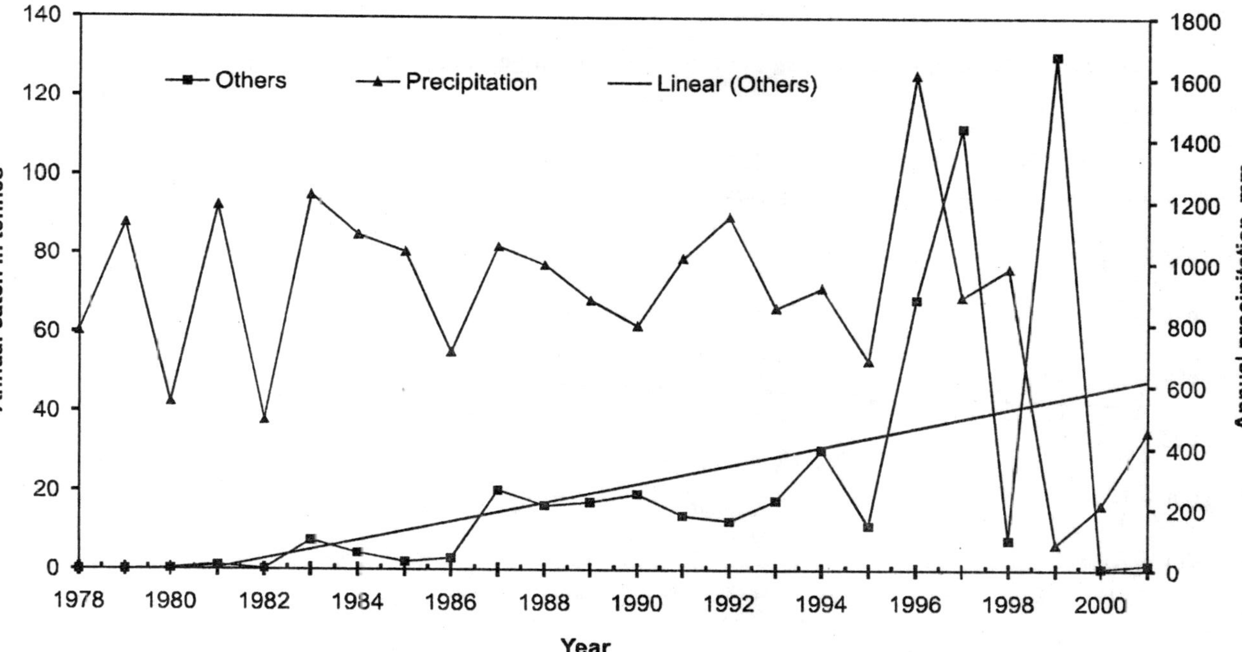

Fig. 8.5: Impact of precipitation on annual fish catch in Sathanur reservoir

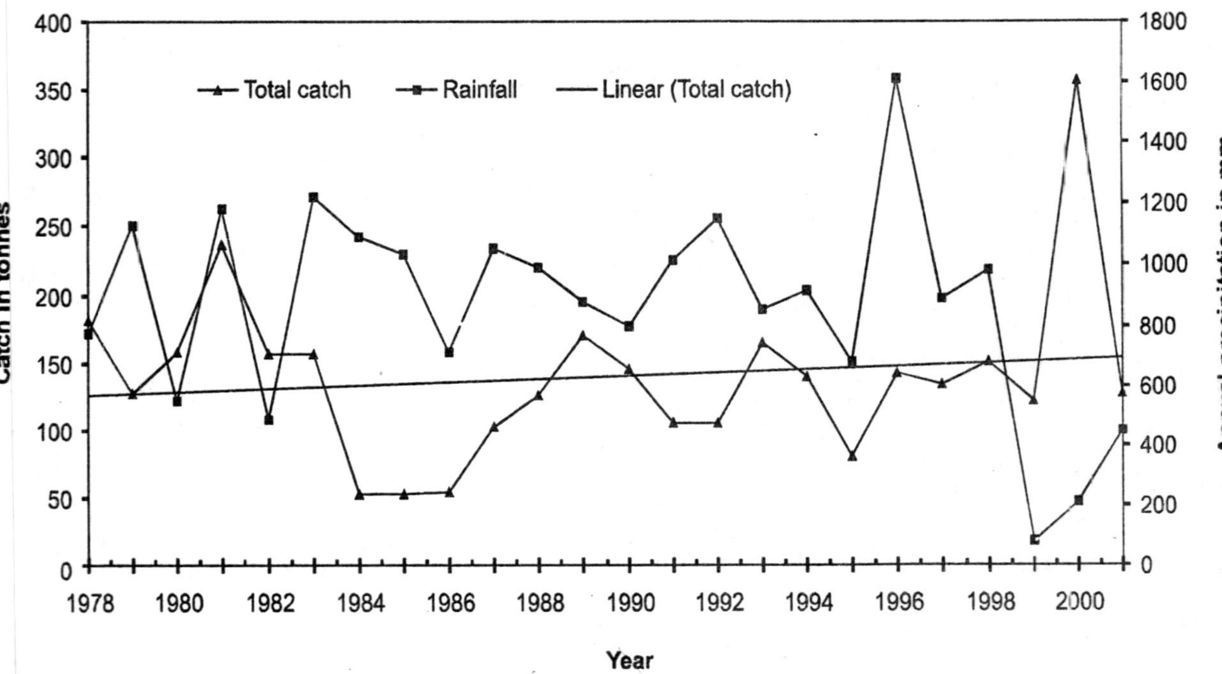

Fig. 8.6: Impact of precipitation on total annual fish catch in Sathanur reservoir

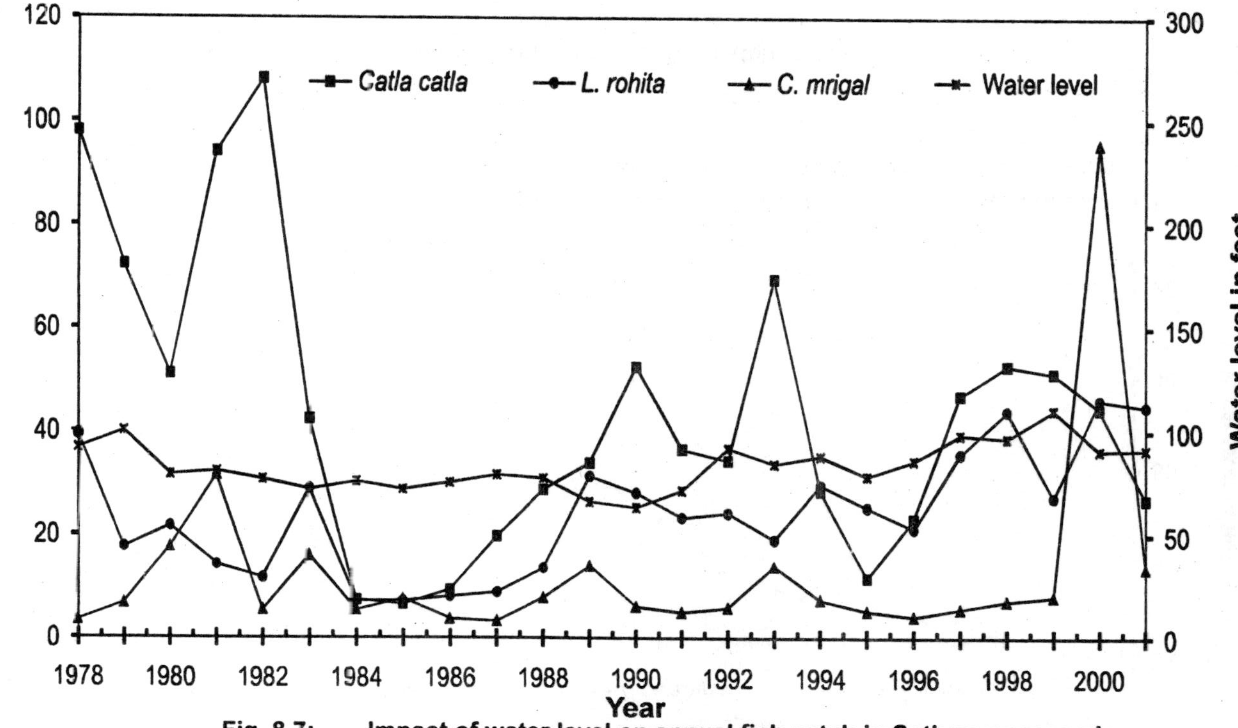

Fig. 8.7: Impact of water level on annual fish catch in Sathanur reservoir

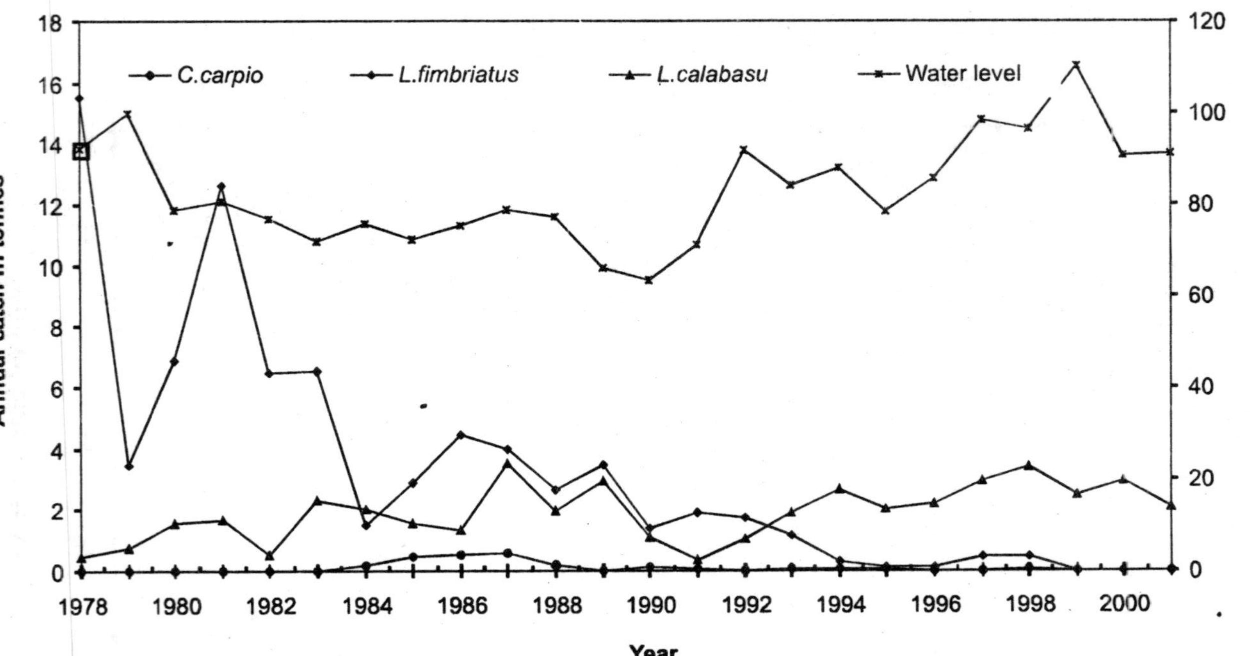

Fig. 8.8: Impact of water level of annual fish catch in Aathanur reservoir

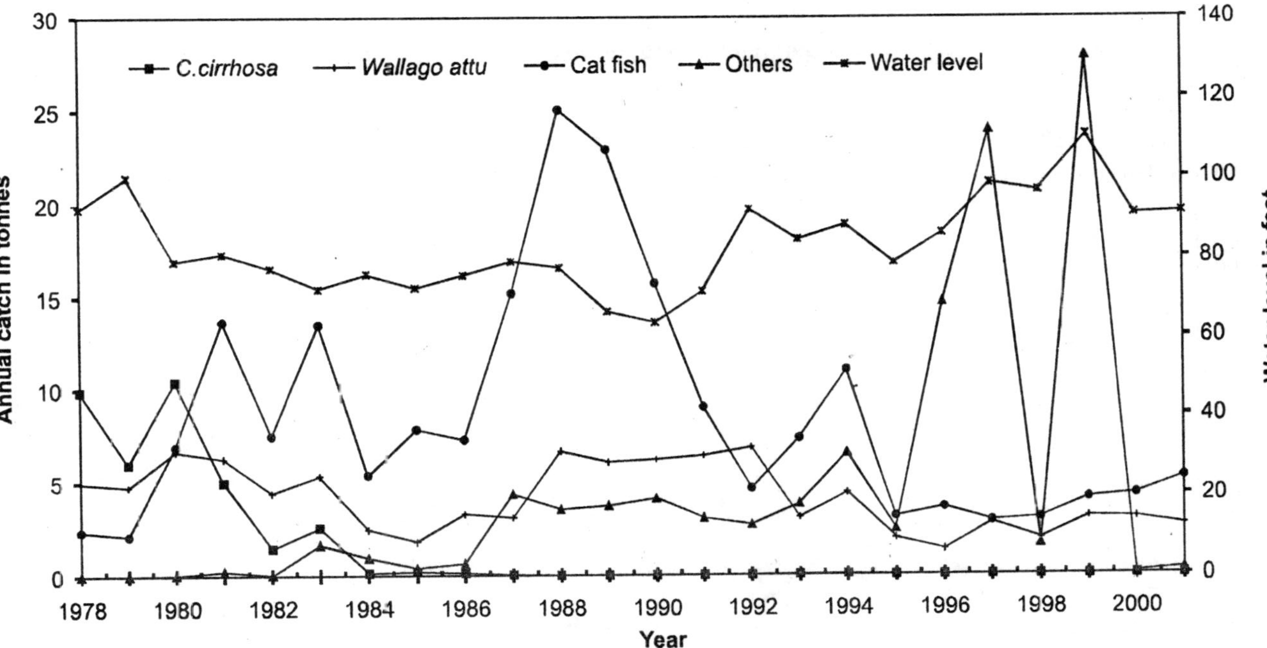

Fig. 8.9: Impact of water level on annual fish catch in Sathanur reservoir

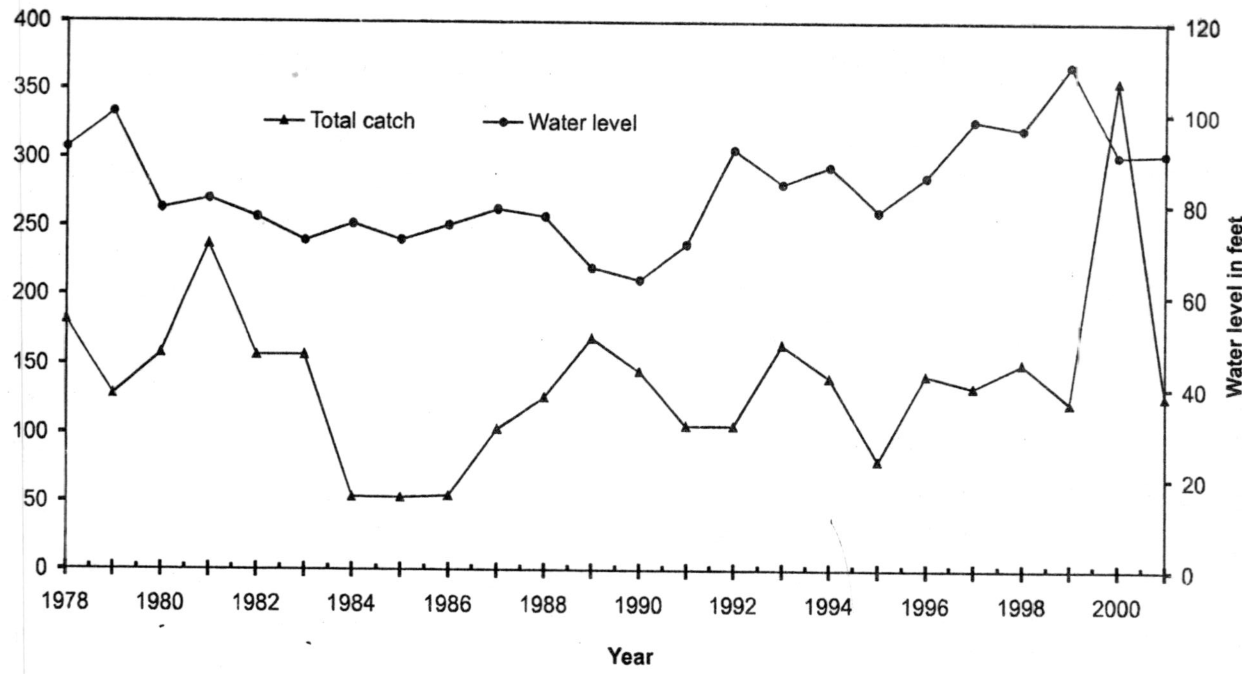

Fig. 8.10: Impact of water level on total annual fish catch in Sathanur reservoir

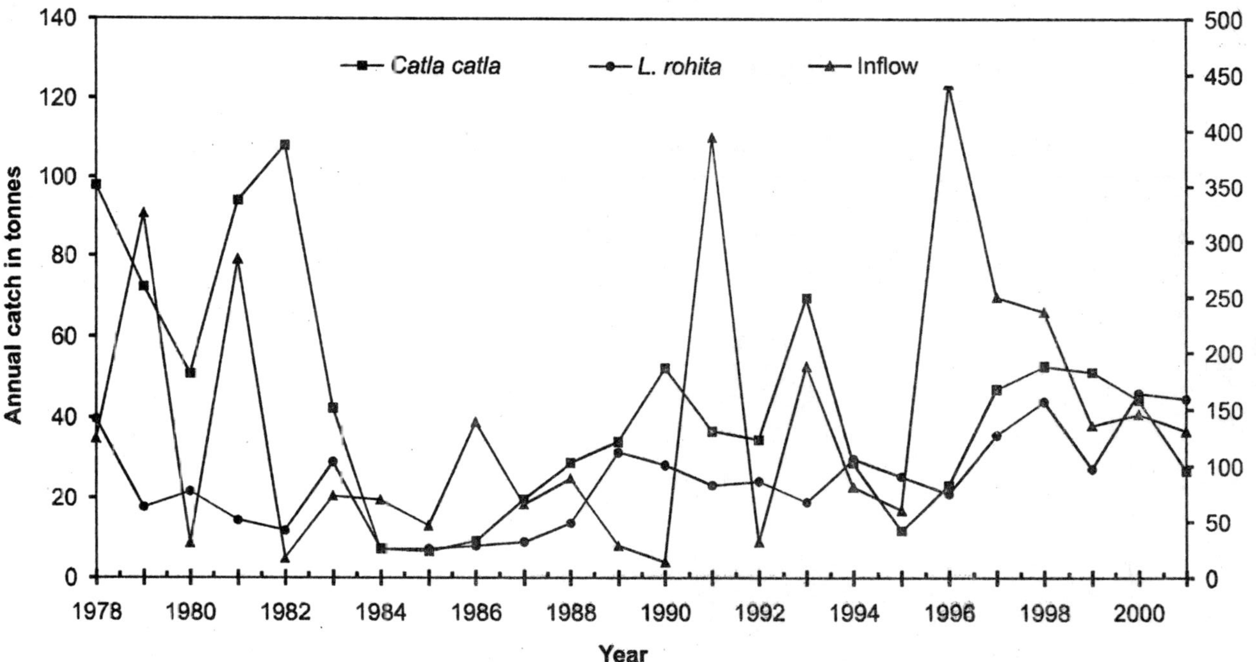

Fig. 8.11: Impact of inflow on annual fish catch in Sathanur reservoir yield

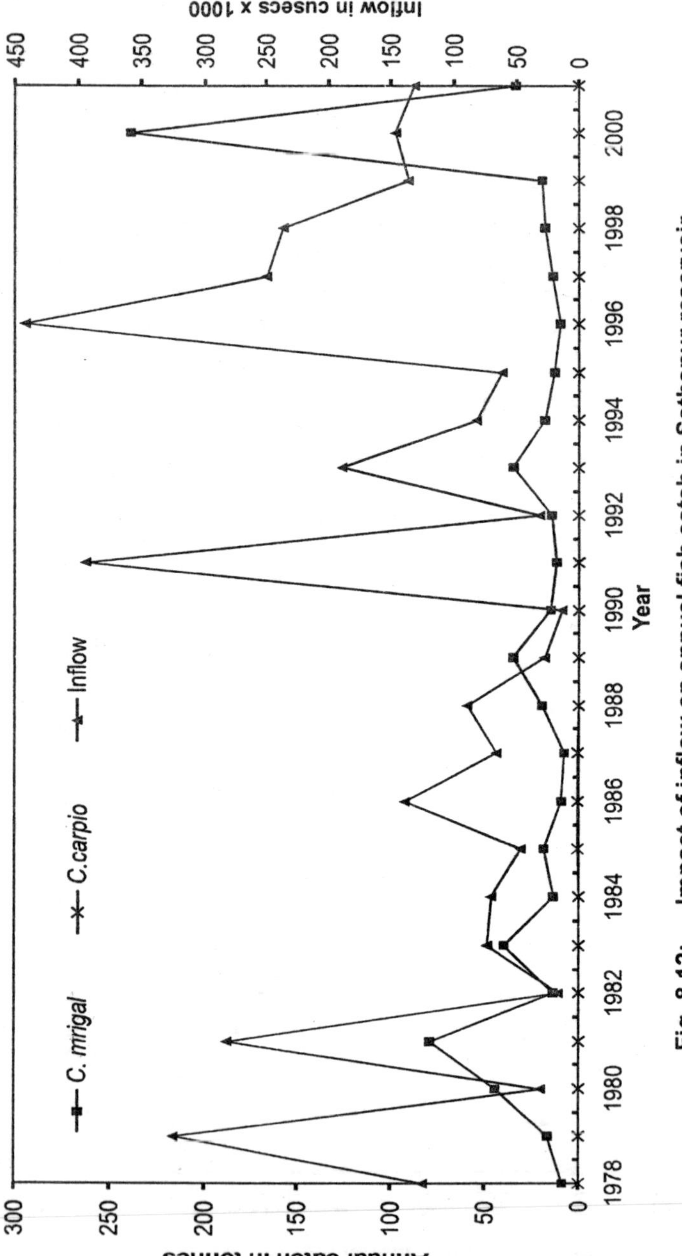

Fig. 8.12: Impact of inflow on annual fish catch in Sathanur reservoir

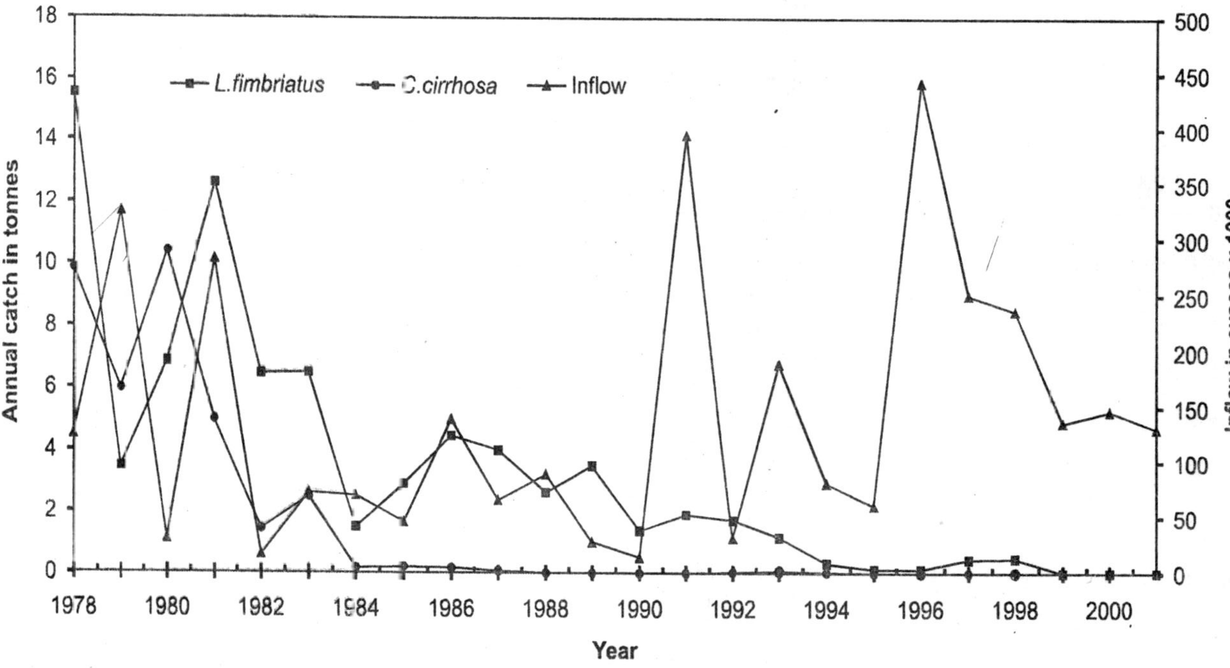

Fig. 8.13: Impact of inflow on annual fish catch in Sathanur reservoir

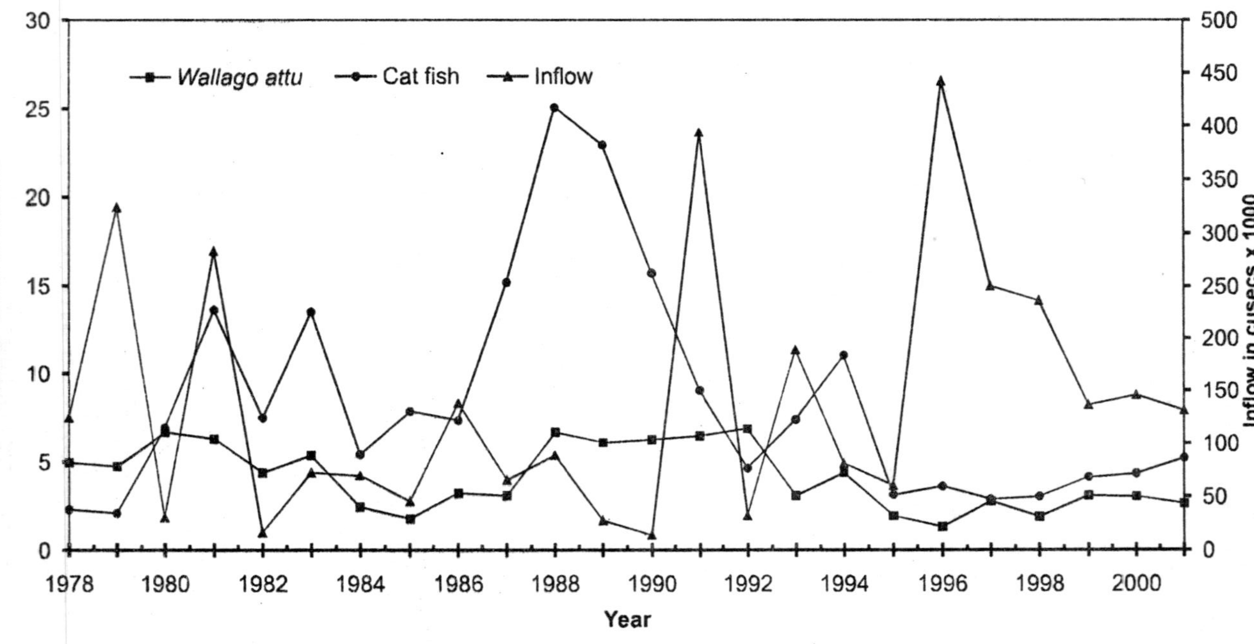

Fig. 8.14: Impact of inflow on annual fish catch in Sathanur reservoir

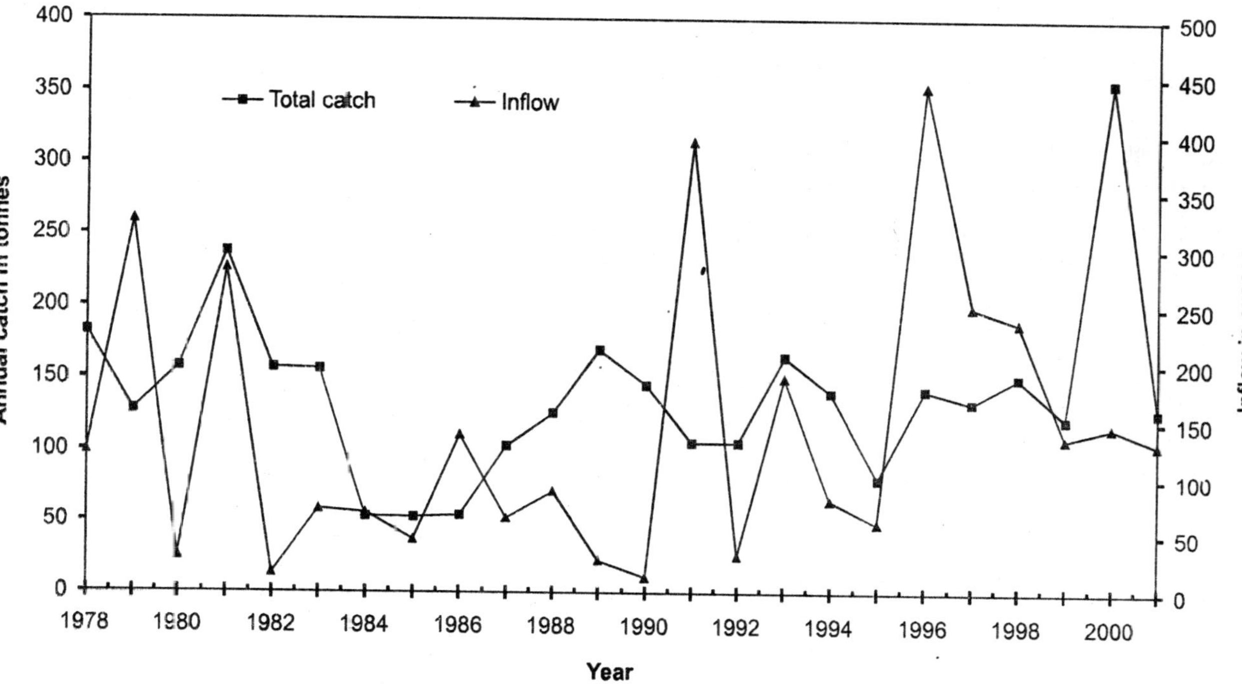

Fig. 8.15: Impact of inflow on total fish catch in Sathanur reservoir

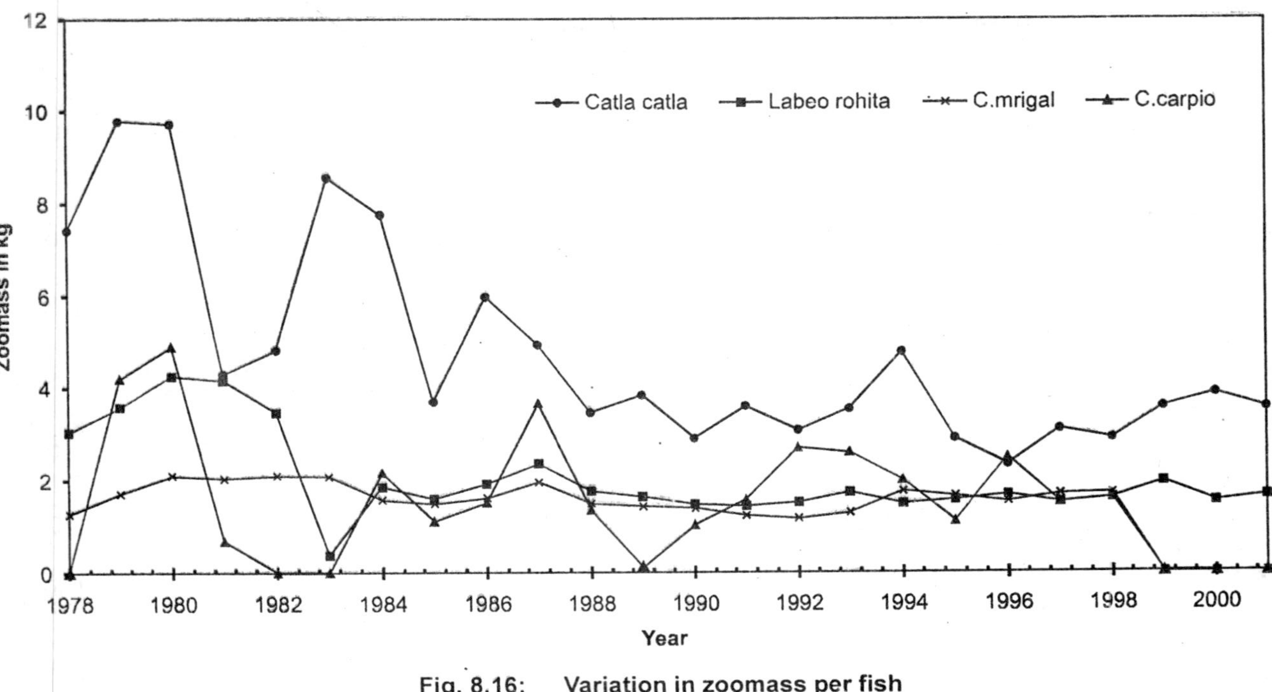

Fig. 8.16: Variation in zoomass per fish

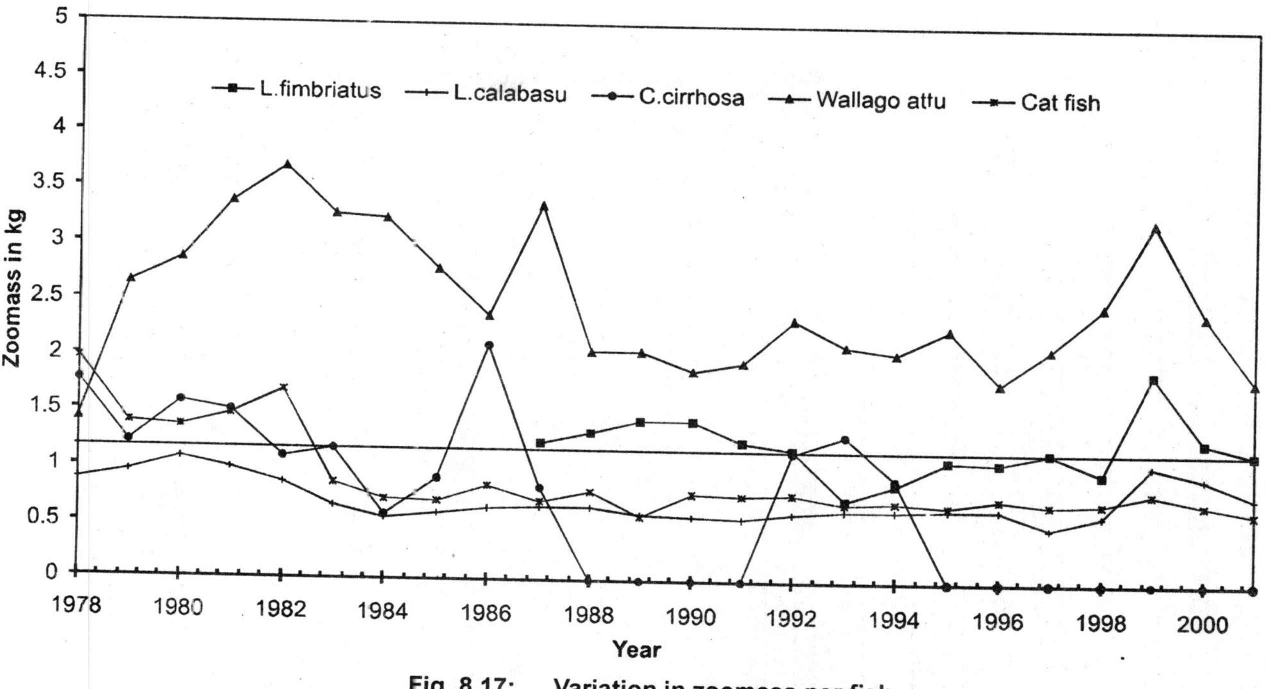

Fig. 8.17: Variation in zoomass per fish

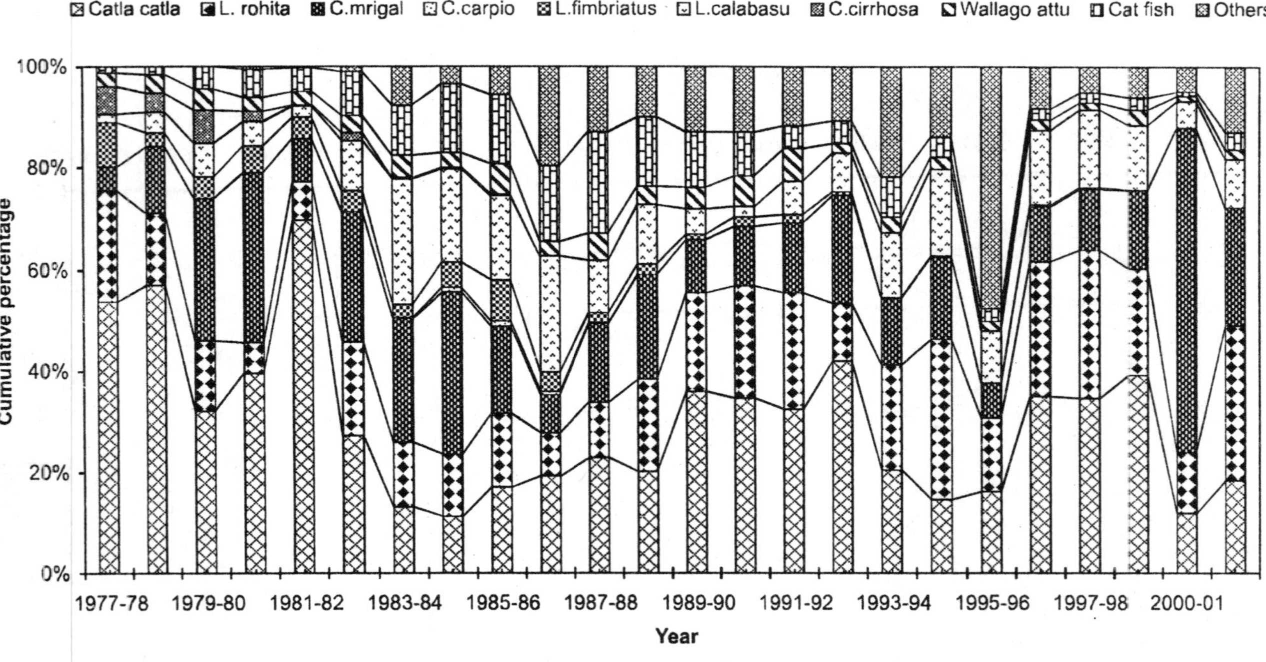

Fig. 8.18: Variation in annual catch of ichthyofauna in Sathanur reservoir

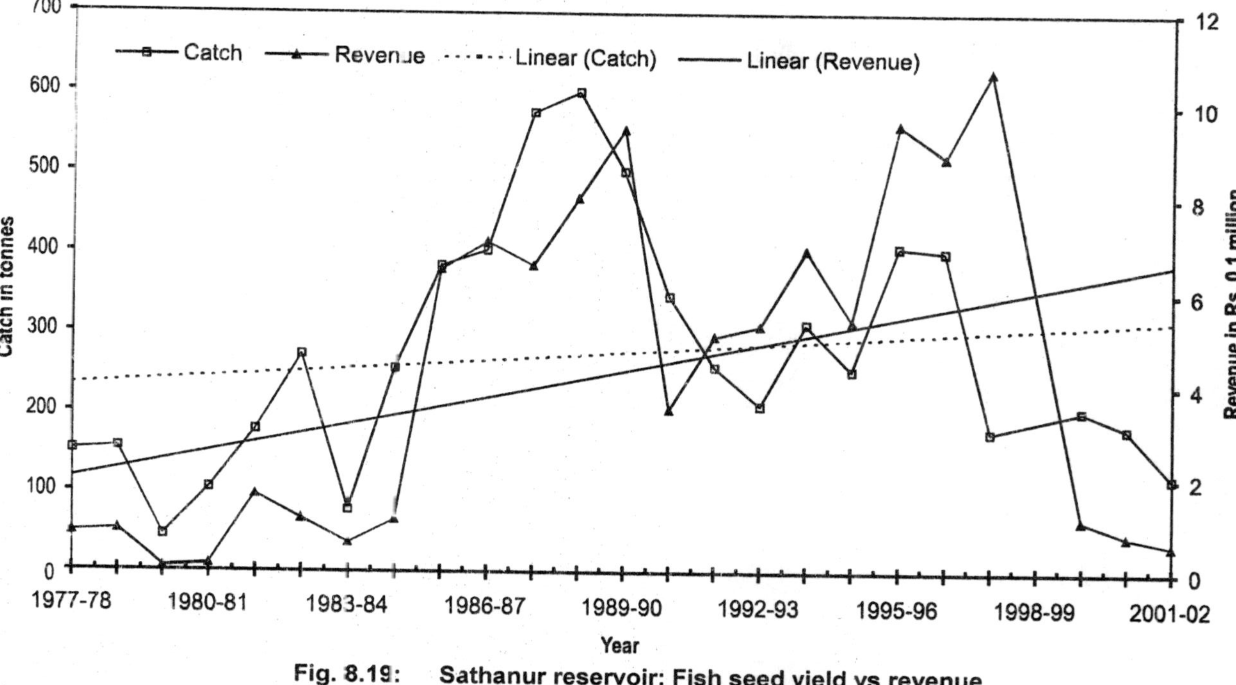

Fig. 8.19: Sathanur reservoir: Fish seed yield vs revenue

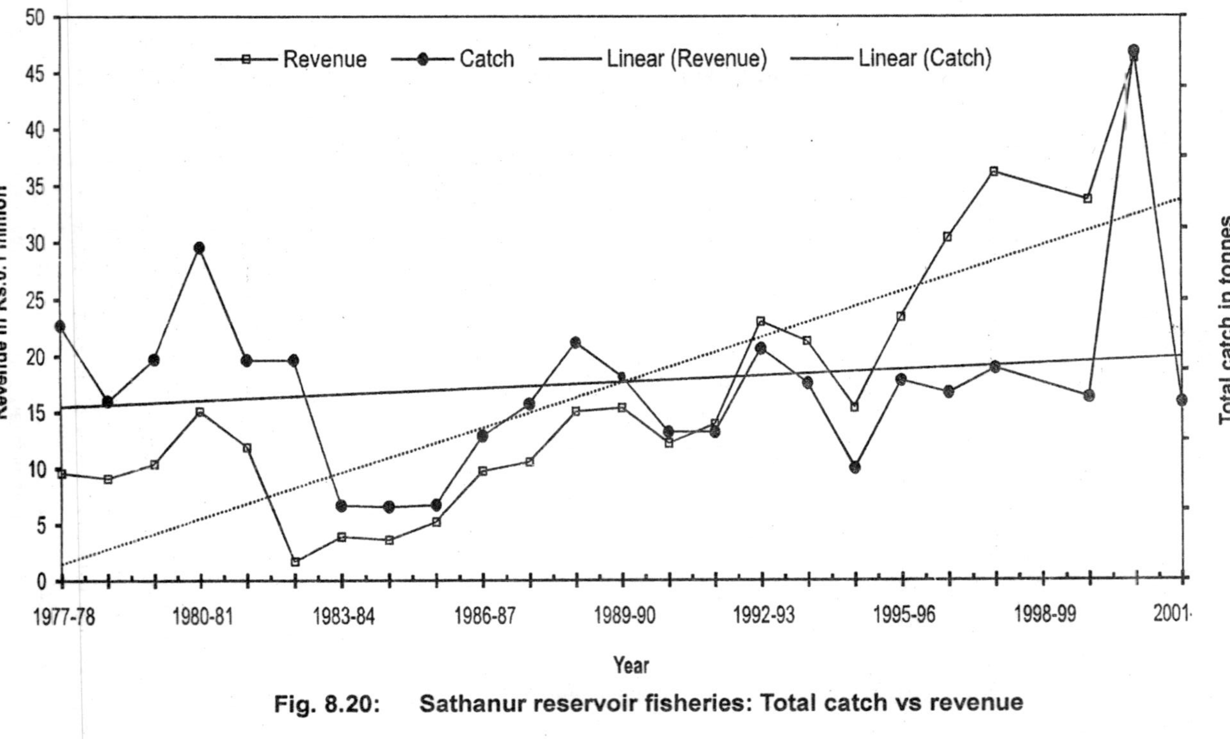

Fig. 8.20: Sathanur reservoir fisheries: Total catch vs revenue

In contrast, the yield of *L.rohita* doesn't display the widely ranging fluctuations as that of *Catla catla* (Figure—8.1). The yield after indicating a downward trend till 1980, picked up in 1983 before drooping down in 1984. The yield was more or less constant till 1987 and picked up in 1988. The curve representing *L.rohita* yield after illustrating minor fluctuations from 1989 to 1996 depicts an upward trend from 1996 onwards. There is a slight increasing trend in the rohu yield as inferred from the trendline (Figure 8.1).

There is no significant correlation between the annual precipitation and the annual catch of *Catla catla* and *L.rohita*, though at some stages rainfall does depict subtle influence over the catch. After 1983 there was a fall in the annual precipitation recorded in the study area, which continued till 1986. Incidentally the annual catch of the *Catla catla* and *L.rohita* also depicted a fall for the above said period but the trend reversed in 1987 when rainfall in the catchment again registered a fall, which continued through 1988 to 1990. In contrast there was a rise in the catch of the *Catla catla* and *L.rohita* for the above said period.

Figure—8.2 depicts the annual catch of *C. mrigal, L.fimbriaties* and *C.carpio* as a function of annual precipitation. The yield of *C.mrigal* registered an upward trend till 1981 when it recorded the maximum yield of 80 tonnes for the entire study period. The yield came down drastically by 83 per cent of its 1981 value in 1982, picked up in 1983 before falling in 1984. These well marked, fluctuations continued till 1998 when the recorded yield was 77 per cent of the 1981 value. Thus overall there is a downward trend in the yield of *C. mrigal* over the years, as is observed from the trend line (Figure 8.2).

The yield pattern of *L.fimbriatus* depicts a downward trend after 1981. The yield did pick up slightly in 1984, which continued till 1986, but thereafter there had been a gradual and steady decline in the yield. The yield in 1996 registered mere 2.8 per cent of its value in 1978, the year that had the maximum yield of *L.fimbriatus*. There was a very slight increase in the annual catch in the years 1997 and 1998. There is a marked declining trend in the annual catch (Figure—8.2). The annual catch of *C.carpio* was very low during the entire study period which paramounts to almost negligible when compared with the catch of other fished (Figure—8.2).

The yield curve of *C.mrigal* emulates precipitation curve during the initial years. Both curves depict a peak in 1981 and fall sharply in 1982. Both the curves illustrate a declining trend till 1986. After 1986 there is no significant correlation between the yield of *C.mrigal* and annual precipitation. A reversed trend is noticed during most of the years afterwards. The yield curve of *L.fimbriatus* emulates rainfall curve and the yield curve of *C.mrigal* during the initial years till 1985 and thereafter it becomes independent (Figure—8.2)

The annual yields of *L.calabasu* and *C.cirrhosa* and the influence of annual precipitation thereupon are depicted in figure 8.3. The yield of *C.cirrhosa* attained its maximum in 1980 and thereafter fell by almost 52 per cent in 1981 and continued thus in 1982 when the yield registered a fall of 86 per cent of its 1980 value. The yield did pick up slightly in 1981 but thereafter it steadily declined registering nil yield from 1988 to 1991 and again from 1995 to 1998. There is a declining trend in the catch as inferred from the trend line (Figure—8.3).

In contrast, the yield pattern of *L.calabasu* illustrates wide variations over the study period. The yield curve depicts an upward trend till 1987 (with minor fluctuation in between) when it registered 86 per cent increase in the yield its 1978 value. Thereafter the yield declined sharply till 1991 when it recorded 90 per cent reduction of its 1987 value. There was a continuous increase in the yield after 1991 till 1994. There was a slight decline in 1992. The yield showed an increasing trend till 1998. There is a marked ascending trend in the catch of *L.calabasu* over the study period as indicated by the trend line (Figure—8.3).

Annual precipitation in the catchments seemed to influence the catches of *C.cirrhosa* and *L. calabasu* during the initial periods as is evident from the Figure. From 1991 to 1996 the yield curves and the precipitation curve depict a similar pattern of rise and fall. After 1996 there is no relationship between *C.cirrhosa* and rainfall while the yield curve of *L.calabasu* seems to emulate the rainfall curve in a subtle manner till 1996 with the exceptions of years 1990 and 1992.

Figure—8.4 illustrates the yield pattern of *Wallago attu* and the cat-fishes and the influence of rainfall on the same. As indicated by the Figure, yield of cat- fishes show wide fluctuations, depicting an upward trend till 1988 and thereafter a declining trend till 1998. The yield of the cat-fishes in 1988-registered increase in their value in 1978 which was the highest recorded over the study period. Thereafter, the yield in 1998 was reduced to 88 per cent of their 1988 value. Yield pattern of *Wallago attu*, in contrast, didn't depict marked variations over the years and was more or less steady. The years 1984 to 1987 registered low yield of *Wallago attu*. The yield was almost steady from 1988 to 1992, came down slightly in 1993, picked up in 1994, reducing again in 1995 and 1996 picking up marginally thereafter. The annual catch of both the fish species register a declining trend for the entire study period (Figure—8.4).

There is no significant correlation between the annual precipitation and the yields of *Wallago attu* and cat-fishes throughout the study period. The yield curve of cat-fishes emulates the rainfall curve from 1980 to 1984 and yet again in 1986-87 and again in 1993-97 (Figure—8.4). The yield curve of *Wallago attu* follows the rainfall pattern from 1981-85 and again from 1991-95.

Figure—8.5 illustrates the yield pattern of fish fauna bracketed as 'others'. There has been a remarkable increase in their yield over the years. The yield which was nil in 1978 increased to sixty eight tonnes in 1994 before coming down to seven tonnes in 1998.There is a sharp increasing trend in their catches as indicated by the trend lines (Figure—8.5). The yield pattern is similar to the pattern of annual precipitation recorded in the catchments. There was an increasing trend from 1978 to 1993 in the rainfall received in the catchment and during this period the yield of the fishes started picking up. The rainfall and fish yield peaked in 1983 as indicated in the Figure. There was a slight downslide in the rainfall from 1993 to 1998 and so also in the yield. From 1989 to 1993 there is no marked similarity in the pattern of rainfall and yield curves (Figure—8.5). Again the patterns of the variations of the curves are similar from 1994 to 1996 and thereafter the patterns are reversed. There was a sharp decline of 31.6 per cent from 1997 to 1998 in the yield of the fishers.

The cumulative catch of the piscifauna is depicted in igure—8.6. The catch was maximum in 1981 and came down sharply in 1982 and the decline continued further till 1984 when the annual yield was reduced by 76 per cent of its 1981 value. The yield was almost constant during 1984-86 periods and thereafter depicted a rising trend till 1989 before falling in 1990-92. The mild fluctuations in the yield continued till 1998.

Figures—8.7 to 8.10 illustrate the relationship between the water level and the piscifauna yield. As evident from the Figures the water level curve doesn't depict wide range of fluctuations over the study period and the water level hasn't gone down below 60 feet during the entire study period. It is evident from Table—8.7 that there is no significant correlation between the water level and the catches of various fish species. Cat-fishes show negative correlation coefficient of -0.6, which suggests the increase in yield when water level was low and vice versa. Figure—8.9 substantiates this fact. The water level curve illustrates a downward trend from 1979 to 1989 while the curve representing cat-fishes depict an upward trend and this trend reverses after 1990 when the water level curve indicates a rising trend and the cat-fishes yield depicts a downward slide. As per the reservoir operating regulations the water level of the reservoir has to be maintained at 46 feet for the aquaculture purpose.

The relationship between the inflow to the reservoir and the yield of the individual species of fishes is indicated in Figures—8.11 to 8.15. The inflow illustrates wide range of fluctuations over the years. The inflow depicted a declining trend from the years 1979 to 1990 and registered a rising trend thereafter till 1996 – year of maximum inflow during the study period. The catches of *Catla catla* and *L.rohita* depict a downward trend till 1986 and thereafter there is an increasing trend till 1993 in case of *Catla catla* and till 1998 in case of *L.rohita* (Figure—8.11). Similarly, the minimum yields in case of *C. mrigal* was observed during the period when, total inflow recorded in the reservoir was minimum (Figure—8.12). Figure—8.13 illustrates the similar phenomenon of gradual decrease in the yield of *L.fimbriatus* and *C.cirrhosa* during the period of decrease in the total inflow. While, after 1990, the total inflow to the reservoir picked up, the catches of the above mentioned species continued sliding down. No significant relationship is observed between the inflow and the catches of *Wallago attu* and cat-fishes.

Table—8.2 presents the correlation coefficients between the commercial catches of piscifauna in the reservoir and the hydrological parameters. There is no significant correlation of the catches with the annual rainfall. Catches of *Catla catla, L.rohita, L. fimbriatus, C. cirrhosa* and *Wallago attu* depict a negative correlations while that of *C.carpio* and cat-fishes have no correlation at all. A mild positive but insignificant correlation is observed between the catches of the fishes categorised as others and C.mirgal, with the annual precipitation. Catches of *C.mrigal, C. carpio, C.cirrhosa, W.attu* and cat-fishes illustrate negative correlation with water level while that of *L.fimbriatus* registers no correlation. The yields of *Catla catla, L.rohita L.calabasu* and others depict mild positive but insignificant correlation with reservoir water level (Table—8.2).

No significant correlation between the catches and annual inflow to the reservoir is discerned. Catches of *C.carpio, L.fimbriatus,* and *L.calabasu* are negatively correlated while those of *C.mrigal* and *C.cirrhosa* are independent of the influence of the inflow. Annual yields of *Catla catla* and *L.rohita* have very slight insignificant correlation while the species categorised as others have slight (0.4) correlation with reservoir inflow.

The coefficients of correlation between the catches of various fish species for the study period are depicted as Table—8.6. The annual yields of *Catla catla* correlates positively with *L.rohita C.mrigal, L.fimbriatus, C.cirrhasu and W.attu* and negatively with *C.carpio, L.calabasu,* cat-fishes and others. The correlations are insignificant except that for with *L.fimbriatus* (0.6) and *C.cirrhosa* (0.6). Catch of *L.rohita* does not depict any significant correlation with the catch of other species except for *C.carpio* (-0.6). Yield of *C.mrigal* correlates positively with *L.fimbriatus L.calabasu, C.cirrhosa, W.attu* and cat-fishes. The correlations are slight and insignificant. Catch of *C.carpio* does not illustrate any significant correlation with the catches of other species. The yield of *L.fimbriatus* shows strong correlation with that of *C.cirrhosa*. Yield pattern of other fish species (others) correlates negatively with most of the fishes except with that of *L.rohita, L.calabasu* and cat-fishes but the correlations are insignificant.

The average zoomass perfish of the various species of fish is represented in Figures—8.16 and 8.17. The zoomass of *Catla catla* followed a steep downslide from 9.7 kg in 1979-80 to 2.9kg in 1998. *L.rohita* illustrated a steady zoomass of 1.5-1.7kg after a sharp fall in 1983 to 0.3kg. *C. mrigal* also had a steady zoomass over the years while the zoomass of *C.carpio* was, gradually reduced to 1.5–1.6 kg in 1997–1998 from 4.9 kg in 1979– 80 (Figure 8.16), *Wallago attu* species featured a slight increase in its zoomass during the initial periods till 1982 and thereafter marked a slight reduction before picking in 1987 and falling back in 1988. The zoomass was almost steady thereafter. *L.fimbriatus* marked a steady zoomass over the years (Figure 8.17). *L.calabasu* registered a steady decline in its zoomass from 1kg in 1980 to 0.6 kg in 1998. Zoomass of cat-fishes was steady over the years while that of *C.cirrhosa* illustrated slight fluctuations reducing to 0.9 kg in 1994 from 2 kg in 1986. With the exception of *Catla catla, L.rohita, C.carpio* and *Wallago attu* rest of the piscifauna illustrated more or less constant zoomass per fish.

Figure—8.18 depicts the distribution patterns of the piscifauna catch from the reservoir over the years. Years 1978 and 1979 were marked by the dominance of *Catla catla*, which comprised of more than 50 per cent of the total piscifauna catch followed by *L.rohita*. Years 1980-81 featured an increase in the *C. mrigal* catch but the pattern reverted in 1982 when again the *Catla catla* yield was the highest (69%). There was an even distribution of the catches in 1983, when *Catla catla* yield fell to 27 per cent and *C. mrigal* and *L.rohita* yields increased to 25.4 per cent and 18.6 per cent respectively. *L.calabasu* also had an appreciable yield of 9.8 per cent of the total catch in the same year.

From 1984-87 there was an increase in the catch of *L.calabasu* and decrease in *Catla catla* yield. Cat-fishes also recorded increased yield during these periods. In 1988 cat-fishes had 19.8 per cent of the share in total catch. After 1989 again there was increase in *Catla catla* yield and the cat-fishes declined. Other unidentified, small fishes like Tilapia and some cat-fishes also illustrated a slight increase in their catch and became predominant in 1996 when the catches of the other fishes were very low. *L.rohita* and *C. mrigal* maintained appreciable catch during the entire study period. Catch of *L.rohita* was highest in 1995 (31.6%). With all these variations in catch, the pattern in 1997–98 is marked with the dominance of *Catla catla*, followed by *L.rohita*, *C.mrigal* and *L.calabasu*. Yields of other species have been reduced to almost nil, thus highlighting the fact that these four species have established themselves well in the reservoir.

Introduction of exotic species is an important casual factor in threats to some native or endemic species because the invaders usually have fast growth rate, strong resistance to extreme environments, a wide food spectrums and high reproduction rate (Xie and Chen, 1999; Diamond and Case, 1986).

The variation in the production the fishes over the years can also be explained due to alteration in the physical and chemical environment of the habitat which may lead to tertiary impacts on fishes through tropic, host parasite, food, habitat competition or in other ways (Ruggles & Watt, 1975; Jessop, 1990b:) Physiological changes in biota may be induced by fluctuations in water level, sediment loading, transparency, draw down, precipitation etc (Baxter and Glaude, 1980; Bernacsek 1984).

From the preceding impact studies it is clear that these hydrological conditions are not the only factors influencing the fish yield. Water quality plays a major role in determining the production of fisher in a particular area. The sediment released in to the reservoir increases turbidity and can settle out on the eggs and can be detrimental to embryonic development as confirmed by other authors too (Zhong and Power, 1996; Fudge and Bodily, 1984). The increase in the total solids over the years as depicted in Figure—9.6 may thus have a profound impact on fish production. Rise in pH, alkalinity, hardness, total dissolved solids and other ions also affect the fish yield. The pH and alkalinity concentrations of the reservoir water exceed the standards for permissible limits specified for fish culture (Table—9.23).

Depletion of dissolved oxygen is the main parameter governing the survival of the fishes in a lake (Bodaly and Rosenberg, 1990; Ruggles and Watt, 1975). The vertical gradient of the dissolved oxygen during the survey period is depicted in Figures—9.16 and 9.17. Tables—9.24 and 9.26 present an account of the D.O concentration various sites and limnions during the study period. The maximum epilimnetic D.O recorded was 6.2 mgl^{-1} in June and minimum hypolimnetic D.O was recorded as 1.3 mgl^{-1} in July. An ambient D.O level of 3 mgl^{-1} is recommended for the survival of fisher (Table—9.23). Fish kills generally over due to the shortage of oxygen and this was confirmed by some local fishermen that during the past four-five years many fishes and fingerlings were found dead and some were suffering from some mysterious disease. Over fishing also contributes greatly to the changes in the fish community (Xie and Chen, 1999; Zhang et al, 1991; Chang et al 1995, Nikolskii, 1969). Though the officials vehemently denied this phenomenon taking place in Sathanur reservoir, the talks with the local fishermen did reveal that illegal fishing does take place especially in the upper reaches of the reservoir.

Table—8.5 presents the average, standard deviation and coefficient of variation for the yield of individual fish species. *C.carpio* has the maximum coefficient of variation followed by *L.fimbriatus* and *C.cirrhosa* referring to maximum fluctuations in their yields. *C.carpio* and *L.cirrhosa* registered nil yields during the study period. *Wallago attu* and *L.rohita* depict the least fluctuations signifying more or less steady yield. In the hydrological parameters, water level depicts least fluctuation followed by rainfall. Inflow registers the maximum C.V. of 91.4 per cent.

Catch per hectare of the reservoir catchment (in kg) of the piscifuana is represented in Table—8.8. The total catch per hectare in 1998 stood at 84 kg. The contribution is dominantly from *Catla catla* (29.3 kg), followed by *L.rohita* (24.4 kg), *C. mrigal* (99 kg) and *L.calabasu* (12.6 kg) (Table—8.8).

The details of the fish seed stocking in Sathanur reservoir for the years 1999-2000, 2000-2001 and 2001-2002 have been presented in Tables—8.13 and 8.14. Table—8.15 presents an account of the expenses and profit incurred from the reservoir fishery. Figure—8.19 illustrates the trends between the fish seed production and revenue earned. Revenue marks a higher trend line compared to the fish seed production, which came down heavily after 1989 and after displaying a slight rise after 1993 again fell sharply in 1998. In contrast revenue kept climbing up after a steep decline in 1990.

Figure 8.20 illustrates the relationship between total fish catch and revenue generated. The trend line for fish catches registers a downslide while that for revenue marks an ascent. Though the total catch of the reservoir has come down in the recent years there has been a steady rise in the revenue due to increased price of the produce. The details are represented in Tables—8.9 Tables 8.9, 8.10, 8.11, 8.12 furnishes the details regarding the expenses incurred and profit earned from the Sathanur reservoir fisheries. There has been remarkable revenue of Rs.1.3 millions.

Table 8.13 Details of fish seed stocking of Sathanur Reservoir 1999–2000

Months	*C.catla*	*L.rohita*	*C.mrigal*	*C.carpio*
April	–	–	–	–
May	–	–	–	–
June	2.10	–	–	–
July	0.80	0.165	–	–
August	0.82	–	–	–
September	0.58	1.26	–	–
October	0.80	0.86	1.34	–
November	0.40	0.24	0.20	–
December	–	–	–	–
January	–	–	–	–
February	–	–	–	–
March	–		–	–
Total	5.50	2.53	1.54	–

Table 8.14 Details of fish seed stocking of Sathanur Reservoir, 2001-2002

Months	*C.catla*	*L.rohita*	*C.mrigal*	*C.carpio*
April	–	–	–	–
May	–	–	–	–
June	0.88	–	–	–
July	0.45	–	–	–
August	–	1.53	–	–
September	0.76	0.53	0.31	–
October	0.075	–	0.90	–
November	0.49	–	–	–
December	–	–	–	–
January	–	–	–	–
February	–	–	–	–
March	–	–	–	–
Total	2.655	2.06	1.21	–

Table 8.15 Expenses and profit details

	Scheme	Income (in million rupees)	Expenditure (in million rupees)	Operating profit (in million rupees)
1.	Fish seed production	1.0	0.8	0.3
2.	Fish exploitation	2.8	2.1	0.7
3.	Fish marketing	2.7	2.3	0.4
	Total	6.6	5.2	1.3

9

Reservoir Water Quality

INTRODUCTION

Artificial lakes, which result from the impoundment of water behind a dam, are limnologically different from natural lakes in several respects. Firstly, in natural lakes water goes out as overflow of the epilimnic layers but in reservoirs made for irrigation it is the hypolimnic water that is generally released from the sluices to the irrigation canals. Secondly, the water level fluctuations in such reservoirs are incessant and sharp compared to slow and steady changes in water levels that occur in natural lakes. In terms of chemistry and biology, too, man-made reservoirs are different from natural lakes. In the former an essentially lotic habitat is suddenly changed to a lentic one. This has far reaching impacts on the chemistry the water and the biota.

Studies on the water quality of man-made reservoirs are important because they reflect the impacts of the hydrologic regime associated with the operation of the reservoir. They also reflect the impacts of what is happening to the feeder river upstream, and of the land-use in the catchment. Above all, a study of the water quality of the reservoir is essential to assess the fitness of the water for various purposes-irrigation, drinking, other domestic use, and industry.

In this chapter we present a study of the water quality of Sathanur reservoir on the basis of our analysis of the historic data obtained from Public Works Department (PWD) as well as extensive experimentation undertaken by us.

STATUS

Data pertaining to the water quality of the Sathanur reservoir, for a period of eleven years (1990-2001), was collected from the SRP authorities and is presented as Table—9.1. The water quality data for the Pick-up dam was sought from the office of Thiruvannamalai Municipality and is presented as Table—9.2. The minimum and maximum values of various limnological parameters for the study periods for Sathanur reservoir have been given in Tables—9.3 to 9.13, and that for Pick–up dam in Tables—9.14 to 9.21. Table—9.22 presents the details of various preservation techniques for physico-chemical parameters. Table—9.23 presents a comparative account of the water quality of reservoir and Pick–up dam with that of the permissible limits of various physico-chemical parameters as specified in Indian standards of water for various usages.

Stratification studies for the reservoir were conducted during the months of June '99 and July'99 when the outflow from the reservoir was nil. Normally the water is released downstream from the reservoir till the middle of the May. Thus, June and July were considered the ideal months to study the stratification pattern of the reservoir along with other limnological parameters. Primary productivity studies were also undertaken during this time. The results are represented in Tables—9.24 to 9.27.

Figures—9.1 to 9.9 depict the variations in the limnological parameters of the reservoir. The impact of precipitation, if any, on the concentration of the parameters, has also been assessed. Figure—9.1 indicates the relationship between the outflow, inflow and precipitation for the reservoir for the study period. The physico–chemical parameters of Pick-up dam as function of outflow from the reservoir, over a period of time, have been plotted as Figures 9.10 to 9.15.

Table—9.1 Physico-chemical characteristics of water in the Sathanur reservoir

Month/ year	Colour	pH	Electrical conductivity (μ mho cm^{-1})	Alkalinity (mg l^{-1})	Total solids (mg l^{-1})	TDS (mg l^{-1})	Suspended solid (mg l^{-1})	Total hardness (mg l^{-1})	Calcium hardness (mg l^{-1})	Magnesium (mg l^{-1})	Sulphate (mg l^{-1})	Chloride (mg l^{-1})	Level (m)
1/90	Clear	8.04	399	190	308	223	85	125	32	10.80	10.50	64	54.65
11/90	Clear	7.67	350	160	256	196	60	145	32	16.8	17.3	53.20	–
12/91	Clear	8.77	475	185	274	252	22	175	28	25.2	20.43	42.48	118.95
1/92	Clear	8.25	505	–	436	268	168	175	8	24.0	9.85	46.08	117.74
6/92	Clear	7.95	615	–	408	326	82	180	22	30.0	19.22	63.72	–
7/92	Clear	7.93	618	–	400	328	72	190	28	28.8	21.87	60.18	–
11/92	Clear	8.25	336	140	238	175	63	125	26	14.4	10.6	28.4	–
1/93	Clear	7.82	399	165	243	207	36	150	28	19.20	10.6	24.8	–
3/94	Clear	7.36	522	210	335	277	58	165	26	24	13.44	60.0	31.4
4/94	Clear	8.27	526	195	376	279	97	250	26	44.40	14.40	56.0	27.4
5/94	Clear	8.63	635	200	499	337	162	205	30	31.2	24	60.3	72.93
6/94	Clear	8.79	560	150	299	297	2	165	26	24	18.2	63.8	69.81
8/94	Clear	8.76	553	83	400	293	107	195	34	26.4	18.2	49.6	21.62
9/94	Clear	8.58	522	165	377	282	95	175	26	26.4	21.1	56.7	22
10/94	Clear	7.5	496	180	295	268	27	160	30	20.4	6.7	49.6	22.9
11/94	Clear	7.65	504	165	304	272	32	160	30	20.4	8.6	49.6	32.1
12/94	Clear	7.22	520	195	285	281	4	170	30	34.8	7.7	40.1	34.9
1/95	Clear	8.03	542	190	304	293	11	175	26	25.2	10.6	53.2	31.3
4/95	Clear	7.97	604	185	326	326	–	175	26	26.4	15.4	74.4	19.35
6/95	Clear	7.97	633	235	414	342	72	195	32	27.6	32.6	148.9	19.1
10/95	Slightly turbid	8.42	703	155	575	380	195	210	32	31.2	11.5	145.3	–
11/95	Clear	8.56	622	150	412	336	76	205	38	26.4	9.6	–	–

(Table Contd...)

Month/ year	Colour	pH	Electrical conductivity (µ mho cm^{-1})	Alkalinity (mg l^{-1})	Total solids (mg l^{-1})	TDS (mg l^{-1})	Suspended solid (mg l^{-1})	Total hardness (mg l^{-1})	Calcium hardness (mg l^{-1})	Magnesium (mg l^{-1})	Sulphate (mg l^{-1})	Chloride (mg l^{-1})
8/96	Clear	7.9	711	250	482	384	98	27 5	32	46.8	7.2	74.4
9/96	Clear	8.02	701	255	464	378	86	275	26	50.4	12.0	–
10/96	Clear	7.7	713	105	472	384	88	225	36	36	7.2	–
11/96	Clear	7.6	716	420	490	386	104	230	24	40.8	7.2	–
12/96	Clear	7.66	719	26	511	388	123	245	28	42	12.0	–
1/97	Clear	7.98	700	110	490	378	112	225	36	32.4	7.2	74.4
2/97	Clear	7.91	706	110	482	381	101	240	26	42	7.2	74.4
3/97	Clear	7.87	703	115	481	380	81	235	28	39.6	19.2	74.4
4/97	Clear	8.46	696	105	468	376	88	225	28	37.2	19.2	74.4
5/97	Clear	8.04	724	260	456	390	66	205	28	32.4	13.44	70.9
6/97	Clear	8.5	812	285	513	438	74	250	30	42	25.88	148.89
7/97	Clear	8.48	802	70	543	432	116	230	30	37.2	26.88	152.44
8/97	Clear	7.97	625	152	–	340	–	195	32	27.6	17.5	125
9/97	Clear	8.12	623	158	–	336	–	190	28	22.8	15.4	141.3
10/97	Clear	8.50	625	185	–	337	–	185	30	25.4	17.3	141.8
11/97	Clear	7.64	625	235	422	333	84	175	32	18	17.3	60.3
12/97	Clear	8.12	620	170	399	335	64	185	26	28.8	11.64	86.7
1/98	Clear	8.67	684	100	432	369	63	190	28	28.8	13.4	63.8
2/98	Clear	8.36	760	60	450	410	40	175	34	21.6	22.08	85.08
3/98	Clear	7.80	687	Nil	452	370	82	175	26	26.4	15.4	70
4/98	Clear	8.41	704	30	390	380	10	200	22	34.8	13.4	67.4
5/98	Clear	8.33	705	Nil	346	381	–	190	26	30	15.4	67.4
6/98	Turbid	8.35	665	40	626	359	267	200	26	32.4	13.4	67.4
7/98	Turbid	8.88	657	80	550	355	195	215	26	36.0	18.2	74.4
8/98	Clear	8.66	616	50	480	333	147	210	22	34.8	15.4	74.4

(Table Contd...)

Month/ year	Colour	pH	Electrical conductivity (μ mho cm^{-1})	Alkalinity (mg l^{-1})	Total solids (mg l^{-1})	TDS (mg l^{-1})	Suspended solid (mg l^{-1})	Total hardness (mg l^{-1})	Calcium hardness (mg l^{-1})	Magnesium (mg l^{-1})	Sulphate (mg l^{-1})	Chloride (mg l^{-1})
7/99	Clear	8.53	662	80	416	357	59	150	18	25.2	15.4	81.5
8/99	Clear	7.2	646	–	386	349	37	200	20	36	12	88.6
9/99	Clear	7.87	561	–	311	303	18	170	26	25.2	5.8	56.7
10/99	Clear	7.9	562	–	343	303	30	170	22	27.6	5.8	60.3
11/99	Clear	7.94	56	–	325	3030	22	175	27.6	27.6	56.7	–
12/99	Clear	7.7	562	–	310	303	7	165	26.4	26.4	56.7	–
05/01	Clear	7.45	601	220	380	325	55	170	22	27.6	19.2	88.6
06/01	Clear	7.75	599	165	345	325	20	165	32	20.4	19.2	88.6
07/01	Clear	7.81	601	205	448	325	123	165	26	24	11.5	92.2
08/01	Clear	7.64	602	150	586	325	65	170	18	36	15.4	85.1
09/01	Clear	7.8	516	353	200	279	74	140	20	21.6	14.4	85.08
10/01	Clear	7.99	567	382	220	306	76	150	22	22.8	14.4	85.08
11/01	Clear	8.0	545	360	205	299	66	145	22	21.6	14.4	88.7
12/01	Clear	7.79	443	373	210	293	79	145	20	22.8	14.4	81.5
01/02	Clear	7.93	541	370	200	291	79	145	22.8	22.8	14.4	74.4

Table—9.2 Physico-chemical characteristics of water in the Pick-up dam

Month/ Year	Turbidity	pH	Electrical conductivity (μ mho cm^{-1})	Total solids (mg l^{-1})	$CaCO_3 + HCO_3$ Hardness (mg l^{-1})	Total Hardness $CaCO_3$ (mg l^{-1})	Chloride (mg l^{-1})	Alkalinity (mg l^{-1})	Fluoride (mg l^{-1})	Iron (mg l^{-1})	Ammoniac Nitrogen (mg l^{-1})	Albuminoidal Nitrogen (mg l^{-1})	Nitrate Nitrogen (mg l^{-1})
3/85	6	8.0	530	370	154	154	54	198	0.08	0.15	Trace	0.02	0.25
6/85	4	8.5	530	370	148	148	58	196	0.8	–	0.10	0.1	1.5
7/85	20	8.5	530	370	140	140	62	196	0.6	0.4	0.04	0.16	–
11/85	8	8.4	380	270	134	134	33	170	0.6	0.25	Trace	Trace	–
5/86	6	8.3	600	460	176	176	56	234	0.8	0.250	Trace	Trace	–
10/86	40	8.3	365	260	112	112	34	144	0.6	1.60	0.01	0.02	–
4/87	10	8.2	560	400	130	130	52	176	0.7	0.5	0.02	0.05	0.25
1/88	10	8.2	400	290	126	126	40	159	0.6	0.2	0.02	0.40	0.25
10/88	5	8.4	430	210	120	120	51	140	0.5	0.05	Trace	0.20	–
4/89	5	8.5	510	360	126	126	58	166	0.6	–	0.02	0.06	–
8/89	16	6.8	420	290	124	124	40	40	0.3	0.2	0.02	0.08	0.25
11/89	10	8.0	420	300	144	144	50	154	0.7	0.3	0.10	0.20	–
8/90	7	8.6	570	400	152	152	66	178	0.7	0.15	0.02	0.20	2.0
12/90	7	7.6	340	240	116	116	33	122	0.3	0.15	0.03	0.16	–
8/91	10	8.5	420	290	137	137	43	154	0.5	0.2	0.02	0.20	–
11/91	15	7.9	340	240	102	102	36	112	0.3	0.4	0.04	0.20	2.00
10/92	8	8.2	380	280	127	127	32	141	0.5	0.2	0.02	0.20	0.50
12/92	10	8.2	310	220	125	125	31	141	0.6	0.3	0.12	0.20	–
6/93	7	7.9	430	300	145	145	33	168	0.2	0.2	Trace	0.16	–
11/93	7	8.3	490	340	141	141	64	172	0.5	0.05	0.02	0.4	–
8/94	5	8.2	510	350	114	114	54	196	0.7	0.05	0.16	0.32	–
11/94	10	7.8	410	380	162	162	31	204	0.4	0.3	Trace	0.16	–
9/95	7	8.4	530	370	150	150	68	196	0.5	0.15	Trace	0.40	–
12/95	6	7.8	680	380	156	156	92	240	0.6	0.05	Trace	0.32	–

(Table Contd...)

Month/ Year	Turbidity	pH	Electrical conductivity (μ mho cm^{-1})	Total solids (mg l^{-1})	$CaCO_3$ + HCO_3 Hardness (mg l^{-1})	Total Hardness $CaCO_3$ (mg l^{-1})	Chloride (mg l^{-1})	Alkalinity (mg l^{-1})	Fluoride (mg l^{-1})	Iron (mg l^{-1})	Ammoniac Nitrogen (mg l^{-1})	Albuminoidal Nitrogen (mg l^{-1})	Nitrate Nitrogen (mg l^{-1})
5/96	15	8.3	740	520	166	166	106	244	0.6	0.15	0.08	0.40	–
10/96	10	7.5	560	400	160	160	54	210	0.6	0.3	0.08	0.40	–
11/96	15	7.3	510	360	170	170	50	220	0.7	0.3	–	–	–
6/97	10	8.6	640	450	182	182	72	266	0.5	0.3	–	–	–
11/97	10	8.1	560	390	170	170	68	216	0.7	0.15	Trace	0.16	–
6/98	10	8.6	570	400	156	156	72	266	0.8	0.3	Trace	0.16	–
12/98	7	8.3	510	360	168	168	48	250	0.3	8.3	0.04	0.24	–
6/99	7	8.3	640	450	198	198	70	250	0.6	0.15	–	–	–
11/99	5	7.6	520	360	146	146	60	202	0.4	0.10	0.01	0.28	–
3/00	7	7.7	540	380	136	136	64	206	0.5	0.15	Trace	0.08	–
11/00	10	7.4	500	410	164	164	52	200	0.6	0.4	Trace	0.08	–
6/01	7	8.8	580	410	148	148	72	202	0.4	0.1	Trace	0.08	–
11/01	7	7.8	460	320	148	148	58	162	0.5	0.1	0.02	0.16	–

Table—9.3 Total alkalinity values in the Sathanur reservoir

Year	Maximum value (in mg l^{-1})	Minimum value (in mg l^{-1})
1990	190	160
1991	185	–
1992	140	–
1993	165	-
1994	210	83
1995	235	150
1996	420	26
1997	285	105
1998	280	230
1999	240	200
2001	373	150

Table—9.4 pH values in the Sathanur reservoir

Year	Maximum value	Minimum value
1990	8.04	7.6
1991	8.7	–
1992	8.3	–
1993	7.8	–
1994	8.8	7.2
1995	8.6	7.9
1996	8.2	7.6
1997	8.5	7.6
1998	8.9	7.8
1999	8.5	7.2
2001	8.0	7.4

Table—9.5 Electrical conductivity values in the Sathanur reservoir

Year	Maximum value (in μ mho cm^{-1})	Minimum value (in μ mho cm^{-1})
1990	399	350
1991	475	–
1992	618	336
1993	399	–
1994	635	496
1995	703	542
1996	719	701
1997	812	620
1998	760	616
1999	662	56
2001	602	443

Table—9.6 Values of suspended solids in the Sathanur reservoir

Year	Maximum value (in mg l^{-1})	Minimum value (in mg l^{-1})
1990	85	60
1991	22	–
1992	168	63
1993	36	–
1994	162	2
1995	195	11.7
1996	123	86
1997	116	66
1998	267	10
1999	175	18
2001	123	20

Table—9.7 Values of total dissolved solids in the Sathanur reservoir

Year	Maximum value (in mg l^{-1})	Minimum value (in mg l^{-1})
1990	223	196
1991	252	–
1992	328	175
1993	207	-
1994	337	268
1995	380	293
1996	386	378
1997	438	333
1998	410	333
1999	357	303
2001	325	279

Table—9.8 Calcium hardness values in the Sathanur reservoir

Year	Maximum value (in mg l^{-1})	Minimum value (in mg l^{-1})
1990	32	–
1991	28	–
1992	28	8
1993	28	–
1994	31	26
1995	38	26
1996	36	24
1997	32	26
1998	34	22
1999	28	18
2001	32	20

Table—9.9 Chloride values in the Sathanur reservoir

Year	Maximum value (in mg l^{-1})	Minimum Value (in mg l^{-1})
1990	64	53.2
1991	42.5	–
1992	63.7	28.4
1993	24.8	–
1994	63.8	40.1
1995	148.9	53.2
1996	74.4	–
1997	152.4	60.3
1998	85.1	63.8
1999	88.6	56.7
2001	92.2	81.5

Table—9.10 Total solid values in the Sathanur reservoir

Year	Maximum value (in mg l^{-1})	Minimum value (in mg l^{-1})
1990	308	256
1991	274	–
1992	436	238
1993	243	–
1994	499	285
1995	575	304
1996	511	464
1997	543	376
1998	626	346
1999	416	310
2001	586	200

Table—9.11 Total hardness values in the Sathanur reservoir

Year	Maximum value (in mg l^{-1})	Minimum value (in mg l^{-1})
1990	145	125
1991	175	-
1992	190	125
1993	150	-
1994	250	160
1995	210	175
1996	275	225
1997	250	175
1998	215	175
1999	200	2
2001	170	140

Table—9.12 Magnesium values in the Sathanur reservoir

Year	Maximum value (in mg l^{-1})	Minimum value (in mg l^{-1})
1990	16.8	10.8
1991	25.2	-
1992	30	14
1993	19.2	-
1994	44.4	20.4
1995	31.2	25.2
1996	50.4	36
1997	42	18
1998	36	21.6
1999	36	6
2001	23	22

Table—9.13 Sulphate values in the Sathanur reservoir

Year	Maximum value (in mg l^{-1})	Minimum value (in mg l^{-1})
1990	17.3	10.5
1991	20.4	–
1992	21.9	9.9
1993	10.6	–
1994	24	6.7
1995	32.6	9.6
1996	12	7.2
1997	26.9	7.2
1998	22.1	13.4
1999	12	5.8
2001	19.2	11.5

Table—9.14 Turbidity of water in the Pick-up dam

Year	Maximum value (in N.T.U)	Minimum value (in N.T.U)
1985	20	4
1986	40	6
1987	10	–
1988	10	5
1989	16	5
1990	7	–
1991	15	10
1992	10	8
1993	7	–
1994	10	5
1995	7	6
1996	15	10
1997	10	7
1998	10	7
1999	10	5
2000	10	7
2001	7	–

Table—9.15 pH of the water in the Pick-up dam

Year	Maximum value	Minimum value
1985	8.5	8
1986	8.3	–
1987	8.2	–
1988	8.4	8.2
1989	8.5	6.8
1990	8.6	7.6
1991	8.5	7.9
1992	8.2	–
1993	8.3	7.9
1994	8.2	7.8
1995	8.4	7.8
1996	8.3	7.3
1997	8.6	8.0
1998	8.5	8.3
1999	8.3	7.6
2000	7.7	7.4
2001	8.8	7.8

Table—9.16 Chloride values in the Pick-up dam

Year	Maximum value (in mg l^{-1})	Minimum value (in mg l^{-1})
1985	62	33
1986	56	34
1987	52	–
1988	51	40
1989	58	40
1990	66	33
1991	43	36
1992	32	31
1993	64	33
1994	54	31
1995	92	68
1996	106	50
1997	86	68
1998	72	48
1999	70	60
2000	64	52
2001	72	58

Table—9.17 Fluoride values in the Pick-up dam

Year	Maximum value (in mg l^{-1})	Minimum value (in mg l^{-1})
1985	0.8	0.6
1986	0.8	0.6
1987	0.7	–
1988	0.6	0.5
1989	0.7	0.3
1990	0.7	0.3
1991	0.5	0.3
1992	0.6	0.5
1993	0.5	0.2
1994	0.7	0.4
1995	0.6	0.5
1996	0.7	0.6
1997	0.7	0.5
1998	0.8	0.3
1999	0.6	0.4
2000	0.6	0.5
2001	0.5	0.4

Table—9.18 Electrical conductivity values in the Pick-up dam

Year	Maximum value (in μ mho cm^{-1})	Minimum value (in μ mho cm^{-1})
1985	530	380
1986	600	365
1987	560	–
1988	430	400
1989	420	–
1990	570	340
1991	420	340
1992	380	310
1993	490	430
1994	510	410
1995	680	530
1996	740	510
1997	730	520
1998	570	510
1999	640	500
2000	540	500
2001	580	460

Table—9.19 Alkalinity values in the Pick-up dam

Year	Maximum value (in mg l^{-1})	Minimum value (in mg l^{-1})
1985	198	170
1986	234	144
1987	176	–
1988	159	140
1989	166	140
1990	178	122
1991	112	154
1992	141	–
1993	172	168
1994	204	196
1995	240	196
1996	244	220
1997	280	216
1998	266	250
1999	250	202
2000	206	200
2001	202	162

Table—9.20 Total solids values in the Pick-up dam

Year	Maximum value (in mg l^{-1})	Minimum value (in mg l^{-1})
1985	370	–
1986	460	260
1987	400	–
1988	290	210
1989	360	290
1990	400	240
1991	290	240
1992	280	220
1993	340	300
1994	380	350
1995	380	370
1996	520	360
1997	510	390
1998	400	360
1999	450	350
2000	410	380
2001	410	320

Table—9.21 Total hardness values in the Pick-up dam

Year	Maximum value (in mg l^{-1})	Minimum value (in mg l^{-1})
1985	154	134
1986	176	112
1987	130	–
1988	126	120
1989	144	124
1990	152	116
1991	137	102
1992	127	125
1993	145	141
1994	162	114
1995	156	150
1996	170	160
1997	186	166
1998	168	156
1999	198	146
2000	164	136
2001	148	–

Table—9.22 Preservation techniques for water sample

Sl. No	Parameter	Recommended sample volume	Type of container	Preservation	Allowable holding periods
1.	Acidity	100 ml	P, G	4°C refrigeration	24 hrs
2.	Alkalinity	100 ml	P, G	4°C refrigeration	24 hrs
3.	Chloride	50 ml	P, G	Not required	24 hrs
4.	DO: Winkler	300 ml	P, G	Fix on site	
5.	Hardness	100 ml	P, G	4°C refrigeration	6 hrs
6.	pH	100 ml	P, G	Determine on site	Unstable
7.	Phosphorous	100 ml	P, G	40mg H_gCl_2 1- 4°C refrigeration	7 days
8.	Electrical conductivity	100 ml	P, G	4°C refrigeration	24hrs
9.	Solids (TS, TDS & TSS)	100 ml	P, G	4°C refrigeration	24 hrs
10.	Sulphate	50 ml	P, G	4°C refrigeration	7days
11.	Heavy metals	200 ml	P (A), (A)	HNo_3 5ml/litre PH=2	6 months

P—Plastic, G—Glass P (A) / G (A); rinsed with 1 + 1 HNO_3

Table—9.23 Comparison of water quality of Sathanur reservoir and Pick-up dam with Indian standards for different usages

Sl. No.	Characteristic	For use as raw water for public water supply and bathing ghats IS: 2296–P 1974	Drinking water IS 10500	For fish culture IS 13891: 1994	For irrigation IS 2296-1974	For swimming pool IS 3328: 1993	Water quality of reservoir	Water quality of pick-up dam
1.	Coliform organisms (Monthly average MPN/100 ml)	Not more than 5000 with less than 5% of the samples with value > 20,000 and less than 20% of the sample within value > 5000	-	-	-	-	-	-
2.	PH	6.0-9.0	6.5-8.5	6.5-8.5	5.5-9.0	7.5-8.5	7.2-8.9	6.8-8.6
3.	Fluorides (as F), (mg l^{-1}, max)	1.5	-	-	-	-	-	0.3-0.8
4.	Chlorides (as CI), (mg l^{-1}, max)	600	250	1000	600	500	28.4-152.4	31-92
5.	Cyanides (mg l^{-1}, max)	0.01	-	0.025	-	-	-	-
6.	Selenium (as Se) (mg l^{-1}, max)	0.05	-	-	-	-	-	-
7.	Lead (as Pb) (mg l^{-1}, Max)	0.1	-	0.05	-	-	-	-
8.	Total Chromium (as Cr) (mg l^{-1}, max)	0.05	.05	0.05	-	-	-	-
9.	Arsenic (as As.) (mg l^{-1}, max.)	0.2	0.05	0.05	-	-	-	-

(Table Contd...)

1	2	3	4	5	6	7	8	9
10.	Dissolved Oxygen, (mg l^{-1}, min)	40% saturation value or 3 Mg l^{-1} whichever is higher	–	4	–	–	–	–
11.	BOD_5 ^{20} (Mg l^{-1}, Max)	3	–	–	–	–	–	–
12.	Phenolic Compounds (as C_6H_5OH (mg l^{-1}, max)	0.005	–	0.001	–	–	–	–
13.	Alpha emitters (μc/ml. max)	10^{-8}	–	10^{-9}	10^{-9}	–	–	–
14.	Beta emitters, (μc/ml. max)	10^{-8}	–	10^{-8}	10^{-8}	–	–	–
15.	Nitrates (as NO_3) (mg l^{-1} max)	50	–	0	0	–	–	0-.2
16.	Oils and grease, (mg l^{-1} max)	0.1	–	No visible colour film on the surface and no oily odour	–	–	–	–
17.	Electrical conductance at 25°C, (max mhos.)	-	–	1500 x 10^{-6}	3000 x 10^{-6}		336-812	310-680
18.	Insecticides	Absent	–	Absent	–	–	–	–
19.	Free Carbon – Dioxide (as Co_2) (mg l^{-1} max)	–	–	12 (at sunrise)	–	–	–	–
20.	Free ammonia (as N) (mg l^{-1} max)	–	–	1.5	–	–	–	–

(Table Contd...)

1	2	3	4	5	6	7	8	9
21.	Total dissolved solids (inorganic) (mg l^{-1} max)	–	500	–	2100		175-438	–
22.	Sulphates (as So_4). (mg l^{-1} max)	–	200	–	1000	500	6.7-32.6	–
23.	Boron (as B), (mg l^{-1} max)	–	–	–	60	–	–	–
24.	Percent Sodium, (mg l^{-1} max)	–	–	–	–	–	–	–
25.	Turbidity, NTU, (mg l^{-1} max)	0	–	10	–	7-15	–	4.40
26.	Total Iron (as Fe), (mg l^{-1} max)	–	0.3	2.0	–	0.1	–	.05-.4
27.	Temperature, °C, range	–	–	2-35	–	–	–	–
28.	Salinity, (mg l^{-1} max)	–	250	1000	–	–	–	–
29.	Alkalinity (as $CaCo_3$), (mg l^{-1})	–	200	100-300	–	50-500	26-420	40-240
30.	Hardness (mg $CaCo_3$) (l^{-1})	–	300	–	–	–	125-275	102-176
31.	Zinc (ppm)	–	5					
32.	Calcium (mg l^{-1})		75					

Table—9.24 Physico-chemical characteristics of water in the Sathanur reservoir, June '99

	Site A			Site B			Site C			Site D			Site E									
	S	C	B	S	C	B	S	C	B	S	C	B	E1	E2	E3	E4	E5	E6	E7	E8	E9	E10
Water depth (m)	0.5	5.0	10	0.5	10	20	0.5	7.5	15	0.5	8	16	0.5	2	4	6	8	10	12	14	16	18
PH	8.7	8.7	8.7	8.7	8.6	8.6	8.7	8.7	8.7	8.7	8.6	8.5	8.6	8.7	8.6	8.7	8.6	8.6	8.6	8.6	8.6	8.2
EC (μ mho cm)	670	780	690	680	690	700	680	670	690	670	690	690	710	710	710	720	700	710	700	700	700	740
Temperature °C	23	19	17.5	24.5	18	16	23	18	16.5	23.5	20	16.5	24.5	24	24	23	20.5	18	17.5	17	16	15.5
D.O	5.6	4.8	4.3	5.9	4.7	3.8	5.5	4.7	3.4	5.7	5	4.1	5.9	5.8	5.8	5.1	5.0	4.7	4.5	4.3	4.1	3.2
Alkalinity (mg^{-1})	264	276	280	284	280	290	288	240	296	260	280	296	280	276	300	284	284	276	284	280	284	296
Acidity (mg^{-1})	24	20	32	28	32	34	36	26	32	32	28	36	38	40	32	40	36	40	40	32	42	36
Hardness (mg^{-1})	184	180	188	180	178	188	184	172	186	180	176	180	180	178	184	184	184	172	172	180	184	196
Calcium hardness (mg^{-1})	136	156	156	136	136	188	164	152	180	140	140	148	152	148	152	152	154	128	156	156	144	160
Calcium (mg^{-1})	54.4	62.4	62.4	54.4	54.4	75.2	65.6	60.8	72	56	56	59.2	60.8	59.2	60.8	60.8	61.6	51.2	544	54.4	57.6	64
Chloride (mg^{-1})	40	52	50	52	48	58	58	50	54	56	54	60	52	52	54	54	54	54	52	52	58	58
T.D.S (mg^{-1})	350	340	350	340	350	360	340	350	330	360	350	380	360	350	340	360	350	360	360	370	380	390
S.S (mg^{-1})	30	20	10	40	26	20	34	66	20	22	36	20	22	36	38	40	30	26	30	26	20	10
Total solids (mg^{-1})	380	360	360	380	376	380	374	416	350	382	386	400	382	386	378	400	380	386	390	396	400	400
Sulphates (mg^{-1})	16.9	19	17	14.6	18	20.8	14.4	16.7	18.9	18.3	14.5	23.4	15.7	13.8	13.2	16.5	18.7	18.6	20.4	18.7	20.8	37.2
Nitrates (mg^{-1})	0.6	0.9	1.1	0.6	0.8	1.8	0.7	1.1	1.4	0.8	0.8	1.1	0.09	0.2	0.3	0.2	0.4	0.2	0.3	0.6	0.9	1.9

(Table Contd...)

1	2	3	4		5	6	7	8	9	10	11	12	13	14	15	16	17	18	19	20	21	22
Phosphates (mg^{-1})	0.1	0.2	0.3	0.3	0.4	1.6	0.09	0.2	0.4	0.7	0.3	1.9	0.3	0.4	0.1	0.1	0.2	0.3	0.4	0.4	0.7	1.8
Potassium (ppm)	–	–	–	2.7	2.8	2.9	–	–	–	–	–	–	2.9	3	3	2.9	3	2.9	2.8	2.7	2.7	2.9
Sodium (ppm)	–	–	–	24.5	24.9	25.2	–	–	–	–	–	–	23	24	25	24.5	25	25	24	23.2	23.2	24.2
Copper (ppb)	–	–	–	4	7	4	–	–	–	–	–	–	8	9	10	7	8	8	7	8	14	11
Chromium (ppb)	–	–	–	2	7	6	–	–	–	–	–	–	< 1	1	1	2	3	4	6	8	< 1	9
Zinc (ppb)	–	–	–	48	50	48	–	–	–	–	–	–	87	88	87	88	102	90	89	82	92	89
Cadmium (ppb)	–	–	–	< 1	< 1	< 1	–	–	–	–	–	–	< 1	1	< 1	< 1	2	2	2	2	1	< 1

(Table Contd...)

	Site F			Site G			Site H			Site I		
	S	**C**	**B**	**S**	**C**	**B**	**S**	**C**	**B**	**S**	**C**	**B**
Water depth (m)	0.5	10.5	21	0.5	2.5	5	.5	4	8	0.5	9	18
PH	8.7	8.5	8.5	8.7	8.7	8.7	8.7	8.7	8.7	8.7	8.7	8.7
EC (μ mho cm)	710	710	710	690	690	690	710	710	710	730	710	730
Temperature °C	24	22.5	14.5	24.6	23.5	22	24	23	18	24.5	19.7	16.5
D.O	5.8	4.9	3.1	5.9	5.7	5.1	6.1	5.6	4.4	5.5	4.7	3.7
Alkalinity (mg^{-1})	286	284	290	300	264	276	240	280	288	300	286	320
Acidity (mg^{-1})	28	26	28	16	28	22	34	24	36	22	28	30
Hardness (mg^{-1})	184	188	190	186	172	188	184	172	186	186	190	198
Calcium hardness (mg^{-1})	156	156	160	100	136	152	128	152	168	152	140	152
Calcium (mg^{-1})	62.4	62.4	64.0	40	54.4	60.8	51.6	60.8	67.2	60.8	56	60.8
Chloride (mg^{-1})	48	56	60	64	50	50	52	54	60	58	48	58
T.D.S (mg^{-1})	380	390	380	360	380	390	360	350	380	360	350	370
S.S (mg^{-1})	26	40	30	38	42	36	18	28	20	20	32	20
Total solids (mg^{-1})	406	430	410	398	422	426	378	378	400	380	382	390
Sulphates (mg^{-1})	18	26	40.8	15	12	18	16.8	22.4	26	10	17.9	27.2
Nitrates (mg^{-1})	0.9	1.6	2.7	0.7	15	0.9	0.7	0.5	0.9	0.9	0.1	1.2
Phosphates (mg^{-1})	0.1	0.8	1.7	0.9	1.7	2	0.1	1.8	2.1	0.3	0.9	1.7

1	2	3	4	5	6	7	8	9	10	11	12	13
Potassium (ppm)	-	-	-	-	-	-	2.8	2.8	3.2	-	-	-
Sodium (ppm)	-	-	-	-	-	-	24.1	24.1	25	-	-	-
Copper (ppb)	-	-	-	-	-	-	9	6	8	-	-	-
Chromium (ppb)	-	-	-	-	-	-	3	2	3	-	-	-
Zinc (ppb)	-	-	-	-	-	-	45	57	56	-	-	-
Cadmium (ppb)	-	-	-	-	-	-	<1	1	1	-	-	-

EC—Electrical Conductivity - Not Analyzed
T.D.S—Total Dissolved Solids S.S—Suspended Solids
D.O—Dissolved Oxygen S—Surface
C—Column B—Bottom

Table—9.25 Productivity of Sathanur reservoir, June '99

Seechi depth-95 cm

Euphotic depth-2.4 m

	Depth	Net primary productivity	Gross primary productivity
Surface	.5 m	84.375 mg C fixed m^{-3} day^{-1}	159.375 mg C fixed m^{-3} day^{-1}
Middle	1m	46.875 mg C fixed m^{-3} day^{-1}	103.125 mg C fixed m^{-3} day^{-1}
Bottom	2.4 m	18.75 mg C fixed m^{-3} day^{-1}	28.125 mg C fixed m^{-3} day^{-1}

Table 9.26 Physico-chemical characteristics of water in the Sathanur reservoir, July '99

Seechi depth-70 cm
Climate-moderately sunny

	Site A			Site B			Site C			Site D			Site E									
	S	C	B	S	C	B	S	C	B	S	C	B	E1	E2	E3	E4	E5	E6	E7	E8	E9	E10
Water depth (m)	0.5	5	10	0.5	10	20	0.5	7.5	15	0.5	8	16	0.5	2	4	6	8	10	12	14	16	18
PH	8.6	8.7	8.6	8.6	8.5	7.9	8.6	8.5	8.3	8.7	8.6	8.2	8.7	8.7	8.7	8.7	8.6	8.5	8.5	8.5	8.3	8.3
EC (μ mho cm^{-1})	700	670	700	720	670	710	670	680	690	740	740	740	700	740	720	750	740	750	720	690	710	740
Temperature °C	27	26	23.2	27	26.2	22.8	27	25	20.2	26.6	22.2	19.5	27	26.8	26.5	24	24	23.5	23	22.4	21	19.5
D.O	5.3	4.7	3.6	5.3	4.6	1.5	5	4.3	2.6	5.1	4.9	1.6	5.3	5.4	5.1	5	4.9	5	4.7	4.4	2.3	1.3
Alkalinity (mg^{-1})	272	264	276	276	260	278	276	256	278	252	240	268	280	280	276	284	280	280	284	282	288	284
Acidity (mg^{-1})	60	68	72	56	64	68	56	64	70	60	64	64	56	56	60	58	58	60	60	64	68	70
Hardness (mg^{-1})	186	170	190	184	178	186	182	176	182	180	174	182	184	180	184	184	180	184	178	184	186	184
Calcium hardness (mg^{-1})	90	100	120	95	100	130	100	115	115	105	120	120	90	120	120	110	100	115	120	130	135	135
Calcium (mg^{-1})	36	40	48	38	40	52	40	46	46	42	48	48	36	48	48	44	40	46	43	52	54	54
Chloride (mg^{-1})	58	70.9	74.0	70.0	69	72	74	72	80	72	72	74	72	67	67	69	70	74	74	76	78	78
T.D.S (mg^{-1})	360	350	360	350	350	360	340	350	360	350	340	360	350	350	340	360	350	350	350	340	350	360
S.S (mg^{-1})	32	36	20	36	42	12	30	38	20	34	27	20	42	36	40	42	38	30	[illegible]	22	20	20
Total solids (mg^{-1})	392	386	380	386	392	372	370	388	380	384	367	380	392	386	380	402	388	380	370	362	370	380

(Table Contd...)

1	2	3	4	5	6	7	8	9	10	11	12	13	14	15	16	17	18	19	20	21	22	23
Sulphates (mg^{-1})	11.8	14.7	20.2	11.3	16	24.8	10.8	12	11.8	10	14.2	21.3	11.2	-	-	-	-	13.7	-	-	-	28.8
Nitrates (mg^{-1})	0.6	0.8	1.1	0.6	0.8	1.1	0.7	1.1	1.2	0.8	0.8	1.2	0.01	-	-	-	-	0.2	-	-	-	0.4
Phosphates (mg^{-1})	0.1	0.2	0.3	0.3	0.4	1.4	0.4	0.3	0.4	0.9	0.4	1.1	0.4	-	-	-	-	0.6	-	-	-	1.8
Potassium (ppm)	-	-	-	-	-	-	-	-	-	-	-	-	3	-	-	-	-	3	-	-	-	3.6
Sodium (ppm)	-	-	-	-	-	-	-	-	-	-	-	-	25.1	-	-	-	-	24.9	-	-	-	25.3
Copper (ppb)	-	-	-	-	-	-	-	-	-	-	-	-	6	-	-	-	-	14	-	-	-	8
Chromium (ppb)	-	-	-	-	-	-	-	-	-	-	-	-	9	-	-	-	-	6	-	-	-	11
Zinc (ppb)	-	-	-	-	-	-	-	-	-	-	-	-	23	-	-	-	-	180	-	-	-	40
Cadmium (ppb)	-	-	-	-	-	-	-	-	-	-	-	-	<1	-	-	-	-	<1	-	-	-	<1

(Table Contd...)

	Site F			Site G			Site H			Site I		
	S	C	B	S	C	B	S	C	B	S	C	B
Water depth (m)	0.5	10.5	21	0.5	2.5	0.5	0.5	4	8	0.5	9	18
PH	8.7	8.6	7.9	8.7	8.7	8.7	8.7	8.7	8.6	8.7	8.7	8.6
EC (μ mho cm^{-1})	730	730	740	740	740	720	730	730	710	700	740	74.0
Temperature °C	26.2	24.7	20	26.2	25.8	24	26.5	24.2	21.1	26.8	22.1	19.4
D.O	5.3	4.2	1.4	5.2	5.0	4.0	5.3	5.0	4.8	5.2	5.1	2.7
Alkalinity (mg^{-1})	268	264	280	250	270	280	268	256	270	256	244	268
Acidity (mg^{-1})	58	60	64	60	66	70	56	58	64	6[illegible]	64	68
Hardness (mg^{-1})	182	186	188	176	174	180	126.2	168	176	160	154	168
Calcium hardness (mg^{-1})	105	120	130	77.5	85	95	80	90	102.5	77.5	77.5	82.5
Calcium (mg^{-1})	42	48	52	31	34	38	32	36	41	31	33	33
Chloride (mg^{-1})	72	72	72	74	69	74	74	74	77	72	70	74
T.D.S (mg^{-1})	350	350	360	380	350	390	360	350	360	350	340	350
S.S (mg^{-1})	36	38	20	20	26	20	20	24	36	20	38	32
Total solids (mg^{-1})	386	388	380	400	376	410	380	374	39.6	370	378	382
Sulphates (mg^{-1})	13.7	16.3	29.2	11.2	14.7	21.2	10.3	13.8	24.3	10.7	16.2	20.3
Nitrates (mg^{-1})	0.5	0.4	1.2	0.7	0.5	0.9	0.9	1.02	1.09	0.9	1.1	1.23
Phosphates (mg^{-1})	1.3	0.6	1.9	0.9	1.2	2.7	1.7	0.8	2.2	1.2	1.6	2.4

(Table Contd...)

1	2	3	4	5	6	7	8	9	10	11	12	13
Potassium (ppm)	-	-	-	-	-	-	-	-	-	-	-	-
Sodium (ppm)	-	-	-	-	-	-	-	-	-	-	-	-
Copper (ppb)	-	-	-	-	-	-	-	-	-	-	-	-
Chromium (ppb)	-	-	-	-	-	-	-	-	-	-	-	-
Zinc (ppb)	-	-	-	-	-	-	-	-	-	-	-	-
Cadmium (ppb)	-	-	-	-	-	-	-	-	-	-	-	-

EC—Electrical Conductivity
T.D.S—Total Dissolved Solids
D.O—Dissolved Oxygen
C—Column

-Not Analyzed
S.S—Suspended Solids
S—Surface
B—Bottom

Table—9.27 Productivity of Sathanur reservoir July '99

Seechi depth-70 cm

Euphotic depth-1.75 m

	Depth	Net primary productivity	Gross primary productivity
Surface	25.5 cm	810 mg C fixed m^{-3} day^{-1}	1.350 mg C fixed m^{-3} day^{-1}
Middle	70 cm	720 mg C fixed m^{-3} day^{-1}	1.170 mg C fixed m^{-3} day^{-1}
Bottom	1.75 cm	18 mg C fixed m^{-3} day^{-1}	360 mg C fixed m^{-3} day^{-1}

The parameters and the time scale thus selected for studying the trend were chosen so as to represent the influence of various hydrological features viz inflow, outflow, water level and precipitation on them. The reservoir operation in terms of impounding inflow and releasing water downstream has been dealt in detail in chapter 10.

IMPACTS

Materials and Methods for the Study of Water Quality Related Impacts

Materials

The materials used for various analytical purposes were as follows:

Reagents

All the reagents used were analytical grade and had an assay of more than 95 per cent unless otherwise mentioned. Deionized and doubly distilled water were used for all the analytical work. All the reagents were prepared fresh and standardized as and when required, except for $SnCl_4$, which has a consistency of over 6 months (APHA, 17th ed.).

Plastic-ware and Glass-ware

The plastic-ware used for sample collection were made up of temperature and scratch resistant, as well as odour-free material. Alkali and temperature resistant vensil and borosil glass-ware were used for estimation purposes.

Equipment

The following equipments were used for all the analytical work:

(i) pH EP-pH electronic paper, Hanna instruments, range: 0-14.0 pH, resolution: 0.1pH, accuracy: $\pm$0.2 pH.

(ii) Systronics Griph 'D' pH meter 327 range: 0-14 pH relative accuracy: $\pm$ 0.05 pH, readability: $\pm$ 0.01 pH.

(iii) Naina digital field conductivity meter, NDC 730, range 100 µ mhos-100 µ mhos in 3 ranges, automatic temperature compensation from 10°C-60°C.

(iv) Systronics dissolved oxygen meter, a battery operated portable instrument for the estimation of DO and temperature measurement. DO is indicated in ppm and temperature in °C. The DO ranges are automatically temperature compensated for the solubility of oxygen in water and permeability of the probe membrane. Salinity compensation is manual. Probe is clark type, membrane covered polarographic sensor. DO accuracy: ±1 per cent of full scale at calibration temperature or 0.1 ppm; temperature accuracy: ±0.5°C.

(v) Anamed Top Pan electronic balance, MX 7301A, capacity: 100gms, resolution: ± 0.001gms.

(vi) Roy top pan balance, KTP-200 CSI; capacity: 2kg, Sensitivity: 0.1gm.

(vii) Systronics UV—visible spectrophotometer 108, photometric range accuracy ± 0.005 absorption at 1 absorption, repeatability 0.002 absorption at 1 absorption, cuvettes: 10 mm.

(viii) Hemco oven ; range : 30°C-50°C.

(ix) Tarsons vaccum filter, 250ml. capacity.

(x) Van dohr sampler

(xi) Seechi disk.

Methods

Water Quality Sampling

The sampling for the stratification studies was done in the months of June and July. There is no outflow during these months. These months were ideal for stratification studies because monsoon breaks only during the last week of July or the first week of August. The water from reservoir is let downstream for irrigation purposes till the middle of the May.

Samples were collected from the nine spots. Three parallel spots at the confluence of river and reservoir (sites A. D and G), three parallel sites at the middle of the reservoir (sites B, E and H) and three parallel sites near the dam structure (sites C, F, and I). Site E (the mid point of the reservoir) was selected for vertical gradient sampling.

A motorboat, courtesy Tamil Nadu Fisheries Development Corporation (TNFDC), Sathanur dam, was employed for collecting samples with the help of a Van-dohr sampler. The surface samples were collected just below sub-surface (0.5m from the surface), at the column depth and just above the bottom of the reservoir avoiding the sediments. At the site E, samples were collected at the depth of every two meters. All the samples were collected in the plastic water cans of 2-litre capacity and preserved for further analyses (Table—9.22). The most changeable and sensitive parameters like temperature, dissolved oxygen and electrical conductivity were analyzed *'in-situ'*

Analytical Methods

For this study, the most advanced methods were utilized. The bulk parameters were analyzed following the latest and internationally accepted techniques (APHA, 1992; Abbasi, 1995). The metal contents of the samples were analyzed by the state-of the art technique of AAS-GF (Graphite furnace) for copper, chromium and cadmium and ASS-Flame for zinc in collaboration with Geological Survey of India, Chennai. Sodium and potassium were analyzed by flame photometer technique. The following methods were used for estimating various water quality parameters.

Estimation of pH

pH was estimated potentiometrically using the pH meter, calibrated with freshly prepared buffer solutions. As pH is a function of temperature, temperature of samples were also recorded. The samples were stirred while recording the results for maintaining the uniformity.

Estimation of Suspended Solids

These were estimated gravimetrically. A well mixed sample was filtered through a previously weighed standard glass fibre filter paper and the residue retained on the filter paper was dried to a constant weight of 103°C-105°C. The increase in weight of the filter paper represented the total suspended solids. Care was taken to exclude large sample sizes to prevent formation of water-entrapping crust on the filter paper due excessive residue.

Estimation of Total Dissolved Solids

These, too, were estimated gravimetrically. A well-mixed sample was filtered through a standard glass fibre filter paper and the filtrate was evaporated to dryness in a pre-weighed porcelain dish and dried to constant weight at 180°C. The increase in dish weight represented the total dissolved solids. Care was taken to exclude large sample sizes to prevent formation of water-entrapping crust on the filter paper due to excessive residue.

Estimation of Acidity

Acidity was estimated by volumetric method using acid-base titration, where the sample was titrated against a strong base, generally, sodium hydroxide (NaOH). The amount of base used to raise the pH of the sample to designated levels was taken as a measure of acidity of the sample.

Estimation of Alkalinity

Alkalinity was estimated by volumetric method using acid-base titration, where the sample was titrated against a strong mineral acid, generally hydrochloric acid (HCl). The amount of acid used to raise though pH of the sample to designated levels, was taken as a measure of alkalinity of the sample. Indicator titration's with coloured or turbid samples was avoided. In such cases, alkalinity was estimated potentiometrically.

Estimation of Hardness

Hardness was estimated by complexometric titration method, where the sample was titrated against di-sodium salt of EDTA with suitable metallochromic indictors like calmagite. Heavy metal ions which interfere during hardness estimation was inhibited using sodium sulphide (Na_2S). Organic matter interference was also removed by drying the sample and then heating in muffle furnace.

Estimation of Chloride (Cl)

Chloride was estimated through argentometric precipitation titration of neutral or slightly alkaline samples against standard silver nitrate solution. On highly coloured samples, where the endpoint could not be identified easily, aluminum hydroxide [Al

(OH_3)] was added and this mixed solution was allowed to settle and then filtered. This filtrate was then used for further estimation.

Estimation of Dissolved Oxygen (DO)

DO was estimated by Winkler's iodometric method based upon redox reaction of manganese salt. It was reacted in the presence of oxygen with potassium iodide and the liberated iodine was titrated against standard sodium thio-sulphate ($NA_2S_2O_3$). In order to remove the effect of interfering materials, the aside modification of the Winkler's method was used.

Estimation of Sulphates (SO_4)

Sulphate in samples was estimated by turbidimetric method, where sulphate ion was precipitated in an acetic acid medium with barium chloride ($BaCl_2$) to form barium sulphate ($BaSo_4$) crystals. The light absorbance of the ($BaSo_4$) suspension was measured by a spectrophotometer at 420 nm and the sulphate ion concentration was then determined by comparing the reading with a standard curve.

Estimation of Phosphorus

Phosphorus was determined as phosphates (PO_4) by stannous chloride method. Dissolved phosphorus was analyzed after filtering the samples through a 0.45 μ membrane filter. The glasswares were rinsed with warm HCl. The membrane filters were soaked for more than 24 hours in 500 ml, distilled water and further their contribution of PO_4 to the samples was checked for and corrections were made for all the determination of dissolved phosphates.

Estimation of Nitrogen

Nitrate nitrogen was determined by phenol-di-sulphonic acid method.

Sathanur Reservoir

In the subsequent paragraphs various important physico-chemical parameters of the reservoir had been discussed across the years, in the months of peak summer (June) and the peak monsoon (Dec). As the reservoir water level stocking would indicate several factors such as inflow, outflow and the evapo-transpiration factors, the change in the water quality parameters had been correlated with the stock water level of the reservoir. The relationship between the reservoir inflow, outflow and the precipitation has been presented as figure—9.0.

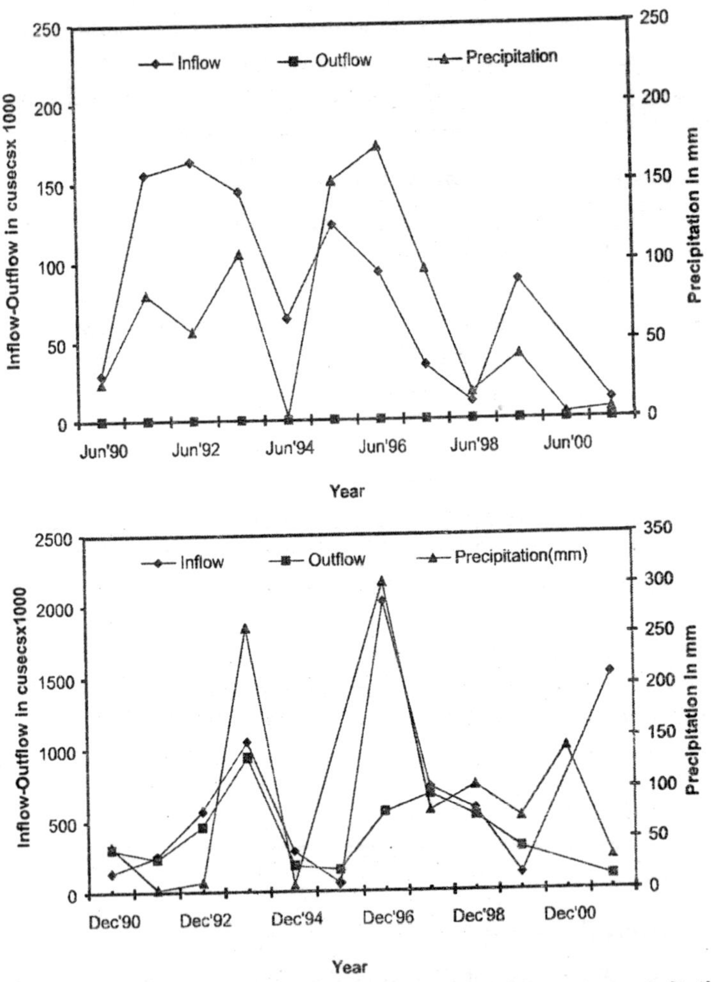

Fig. 9.0: Relationship between inflow, outflow and precipitation in Sathanur reservoir during the peak summer (above) and the peak monsoon (below)

pH

pH of the Sathanur reservoir during the peak summer, from 1990 to 2001, varied between 7.7 (Jun '01) and 8.7 (Jun '97). The fluctuation in the pH is not significant as the value varied between 7.7 to 8.7. Interestingly, though the water level stocking in the reservoir had been on the rise (Figure—9.1), no significant increase in the pH of the reservoir was observed.

pH of the Sathanur reservoir during the months of peak monsoon varied between 7.2 (Dec '94) and 8.7 (Dec '91). This fluctuation in the pH is not significant as happened in the case of peak summer. Though the reservoir water stocking during the monsoon varied widely, this did not affect the pH of the reservoir water much (Figure—9.1). As the pH of the reservoir falls in the range of 7.2 and 8.7, the reservoir can be categorized as alkaline.

Electrical Conductivity

EC of the Sathanur reservoir during the peak summer, from 1990 to 2001, varied between 560 μ mhos (Jun '94) and 812 μ mhos (Jun '97). The EC was on rise at an alarming rate of 100μ mhos per year during the period June '94 till June '99 (Figure 9.2). The trend line of EC indicates that the EC is on rise, which correlates with the increasing water level of the reservoir. This increase in the EC during the summer can be attributed to the evapo-transpiration factor and the increase in concentrations of the physico-chemical parameters such as calcium, chlorine, sulphate and magnesium ions.

During the peak monsoon, EC of the Sathanur reservoir varied between 337μ mhos (Dec '92) and 620 μ mhos (Dec '97). A positive correlation was observed between the stocking water level and the reservoir EC (Figure—9.2). The increase in EC during the peak monsoons can be attributed to the surface run-off. Electrical conductivity is a function of total dissolved solids, and also influenced by the cations and anions present in a water body. Water bodies with EC above 200 μ mhos have been categorized as eutrophic by Rawson (1960) and Berg et al., (1958). Thus, Sathanur reservoir, with EC values in the range of 337 μ mhos and 812 μ mhos can be considered as eutrophic.

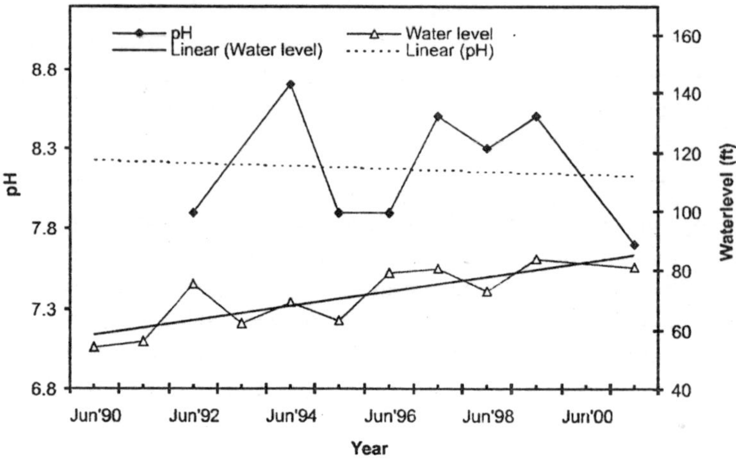

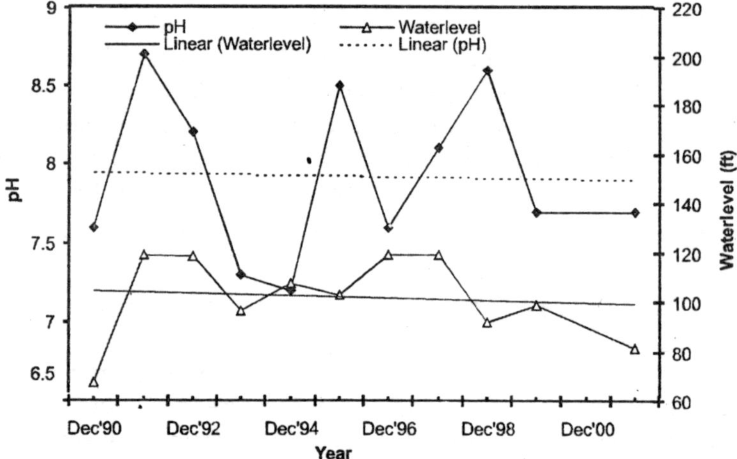

Fig. 9.1 Variation in pH and water level in Sathanur reservoir during the peak summer (above) and the peak monsoon (below)

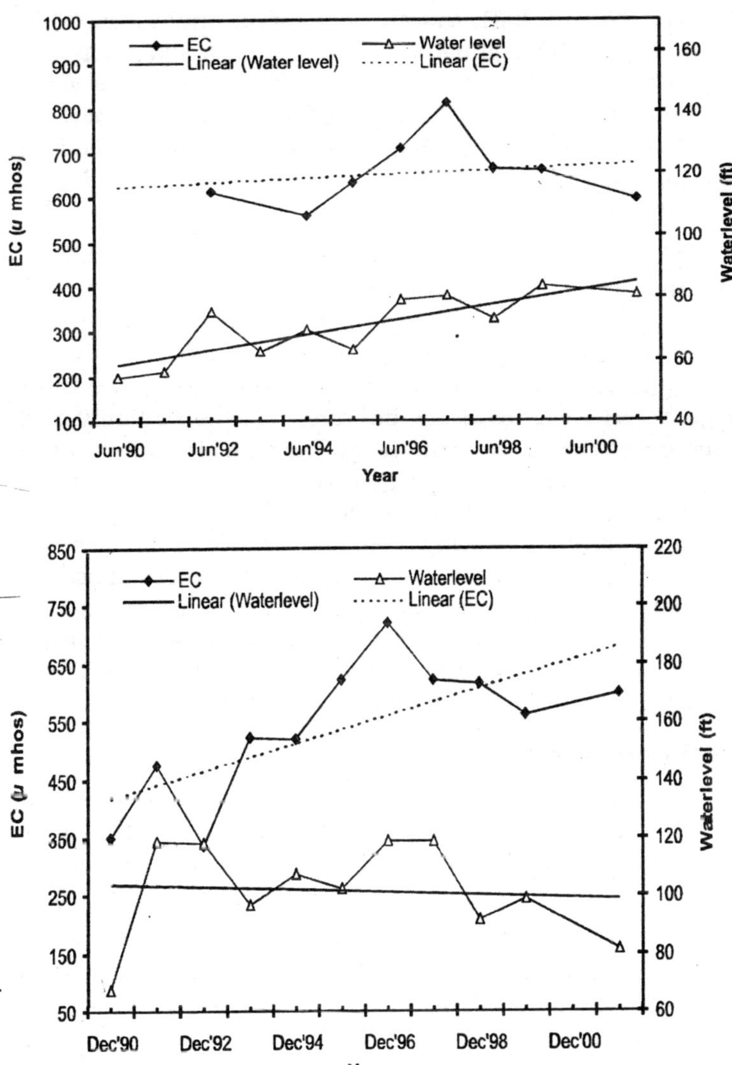

Fig. 9.2: Variation in electrical conductivity and water level in Sathanur reservoir during the peak summer (above) and peak monsoon (below)

Alkalinity

Alkalinity of Sathanur reservoir during the peak summer, during 1990 till 2001, varied between 40 mgl^{-1} (Jun '98) to 285 mgl^{-1} (Jun '97). Figure—9.3 indicates a clear and sharp rise of alkalinity of Sathanur reservoir between June' 94 and June' 97 with the reservoir water level. As observed in the case of EC and total hardness, there was a sudden fall in the alkalinity value in the year June' 98 and the subsequent years again saw a sharp increase in alkalinity.

During the months of peak monsoon, during 1990 till 2001, Sathanur reservoir recorded the alkalinity values between 26 mgl^{-1} (Dec '96) to 373 mgl^{-1} (Dec '01). Also, there was a continuous rise in the alkalinity values from Dec '90 to Dec '93, subsequently the alkalinity values were found to be inconsistent (Figure—9.3). The trend line indicates that the alkalinity of the reservoir increased along the years. The high alkalinity values of the reservoir, above 100 mgl^{-1}, point towards the highly eutrophic nature as per the classification of Moss (1973) and Philipose (1960). As per the criteria of Moyle's (1946), Sathanur reservoir falls under the category of hard water body. Further, the studies indicate the alkalinity of the reservoir is mainly due to the bicarbonates, which could have been constituted by the high concentration of calcium due to calcium rich rocks found in the study area. Similar observations were made by Zutchi et.al (1980).

Total Hardness

Total Hardness of the Sathanur reservoir water during the peak summer, from 1990-2001, varied between 18 mgl^{-1} (Jun '99) and 275 mgl^{-1} (Jun '96). The trend line of total hardness indicates that there is a fall in total hardness along the period of study during the summer months. (Figure—9.4)

During the peak monsoon, there was a steady rise in the TH values of Sathanur Reservoir from Dec '90 to Dec '96 and there was a sudden fall in Dec '99. Similar trend was observed in the case of EC.

Calcium Hardness

Calcium Hardness of the Sathanur reservoir water during the peak summer, June 1990 – June 2001, varied between 22 mgl^{-1}

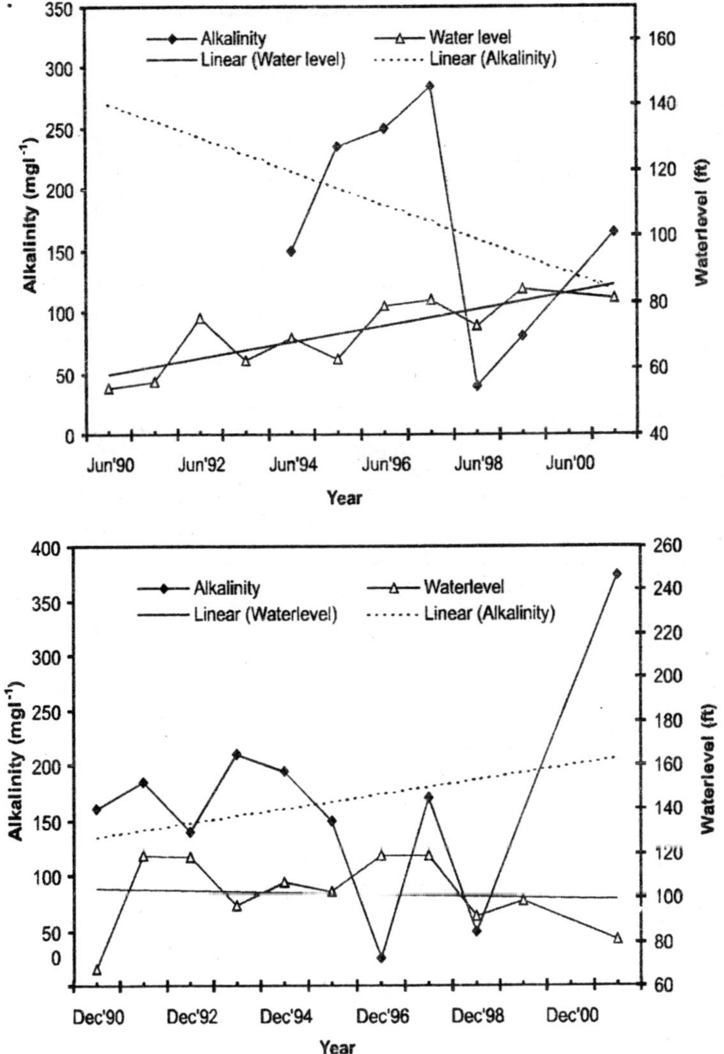

Fig. 9.3: Variation in alkalinity and water level in Sathanur reservoir during the peak summer (above) and peak monsoon (below)

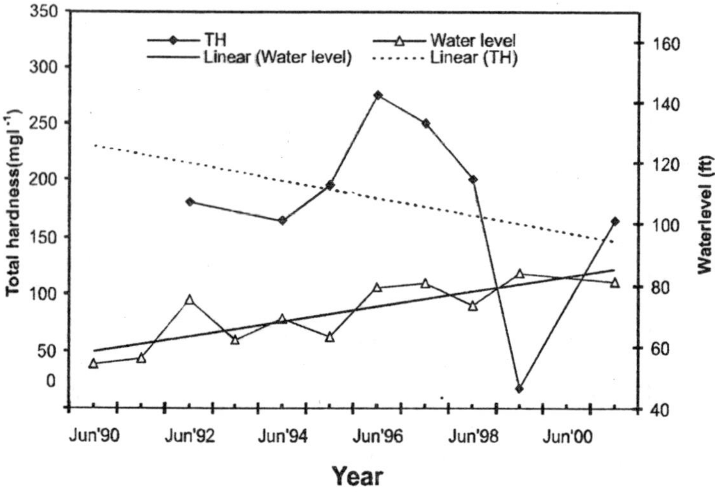

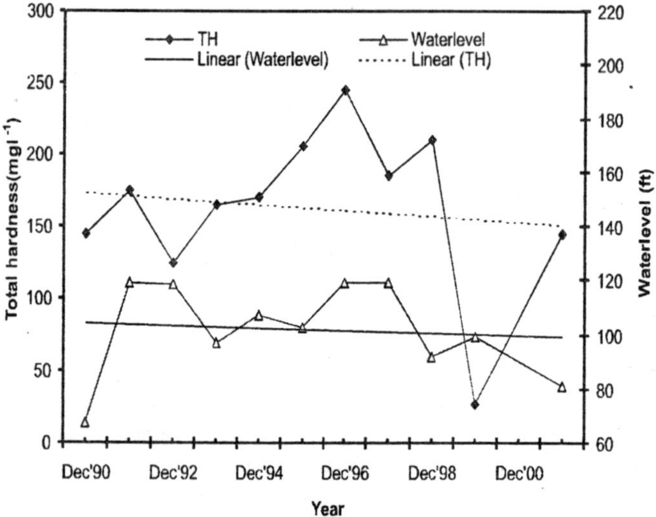

Fig. 9.4: Variation in total hardness and water level of Sathanur reservoir during the peak summer (above) and the peak monsoon (below)

(Jun '92) and 32 mgl^{-1} (Jun '01). A slight increasing trend has been observed in the values of Calcium hardness during the study period (Figure 9.5).

During the months of peak monsoon (Dec '90 – Dec '01), the Calcium hardness of the reservoir water varied between 20 mgl^{-1} (Dec '01) to 38 mgl^{-1} (Dec '95). The trend line indicates that there is a slight fall in Calcium hardness, over the peak monsoon.

Solids

The total dissolved solids, total suspended solids and total solids of the Sathanur reservoir, during the peak summer and peak monsoon months, have been presented in the Figure—9.6. The suspended solids ranged between 2mgl^{-1} (Jun '94) to 267 mgl^{-1} (Jun '98); while the values of Dissolved solids were observed to have a range between 297 mgl^{-1} (Jun '94) to 438 (Jun '97). The values of total solids varied between 299 mgl^{-1} (Jun '94) and 626 mgl^{-1} (Jun '98).

During the peak monsoon from 1990-2001, the Sathanur reservoir recorded the values of suspended solids between 4 mgl^{-1} (Dec'94) and 147 mgl^{-1} (Dec '98). The concentration of the dissolved solids varied between 175 mgl^{-1} (Dec '92) and 388 mgl^{-1} (Dec '96), while the values of total solids were recorded between 21 mgl^{-1} (Dec '01) and 511 mgl^{-1} (Dec '96).

Chloride

Chloride concentration of the Sathanur reservoir during the peak summer from June 1990 till June 2001, varied between 63.8 mgl^{-1} (Jun '94) and 148.9 mgl^{-1} (Jun '98). The trend line shows that there is a slight increase in the concentration of the Chloride during the peak summer (Figure—9.7)

During the peak monsoon period, Dec' 1990 to Dec' 2001, the Chloride concentration of the reservoir varied between 28.4 mgl^{-1} (Dec '92) and 86.7 mgl^{-1} (Dec '97). The trend line indicates that there is a steady rise in the Chloride concentration.

High level of chloride is an index of organic pollution in a water body (Munawar, 1970). Similar concept has been presented by other authors too. An increase in the amount of chloride laid by various aquatic fauna may account for the increase in chloride

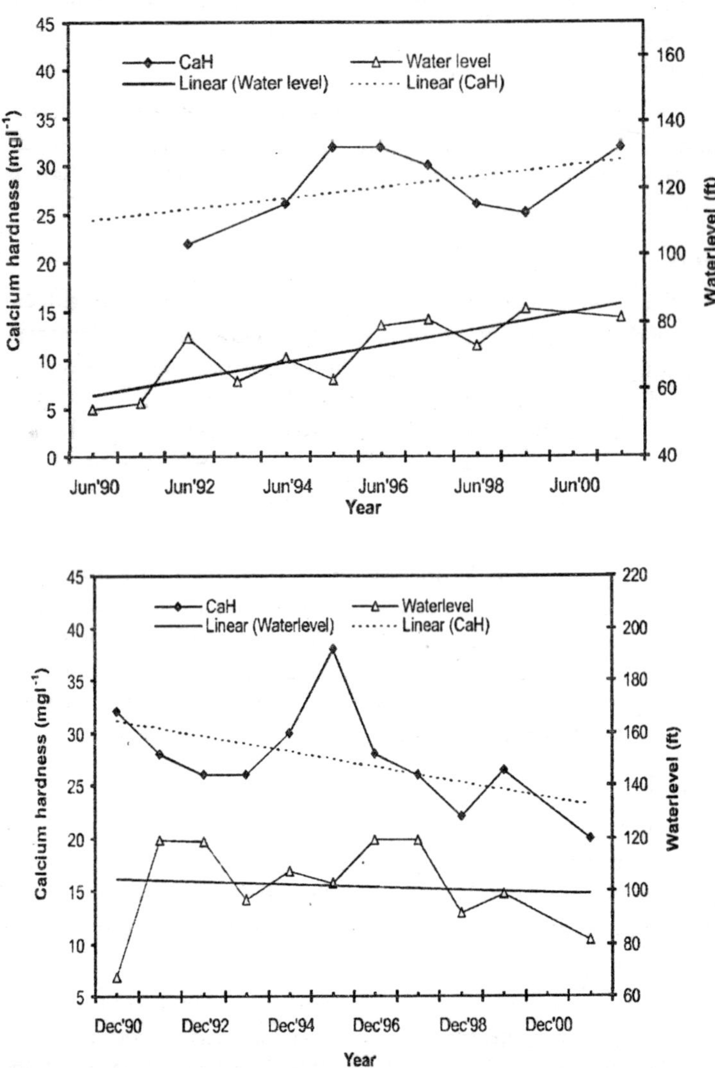

Fig. 9.5: Variation in calcium hardness and water level in Sathanur reservoir during the peak summer (above) and the peak monsoon (below)

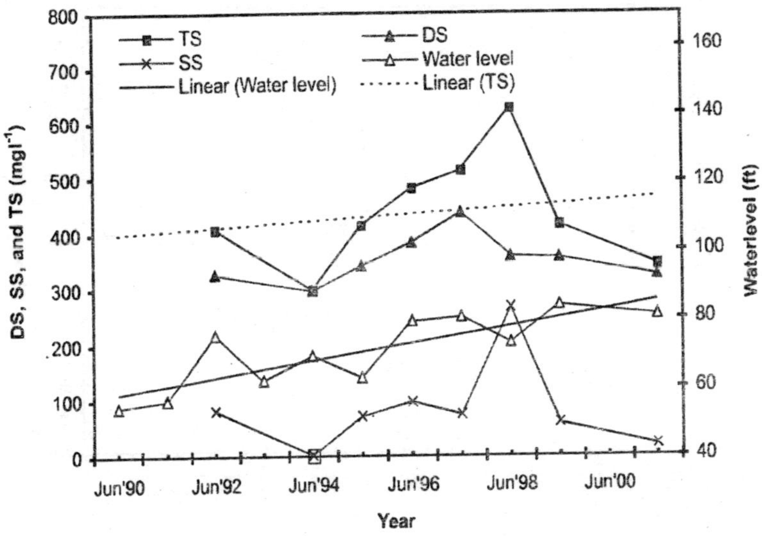

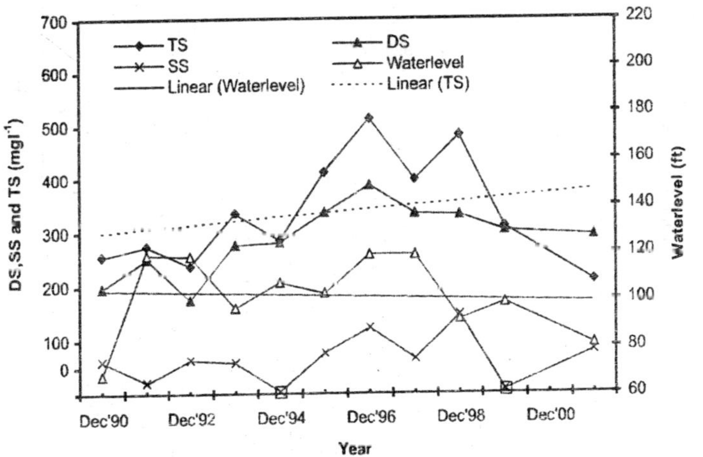

Fig. 9.6: **Variation in dissolved solids, suspended solids, total solids and water in Sathanur reservoir during the peak summer (above) and the peak monsoon (below)**

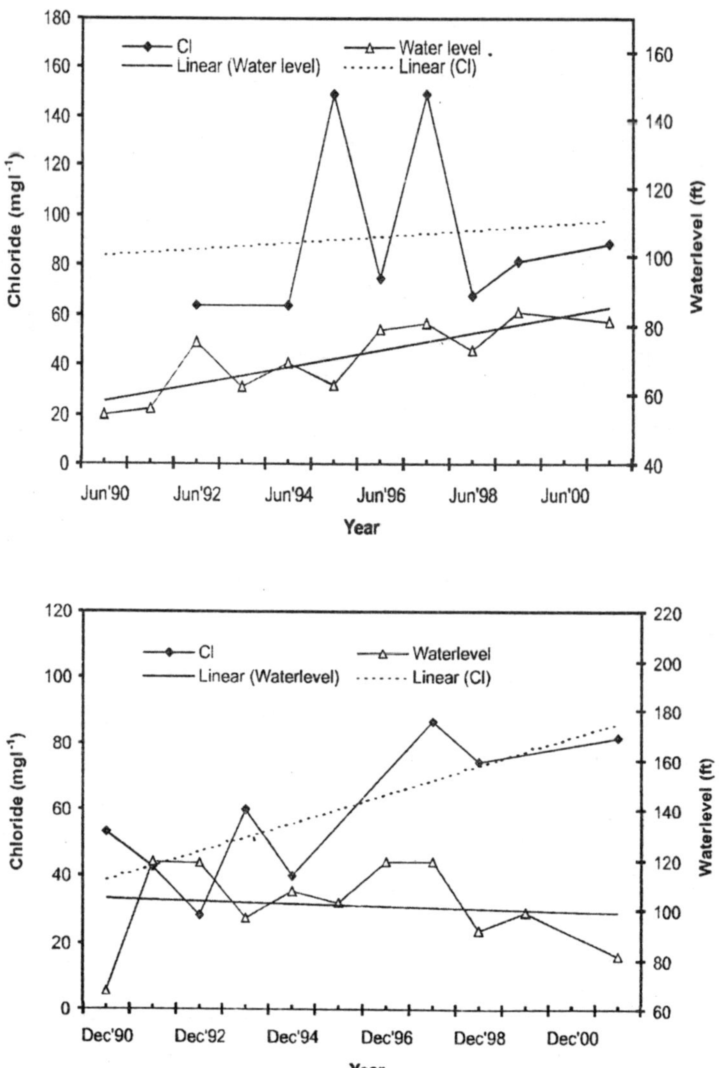

Fig. 9.7: Variation in chloride and water level in Sathanur reservoir during the peak summer (above) and the peak monsoon (below)

concentrations (Cole, 1979; Pandey and Misra, 1991). The role of chloride based fertilizers and chemicals have also been discussed by several workers such as Pawar and Shaikh (1995). Sathanur reservoir is an active site for aquaculture, which is practised mainly in the form of fisheries (Chapter 8). The reservoir also supports crocodiles (*C.palustris*) and several other fauna. The upstream of the reservoir is an agrarian site with many other major and minor projects in operation (Chapter 3). Thus, the presence of high concentration of chlorides and other ions such as sulphates may be a result of agricultural run-off too.

Several factors such as depth, sedimentation, deforestation, agricultural practices, soil erosion and other anthropogenic factors influence the limnitic loading rate of the reservoir as supported by other such studies on the water quality of water bodies (Whitemore et.al., 1997; Jin et.al., 1990; Gotterman et.al., 1983).

Magnesium

Magnesium concentration of the Sathanur reservoir water during the peak summer months (1990-2001) varied between 15.4 mgl^{-1} (Jun '99) and 46.8 (Jun '96). The concentration of Magnesium appears to have decreased across the years during the months of Sathanur. (Figure 9.8)

During the peak monsoon period from Dec '90 to Dec '01, the values of magnesium concentration fell between 47.7 mgl^{-1} (Dec '94) and 56.7 mgl^{-1} (Dec '99). The trend line indicates that there was a slight increase in the Magnesium concentration over the years during peak monsoon.

Sulphate

During the peak summer months from 1990 to 2001, the Sulphate concentration of the Sathanur reservoir varied between 7.2 mgl^{-1} (Jun '96) and 32.6 mgl^{-1} (Jun '95). No specific pattern was observed in the fluctuation of Sulphate concentrations; however, the trend line indicates a slight fall in the Sulphate levels over the year (Figure—9.9)

During the peak monsoon, the Sulphate concentration of the reservoir falls between 7.7 mgl^{-1} (Dec '94) to 56.7 mgl^{-1} (Dec '99). The trend line suggests that there had been a steady rise in the Sulphate concentration during the study period.

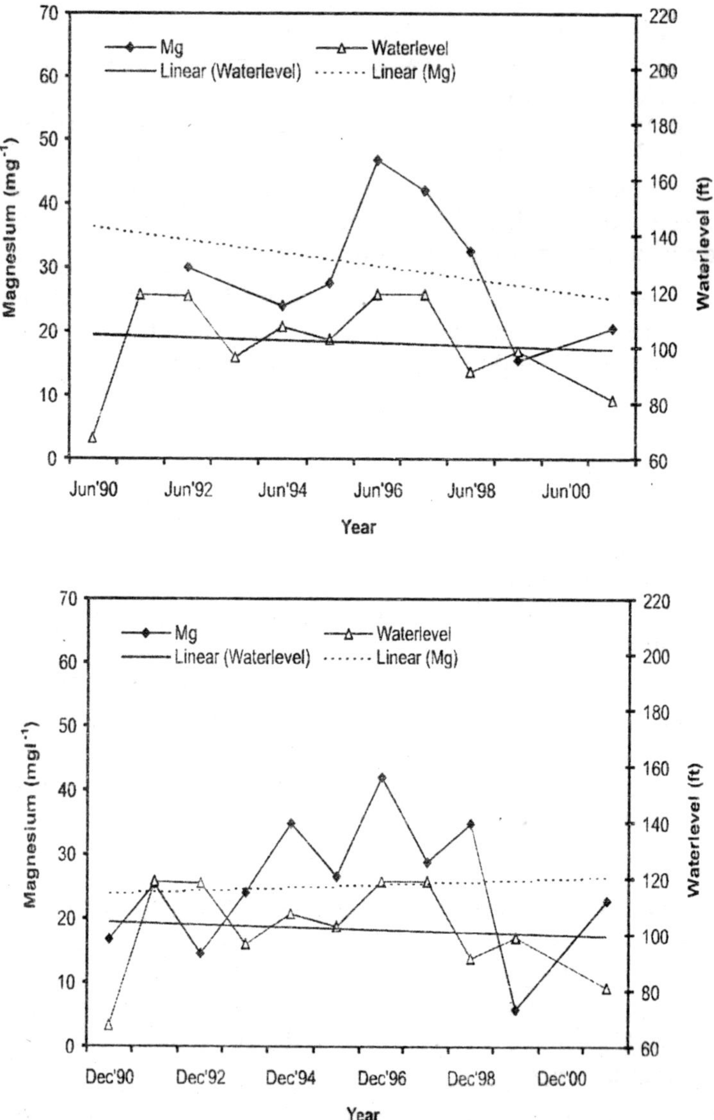

Fig. 9.8: Variation in magnesium and water level in Sathanur reservoir during the peak summer (above) and the peak monsoon (below)

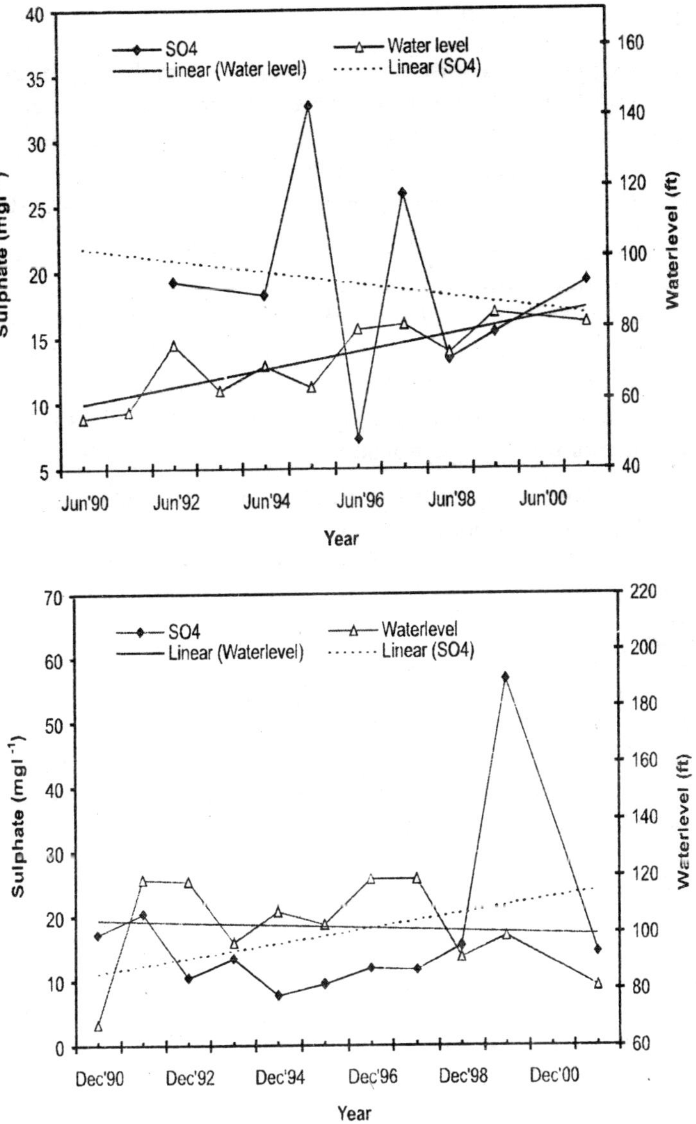

Fig. 9.9: Variation in sulphate and water level in Sathanur reservoir during the peak summer (above) and the peak monsoon (below)

Pick-up Dam

pH

The pH values of the water at pick-up dam during the peak summer, 1991 to 2001, varied between 7.9 (Jun'93) and 8.8 (Jun'01). The trend line shows that there is a slight rise in pH across the years (Figure 9.11 a). No significant difference has been observed between the Sathanur reservoir water quality and the pick-up dam.

During the peak monsoon of 1991 to 2001, the pH values ranged within 7.3 (Dec'96) and 8.3 (Dec'98). The trend line indicates that there is a slight decrease in pH values, over the study period (Figure 9.11 b), as in the case of Sathanur reservoir (Figure 9.1). It is interesting to note that the pH of Sathanur reservoir follows a steady pattern, while the pick-up dam has a record of slight fall over the months of peak monsoon.

On comparing with the BIS water quality standards for irrigation (1974), fisheries (1994), and drinking water (1991), the pH of the pick-up dam falls within the prescribed range of 6.5 – 8.5 pH units.

Alkalinity

Alkalinity of the water at pick-up dam during the peak summer, 1991 till 2001, varied between 124 mgl^{-1} (Jun'91) to 266 mgl^{-1} (Jun'97). A steady and sharp rise of alkalinity values was observed till the year 1995 and subsequently there was a fall in the alkalinity till it reached 202 mgl^{-1} (June '01) The trend line shows a steady increase in alkalinity values, across the years; while in the case of Sathanur reservoir, the trend line showed sharp decline in the alkalinity (Figure—9.11a, Figure—9.3).

During the peak monsoon months, the pick-up dam showed the alkalinity values of the pick-up dam varied between 112 mgl^{-1} (Jun'90) and 250 mgl^{-1} (Jun'97) (Figure—9.11 b). No significant change has been noticed, when trend lines of both Sathanur reservoir and pick-up dam were compared.

As per the criteria of BIS water quality standards for drinking water (1991), the water at pick-up is above the permissible limits of drinking water as the alkalinity value of pick-up exceeds 200 mgl^{-1},

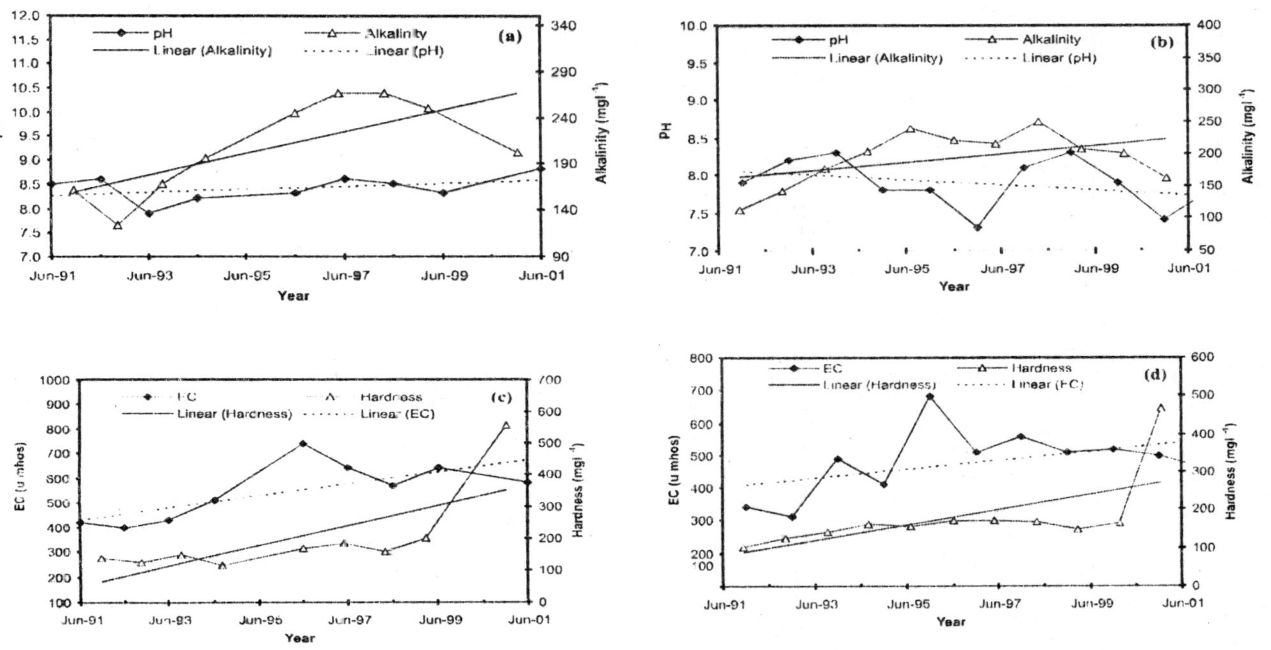

Fig. 9.10: Water quality of the peak-up dam, 1991-2001: (a) pH and alkalinity during peak summer (b) pH and alkalinity during peak monsoon (c) EC and hardness during peak summer (d) EC and hardness during peak monsoon

however, the water quality at pick-up dam is within the recommended levels for fisheries (1994) and swimming purposes (1993).

EC

EC of the pick-up dam during the peak summer of the study period varied between 420 μ mhos (Jun'91) and 740 μ mhos (Jun'96). The trend line indicates that the EC of the pick-up dam is on rise, (Figure—9.11c). A comparison of Sathanur reservoir (Figure—9.2) and the pick-up dam indicates that the later had slightly higher values of EC during the period of study.

During the peak monsoon, EC of the pick-up dam varied between 310 μ mhos (Dec'92) and 680 μ mhos (Dec'95). The trend line shows a steady rise in the values of EC (Figure—9.11d). Similar trend was observed in the case of Sathanur reservoir (Figure—9.2).

The pick-up dam water does not satisfy the criteria of BIS water quality standards for drinking (1991), irrigation (1974) and aquaculture (1994).

Total Hardness

Hardness of the pick-up dam during the peak summer, from 1991-2001, varied between 114 mgl^{-1} (Jun'94) and 558 mgl^{-1} (Jun'01). The trend line suggests that there had been a steady rise in the hardness values across the years (Figure—9.11 c). However, the trend line of Sathanur reservoir indicates that there had been a fall in total hardness across the period of study (Figure—9.4). During the peak monsoon months, the hardness of the pick-up dam varied within a narrow range of 46 mgl^{-1} (Dec'01) and 102 mgl^{-1} (Dec'91). A sudden rise in the values of hardness can be observed from 164 mgl^{-1} (Dec'00) to 468 mgl^{-1} (Dec'01) can be observed in Figure—9.11 (d). The trend line indicates an increase in the hardness of pick-up dam water during the peak monsoon (Figure—9.11 d). In the case of Sathanur reservoir, the trend line indicates a slight fall in the values of hardness along the period (Figure—9.4).

When compared with the BIS water quality standards, for drinking water (1991), the pick-up dam water falls within the permissible limits.

Turbidity

The turbidity values of the pick-up dam ranged from 5 NTU units (Jun'94) to 15 NTU units (Jun'96), during the peak summer months of the study period. There had been a sharp and steady decline in the turbidity values between 1991 and 1994, later the turbidity of the pick-up dam increased to 15 NTU units in the year 1996 (Figure 9.12 a). Subsequently, there had been a gradual decrease in the turbidity values between 1996 and 2001. The trend line suggests that there had been a slight decline in the turbidity values during the peak summer months.

During the peak monsoon months, the pick-up dam recorded the turbidity values between 6 NTU units (Dec'95) and 15 NTU units (Dec'96); following the similar pattern as in the case of summer period. A steady fall in the turbidity values can be observed from the trend line (Figure—9.12 b).

When compared with the BIS water quality standards, the pick-up dam water is suitable for aquaculture (1994) and swimming (1993).

Total Solids

The total solids (TS) present in the pick-up dam during the peak summer ranged between 250 mgl^{-1} (Jun'91) and 520 mgl^{-1}. The trend line indicates that there had been a steady increase in the concentration of total solids present at pick-up dam during the peak summer months between 1991 and 2001 (Figure—9.12 a). The TS of the water at Sathanur reservoir water was slightly more when compared with that of pick-up dam, across the years (Figure—9.6).

During the peak monsoon months of the study period (1991-2001), the pick-up dam has a record of total solids in the range of 220 mgl^{-1} (Dec'92) and 480 mgl^{-1} (Dec'95). A steady and slight increase in the values of total solids can be noticed from the trend line of Figure—9.12(b). Interestingly, the Sathanur reservoir also follows the same trend in the total solids content during the peak monsoon (Figure—9.6).

The total solids of the pick-up dam were within the permissible limits of BIS water quality standards for drinking (1991) and irrigation purposes (1974).

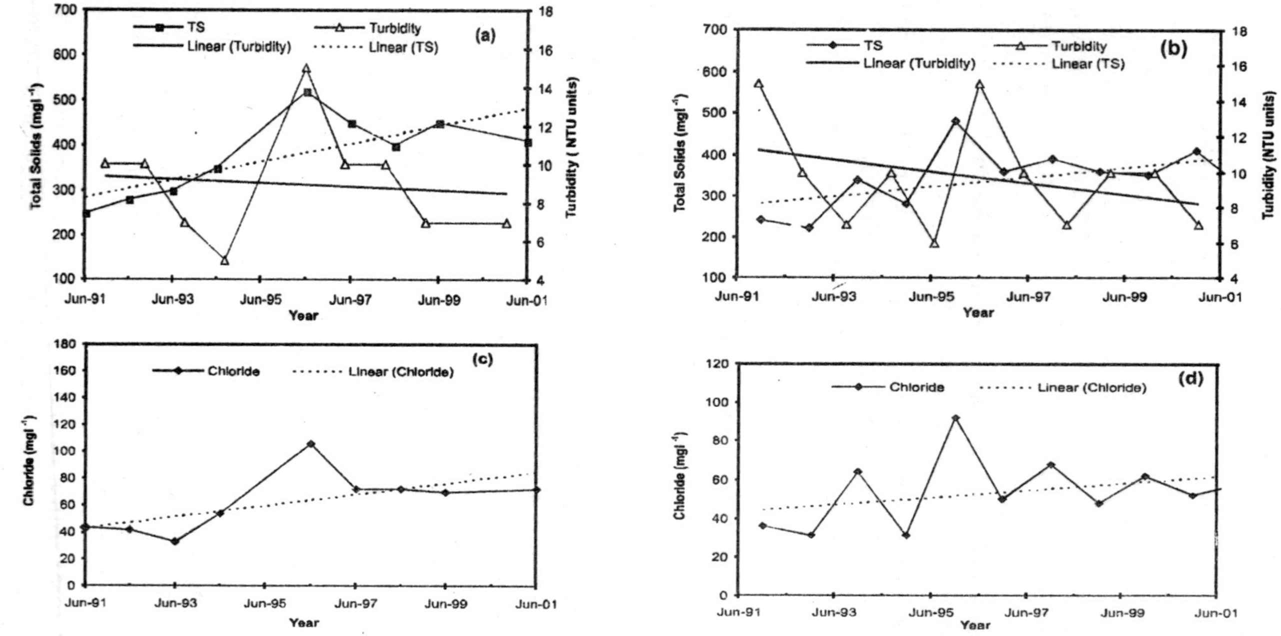

Fig. 9.11: Water quality of the pick-up dam, 1991-2001: (a) Total solids and turbidity during peak summer (b) Total solids and turbidity during peak monsoon (c) Chloride concentration during peak summer (d) Chloride concentration during peak monsoon

Chloride

Chloride concentration of the pick-up dam during the peak summer from June 1991 till June 2001, varied between 33 mgl^{-1} (Jun'93) to 106 mgl^{-1} (Jun'96). The trend line indicates that there had been an increase in the chloride concentration of the pick-up dam (Figure—9.12 c). The chloride concentrations of the Sathanur reservoir were on a higher side when compared with the pick-up dam (Figure—9.7).

The chloride values of pick-up dam during the peak monsoon months, varied between 31 mgl^{-1} (Dec'92) and 92 mgl^{-1} (Dec'95). The trend line indicates that there had been an increase in the chloride concentration of the pick-up dam across the years (Figure—9.12 d).

The chloride content of the pick-up dam falls with in the permissible limits of BIS water quality standards for drinking (1991), fisheries (1994), and irrigation (1974) and for swimming (1993).

Fluoride

The fluoride concentration of the pick-up dam, during the peak summer months from 1991 to 2001 varied between 0.2 mgl^{-1} (Jun'93) to 0.8 mgl^{-1} (Jun'98). The trend line indicates that there had been a slight rise in the fluoride values during the study period (Figure—9.13 a).

During the peak monsoon, the pick-up dam has recorded the fluoride concentration between 0 mgl^{-1} (Dec'98) to 0.15 mgl^{-1} (Dec'00). The trend line indicates that there had been a slight increase in the concentration of fluoride during the period of study (Figure—9.13 b).

As per the water quality criteria of BIS the pick-up dam water is suitable for drinking (1991) and also for swimming (1993), irrigation (1974) and aquaculture (1994).

Iron

The iron content of the pick-up dam, during the peak summer of the study period ranged between 0.05 mgl^{-1} (Dec'94) and 0.3 mgl^{-1} (Dec'98). There was a sharp decline in the iron concentration

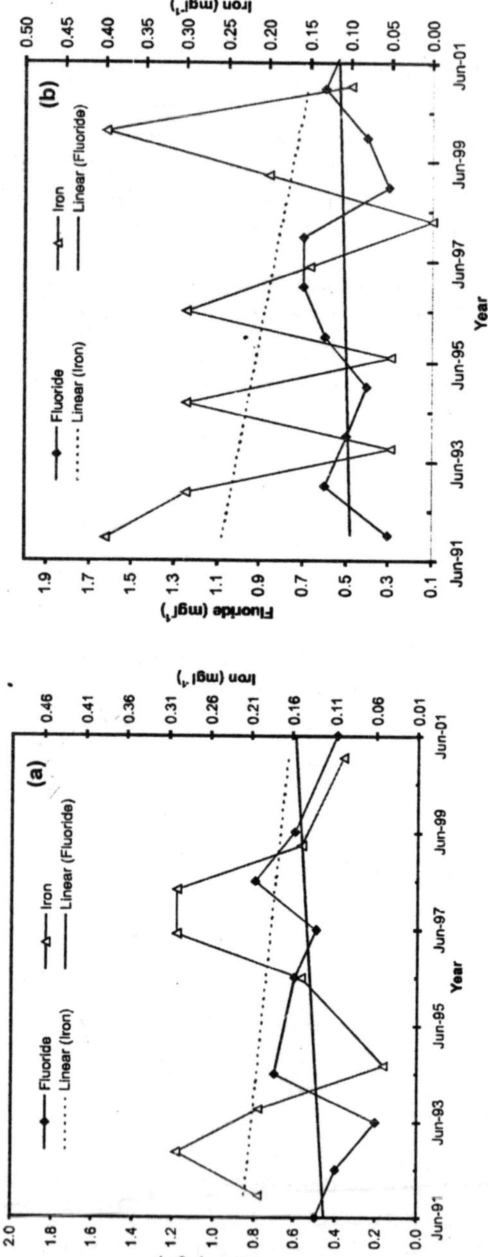

Fig. 9.12: Water quality of the pick-up dam, 1991-2001: (a) fluoride and Iron concentration during peak summer (b) flouride and Iron concentration during peak monsoon

between 1992 and 1994; later on the iron concentration was on a rise till it reached the value of 0.3 mgl^{-1} in the year 1997 (Figure—9.13 a). The trend line shows that there had been a slight fall in the concentration of the iron across the peak summer months.

During the peak monsoon months from 1991 to 2001, the pick-up dam recorded the values between 0 mgl^{-1} (Dec'98) and 0.4 mgl^{-1} (Dec'00). The iron concentration during the study period fluctuated very widely, coupled with a rise and fall in the alternate years. However, the trend line reveals a slight fall in the iron concentration during the study period (Figure—9.13 b).

As per the BIS water quality standards for drinking (1991) irrigation (1974), and swimming (1993), the pick-up dam water falls within the permissible limits.

STRATIFICATION STUDIES OF THE RESERVOIR

Physico-chemical parameters

pH

The pH ranged from 8.2 to 8.7 during June (Table—9.24) and 7.9 to 8.7 during July month (Table—9.26). The maximum gradient in pH was observed at site 'E' in June and site B and F in July. The hypolimnetic zone at most of the sites had less alkaline pH compared to epilimnion but the gradient was just one or two units. At site E, though a fall in pH is observed but it is not uniform at the depth of every two meters. The decline in pH is obvious only after 14m or 16m.

Electrical Conductivity (EC)

The EC values ranged from 670 μ mhos cm^{-1} to 740 μ mhos cm^{-1} in both June and July month (Tables—9.24 and 9.26). Hyplolimnetic EC values are higher than the epilimnion at all the sites, except site G during July where EC was observed to be higher in the epilimnion (740 μ mhos cm^{-1}). The difference between the epilimninon and hypolimnion EC is not significant as it varies only from 10-40 units. There is also no marked uniform rise in the EC values at every two metres depth as observed at site E during June and July. The EC values slightly fall at a depth of 12-14 m before picking up sharply at 18 m depth.

Temperature

The epilimnion temperature ranged from 23° C-24.5 °C and the hypolimnion temperature ranged from 14.5°C-22°C during June month (Table—9.24). During the month of July the eplimnion temperature recorded at various sites varied from 26.2°C-27°C while the hypolimnetic temperature varied from 19.4°C to 24°C (Table 9.26). The highest thermal gradient observed during the June month was at site 'E' followed by site 'F' where the epilimnion and hypolimnion temperature difference was 10°C and 9.5°C for a depth of 18m and 21 m respectively. The minimum thermal gradient, of 2.6°C was at site 'G' (5m deep). The highest gradient of 7.5°C was observed at site E in July followed by site I (7.4°C). The minimum thermal gradient was observed at site.G (2.2°C).

The fall in the temperature with depth from does not follow a uniform pattern as inferred from site 'E' (Table—9.24). The decline in temperature is markedly prominent from 6 to 10 m depth in June (Table—9.24). The decline in temperature observed to be homogenous in July (Table—9.26) when no sharp decline is noticed at any particular depth at site E. Figures—9.16 and 9.17 illustrate the trends of vertical gradient of temperature at site 'E' for June and July months. The temperature was almost constant till the depth of 4 m after which it registered a fall fills the depth of 10 m. Thereafter the temperature was more or less constant (Figure—19.14). In July a slight fall is registered after a depth of 14 m, which continues till 18 m (Figure—9.15).

Dissolved Oxygen

The epilimnion DO was observed to vary between 5.6 mg l^{-1} to 6.1 mg l^{-1} during June (Table 9.24) and the hypolimnion DO was recorded to fluctuate between 3.1 mg l^{-1} to 5.1 mg l^{-1}. The lowest value of DO was recorded at the hypolimnion of site 'F' (3.1 mg l^{-1}) followed by site 'E' (3.2 mg l^{-1}). Highest hypolimnetic DO was recorded at site G (5.1 mg l^{-1}) followed by site A (4.3 mg l^{-1}) and site D (4.1 mg l^{-1}) Maximum epliimnion DO was recorded at site H (6.1 mg l^{-1}) as inferred from Table—9.24.

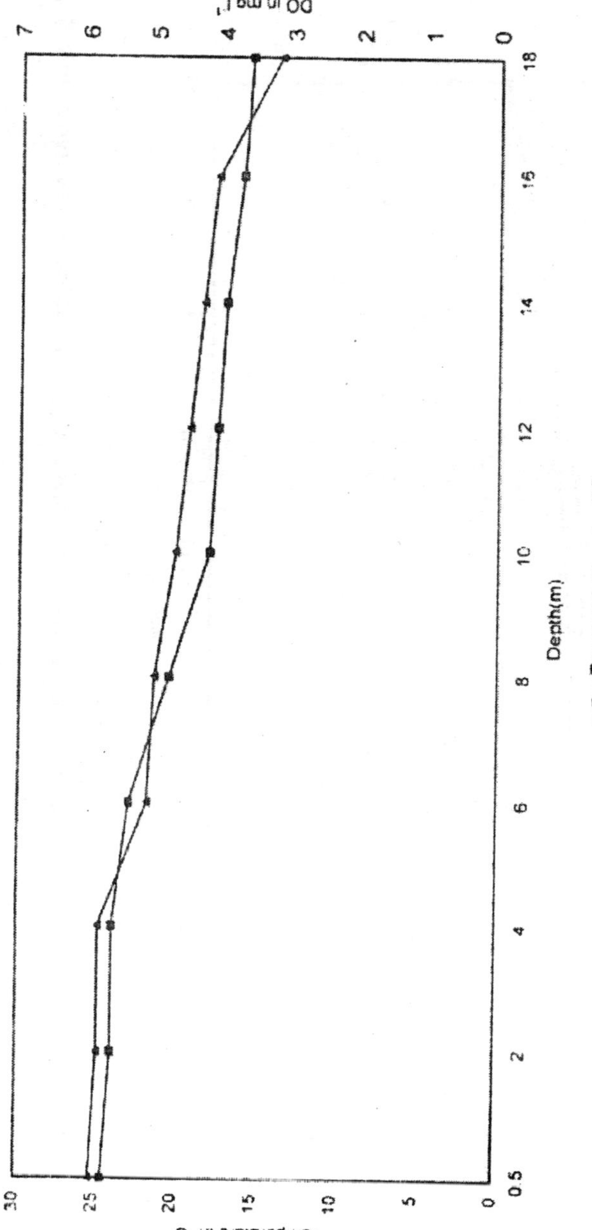

Fig. 9.13: Vertical gradient of temperature and dissolved oxygen in Sathanur reservoir (June '99)

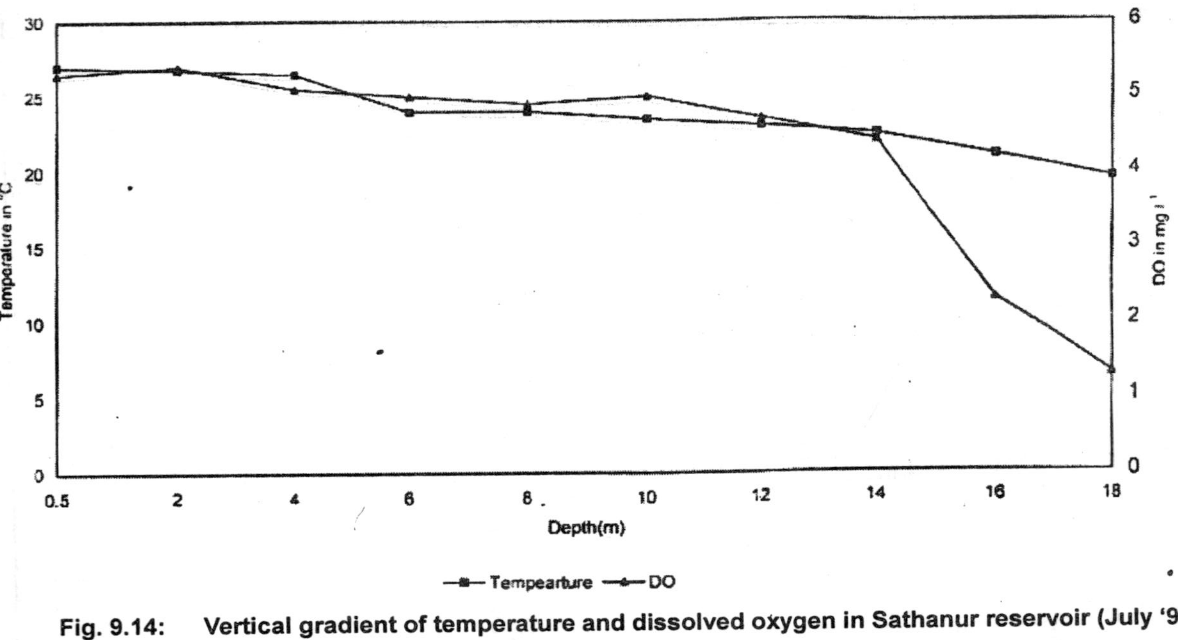

Fig. 9.14: Vertical gradient of temperature and dissolved oxygen in Sathanur reservoir (July '99)

In July the DO values were observed to vary from 5 mg l^{-1} to 5.3 mg l^{-1} in the epilimnion and 1.3 mg l^{-1} to 4.9 mg l^{-1} in the hypolimnion at various sites (Table 9.25). The highest hypolimnetic DO value was recorded at site D and G (4.9 mg l^{-1}). The lowest value of hypolimnetic DO was observed at site E (1.3 mg l^{-1}) followed by site F (1.4 mg l^{-1}). The decline in DO values was uniform till the depth of 14 m-16m after which a sharp decline was observed by a gradient of 1 mgl^{-1} as inferred from the values at site E (Tables—9.24 and 9.26).

The vertical gradients in dissolved oxygen are graphically illustrated as Figure—9.16 and 9.17 for June and July months respectively. In June the DO fell by one unit at a depth of 4 m and sharply again at a depth of 16 m (Figure—9.14). The pattern is uniform otherwise. In July a sharp fall in DO value by more than two units was observed at a depth of 14 m which fell further at a depth of 16 m recording 1.3 mg l^{-1} at 18 m depth (Figure—9.15).

Alkalinity

Alkalinity in the form of bicarbonates illustrated a maximum gradient of 48 mgl^{-1} at site H in June (Table—9.24) and that of 30 mg l^{-1} at site G in July (Table—9.25). The epilimnion alkalinity values in June varied from 260 mg l^{-1} to 300 mg l^{-1}, maximum being at site G. The hypolimnion alkalinity value varied from 280 mg l^{-1} to 320 mg l^{-1} with maximum value observed at site I (Table 9.24).

In July the maximum gradient observed was 16 mg l^{-1} at site D (Table—9.26). The epilimnion values varied from 252 mg l^{-1} to 276 mg l^{-1}, the maximum being at site B and C. The hypolimnetic alkalinity values varied from 268 mg l^{-1} to 284 mg l^{-1}. The hypolimnion values were higher at all the sites. The values don't show a uniform pattern of gradient with depth as inferred from site 'E' (Tables—9.24 and 9.26).

Acidity

The acidity values were less, pointing towards the alkaline nature of the water body. The epilimnion values in June varied from 16 mg l^{-1} to 38 mg l^{-1} the maximum being at site E (Table—9.24). The hypolimnion values ranged from 22 mg l^{-1} to 42 mg l^{-1}, the maximum value observed at site E. The maximum gradient observed was at sites A and I, with 8 mg l^{-1}.

There was a slight increase in the acidity values in July, with the epilimnetic, values in July, with the epilimnetic values ranging from 56 mg l^{-1} to 60 mg l^{-1}. The hypolimnetic values ranged between 64 mg l^{-1} to 72mg l^{-1}, maximum observed at site A (Table—9.26).

There was a pattern of increasing gradient from epilimnion to hypolimnion at all the sites except at site 'C' in June (Table—9.24). There was no uniformity in the pattern of the increase or decrease in the values depth wise as inferred from site E (Tables—9.24 and 9.26).

Hardness

The epilimnetic total hardness concentration in June varied from 180 mg l^{-1} to 186 mg l^{-1} and that of hypolimnion from 180 mg l^{-1} to 198 mg l^{-1}, maximum being, in the hypolimnion site I (Table—9.24). The maximum gradient of 12 mg l^{-1} for a depth of 18m was observed at site I. Similar values of hardness were observed in July as well (Table—9.26). There was an increase in the value vertically from epilimnion to hypolimnion at all the sites.

Calcium hardness values in June ranged from 100 mg l^{-1} to 164 mg l^{-1} in the epilimnion and 148 mg l^{-1} to 188 mg l^{-1} in the hypolimnion (Table—9.24). The high calcium hardness values reflect their contribution in the total hardness values. There is a marked fall in the calcium hardness values in July (Table—9.26). The epilimnion values ranged from 90 mg l^{-1} to 105 mg l^{-1} while the hypolimnion concentration varied from 82.5 mg l^{-1} to 135 mg l^{-1}. An increasing vertical gradient is observed, in the hardness values at all the sites but there is no marked pattern as evident from the depth wise values at site E (Tables 9.24 and 9.26). A vertical gradient of 8 mg l^{-1} observed at site E in June while in July it was 45 mg l^{-1} for the calcium hardness values. For total hardness the gradients were 16 mg l^{-1} and nil at site E in June and July respectively.

Calcium

The epilimnion concentration of calcium varied from 40 mg l^{-1} to 65.6 mg l^{-1} and hypolimnion from 59.2 mg l^{-1} to 75.2 mg l^{-1} in June (Table 9.24). In July the epilimnion calcium values recorded ranged from 31 mg l^{-1} to 42 mg l^{-1} and in hypolimnion the values ranged from 33 mg l^{-1} to 54 mg l^{-1}. The lowest recorded values were at site I (Table—9.26). No marked pattern in the vertical gradient of the calcium values was observed at site 'E' at various depths (Tables—9.24 and 9.26). Hypolimnetic concentration of calcium was more than epilimnion concentration at all the sites.

Chloride

The epilimnetic chloride values ranged from 40 mg l^{-1} to 64 mg l^{-1} in June and that of hypolimnion varied from 50 mg l^{-1} to 60 mg l^{-1}. The values showed a marked increase from epilimnion to hypolimnion at all the sites except for sites C. G and I. The maximum gradient of 14 mg l^{-1} is observed at site G where the values decrease with depth (Table—9.24).

In July, the epilimnetic chloride values were observed to range from 58 mg l^{-1} to 74mg l^{-1} while that of hypolimnion from 72 mg l^{-1} to 80 mg l^{-1} (Table—9.26). There is a rise in the chloride concentration in the reservoir water in July month. The vertical gradients observed this particular month also less compared to the ones in June at various sites. The maximum gradient of 14 mg l^{-1} for a depth of 10 m was observed at site A while at other sites the difference ranged from nil to 6 mg l^{-1} (Table—9.26).

The change in the values from epilimnion to hypolimnion do not occur in a uniform fashion as discerned from the depth wise chloride values at site 'E' (Tables—9.24 and 9.26).

Solids

The total dissolved solids in the epilmnion at various sampling sites ranged from 340 mg l^{-1} to 380 mg l^{-1} during the month of June (Table—9.24). The hypolimnetic values ranged from 330 mg l^{-1} to 390 mg l^{-1} the maximum being at site E. The maximum vertical gradient of 30 mgl^{-1} was also observed at site E. TDS values registered an increase in the hypolimnion except for site C which had higher value (340 mg l^{-1}) in the epilimnion.

In July also more or less similar values were observed (Table—9.26), the TDS values ranging from 350 mg l^{-1} to 380 mg l^{-1}.

Suspended solids have less contribution to the total solid contents of the reservoir as evident from their low values. The epilimnetic values in June varied from 18 mg l^{-1} to 38 mg l^{-1} while the hypolimnetic values varied from 10 mg l^{-1} to 36 mg l^{-1}. (Table—9.24). Similar values were observed in July too (Table—9.26)

There is no strong stratification, in the reservoir in terms of the total solids as observed at site E (Tables—9.24and 9.26). After a depth of 8 m a continuous increase in the total solids at site E, in June, is observed (Table—9.24) but in July there is no such pattern (Table—9.26).

Sulphates

The epilimnion sulphate values at various sites in June varied from 10 mgl^{-1} to 18.3 mg l^{-1} while that in hypolimnion 17 mg l^{-1} to 40.8 mg l^{-1}, the highest hypolimnion value recorded at site F at a depth of 21 m (Table—9.24). There was a slight decrease in the values in July with epilimnetic values 10 mg l^{-1} to 13.7 mg l^{-1} and hypolimnetic values 11.8 mg l^{-1} to 29.2 mg l^{-1} (Table—9.26). The hypolimnion sulphate values are higher at all the sites. There is no pattern in the vertical gradient of the sulphate concentration, as observed from the depth wise values at site E (Table—9.24 and 9.26).

Nitrates and Phosphates

The epilimnion nitrate concentration in June was recorded to range between 0.09 mg l^{-1} to 0.9 mg l^{-1} while the hypolimnion nitrate concentration varied from 1.1 mg l^{-1} to 2.7 mg l^{-1} (Table—9.24). Hypolimnetic concentration was recorded higher at all the sites. Similar values were recorded in July (Table—9.26). There is no marked pattern in the vertical gradient of the nitrate concentration as observed at site E (Table—9.24).

Epiliminion phosphate concentration varied from 0.09 mg l^{-1} (Site C) to 0.9 mg l^{-1} (Site G) during June (Table—9.24) Hypolimnion concentration ranged from 0.3 mg l^{-1} to 2.1mg l^{-1} (Site H). Similar concentrations were observed during July too (Table 9.26). Higher values of hypolimnetic concentrations were observed at site G. H and I (Tables—9.23 and 9.25).

Though the concentration of the hypolimnion phosphate was higher at all the sites still no pattern of vertical gradient at various depths was observed (Site E, Table—9.24).

Potassium and Sodium

Potassium values at the epilimnion of sites B, E and H ranged from 2.7mg l^{-1} to 2.9 mg l^{-1} during the month of June. (Table—9.24). In July the epilimnion value at site E was 3 mg l^{-1} (Table—9.26).

The hypolimnion values in June varied between 2.9 mg l^{-1} to 3.2 mg l^{-1} (Table—9.24) while in July the potassium value obtained at site E was 3.6 mg l^{-1}. No consistency in the pattern of the vertical gradient of potassium values was observed as evident from stie E (Table—9.24).

The epilimnion sodium values at sites B, E and H ranged from 23 mg l^{-1} to 24.5 mg l^{-1} during June and that of hypolmnion varied from 24.2 mg l^{-1} to 25.2 mg l^{-1} (Table—9.26). The values observed at site E in July were 25.1 mg l^{-1} (epilimnion) and 25.3 mg l^{-1} (hypolimnion). There is no much of difference between the epilimnion and hypolimnion concentrations of sodium and potassium at various sites.

Heavy Metals

Heavy metals, with the exception of zinc, were present in a very trace amount. Epilimnion copper values were recorded to range between 4 ppb to 9 ppb, maximum being at sites E and H (Table 9.24). Hypolimnion values were 11 ppb and 8 at sites E and H. Copper concentration was more at a depth of 16 m than at 18 m, as evident from the values of site E (Table 9.24) Chromium concentration varied from 2 ppb to 9 ppb at different limnetic zones (Table—9.24). The highest concentration was recorded at the hypolimnion of site E (9 ppb).

Cadmium was present in trace amounts except for in some depths at site E where 2 ppb concentration of cadmium was recorded (Table—9.24).

The epilimnetic zinc values ranged from 45 ppb to 87 ppb, highest being at site E (Table—9.24). The hypolimnion values varied from 48 ppb to 89 ppb, the highest concentration recorded at site E. At the site E, at 16 m depth 92 ppb and at 8 m depth 102 ppb concentration was recorded.

Primary Productivity Studies

In June the seechi depth was 95 cm and euphotic zone was 2.4 m. Gross primary productivities recorded at epilimnion and euphotic zones were 159.375 and 28.125 mg C fixed m^{-3} day^{-1} (Table—9.25). The climate prevailing on that particular day was windy and cloudy.

The particulars of primary productivity for the month of July are presented in Table—9.27. The seechi depth was only 70 cm and euphotic depth 1.75 m. The productivities recorded at various zones were higher compared to the values in June.

The Sathanur reservoir depicts a very weak stratification as is evident from the foregoing sections. For a true thermocline development a gradient of 3°C per meter has been suggested (Ruttner, 1963). In Sathanur reservoir the highest gradient was 10°C for a depth of 21 m. The gradients reported in other Indian reservoirs vary from 1°C to as much as 8 °C (David et al, 1969, Devasundaran and Ray, 1954).

The gradients in the chemical parameters also are quite narrow, supporting the absence of well-marked stratification. The reservoir water is nutrient rich as inferred from its phosphate and nitrate concentrations. The concentration of Phosphate between 30-100 µ mg l^{-1} point towards the eutrophic and hyper-eutrophic status of the reservoir (Wetzel, 1975). Further seechi depth of less than 1m (95 cm and 70 cm in June and July respectively) also reflects towards the eutrophic condition of the reservoir (Huber et al, 1982). The DO is generally low perhaps due to low photosynthetic activity. The DO values were lower in July compared to June because of the higher water temperature in July. The hypolimnetic DO concentration was only 1.3 mgl^{-1} at a depth of 18 m. Since, there is very weak summer stratification in the reservoir, there is apparently some mixing between the various zones. This mixing prevents the prevalence of complete anoxic conditions in the hypolimnion. The lowest DO value recorded was 1.3 at site E. There have been reports of low productivity and large water volume of deep lakes sustaining mild oxygenated hypolimnion during summer stratification (Moore et al, 1988). The low DO value in the reservoir is alarming in the wake of fisheries being practised in the reservior. For supporting fish life a minimum of 3.0 ppm of DO is essential in the water (Abbasi, 1997).

High concentration of ions in the hypolimnion suggests a nutrient rich sediment in the Sathanur reservoir. Release of major ions from such sediments is enhanced by oxygen deficit in the hypolimnion (Abbasi, 1997) and these maintain a high concentration of nutrients in the water mass that may lead to problematic internal load (Salonen and Varjo, 2000). Studies illustrate that metals at high pH sink down to sediments (Dean et al, 1972).

The presence of high concentration of phosphate, nitrate, chloride, zinc and sulphate may be attributed to the agricultural practices, and land use pattern upstream of the reservoir. The agricultural runoff consists of several fertilizers, pesticides and other chemicals, which are transported downstream to the reservoir. Chlorides, nitrate and phosphate based fertilizers are considered to be the principal source of these ions in the areas under intensive agriculture (Rao et al, 1997).

Section—IV Impacts, Downstream

10

Impacts of Irrigation

INTRODUCTION

The largest demand for world's water comes from the agriculture sector. More than two-thirds of the water withdrawn from the earth's rivers, lakes and underground aquifers is utilized for irrigation. Agriculture has turned out to be not only the largest user in terms of volume, but also relatively low value, low efficiency and highly subsidized water user. These facts are forcing governments and donors to reanalyze the economic, social and environmental implications of large, medium and small irrigation projects operating across the world.

In China, Pakistan and Indonesia irrigation has absorbed over half of agricultural investment. In India, about 30 per cent of all public investment has gone into development of irrigation facilities. Despite such huge investment and subsidies, the irrigation performance indicators have been found to fall short of expectations in terms of yield, area irrigated and technical efficiency.

With this backdrop, we have conducted a study on the Sathanur irrigation project. The project caters to the irrigation demands of 18,000 ha of agricultural land. The study highlights the key features of the irrigation practised in Sathanur Command

Area (SCA) with major thrust on Sathanur canal irrigation. The study also brings out the impacts of the irrigation, beneficial and adverse, on the environment.

STATUS

Irrigation facilities, before the construction of Sathanur reservoir project, were scanty and undependable in the Thiruvannamalai and Villupuram districts, Tamil Nadu, India. The Ponnaiyar river being seasonal and flashy (Chapter 2), there were no means of utilizing its waters. Farming was completely dependent on monsoon in these areas.

Sathanur Reservoir and Canal System

The details regarding the commissioning and operating of the Sathanur project and the network of canal system have been furnished in Chapter 2. The project was designed to irrigate more than 18,000 ha of land downstream through two canal systems: the Sathanur Right Bank Canal (SRBC) and the Sathanur Left Bank Canal (SLBC). The extent of command area under SRBC and SLBC is illustrated in Tables—2.1 to 2.5, Chapter 2. The area under indirect tank irrigation (village wise) in SRBC and SLBC commands is detailed in Tables—2.6 and 2.7, Chapter 2.

Regulation of Release of Water from the Sathanur Reservoir

The regulation of the supply of water from the reservoir is under the control of Sub-Divisional Officer, Public Works Department (PWD), Thiruvannamalai. The person is generally referred as Controlling Officer. The various provisions for scheduling and allocation of water from the reservoir to the downstream command areas are listed below.

Sathanur Left Bank Canal and Old Deltaic Regions

It has been stipulated that inflow in excess of the limit specified hereunder against each month is to be impounded in the reservoir.

1 January -15 April	2.00 cusecs
16 April-15 June	Nil
16 June-30 September	2.00 cusecs
1-31 October	1,500 cusecs
1-30 November	Nil
1-31 December	1500 cusecs

1. During the period 16 June-15 July, if necessary the inflows less than 2.000 cusecs may temporarily be impounded to facilitate the execution of repair works of sluice gates, stilling basin etc. The quantity so impounded shall be correctly assessed and let down immediately depending upon the requirement of lower down command areas.
2. The natural flows between 16 June-30 September that are usually released in small quantities below the limit flows, if impounded temporarily should be released on or before 30th of September.

   (Amendment in G.O. Ms No. 1220, PWD, dt 16th June 1984)
3. Inflow in excess of those specified above thus impounded in the reservoir shall not be claimed at a later date for the benefit of the existing irrigation lower down.
4. Whenever the Sub-Divisional Officer, PWD Cuddalore anticipates good rainfall in the old deltaic area, he may suggest the Controlling Officer to reduce the release from the reservoir. It shall be the responsibility of the Sub-Divisional Officer, PWD, Cuddalore to see to it that no water is wasted at Cuddalore bridge.
5. The Controlling Officer may, in exceptional cases when heavy rains are anticipated in the neighbourhood of the reservoir, reduce or increase the discharge from the reservoir on his own understanding and responsibility.
6. In the event of the failure of monsoon the water requirement, for the period 16 April-15 June and 1-30

November for the existing cultivable land below the Sathanur dam, may be fulfilled from the impounded storage, if any in the reservoir, with the prior approval of the Chief Engineer (Irrigation Department).

7. Water allocation for the command area under the SRP shall normally be allowed from 15 December-30 April. If there is no sufficient storage in the reservoir in any year for irrigating the entire command area under SRP canal system, the Executive Engineer, PWD, Thiruvannamalai, shall, in consultation with the Collectors of North Arcot and South Arcot districts (now Thiruvannamalai and Villupuram respectively) fix the extent that can be irrigated with and the probable date from which the release could be made. The beneficiaries will be informed not later than middle of November.

   The supply for the command irrigation under the SRP canals shall normally be allowed at the duties specified below for each period for the area proposed to be under irrigation. It shall be in addition to the quantities let down for the irrigation of old deltaic regions till 15 April.

15-31 December	100 duty
1 January -14 February	50 duty
15 February-31 March	60 duty
1-30 April	80 duty

   The duties proposed are for the general guidance and the Controlling Officer can alter the duty for the purposes of regulation according to the extent of wet and dry crops raised in the command area. It should however be noted that the total quantity should not normally exceed that provided for as per the duty above.

8. Stabilization of 5000 acres of second crop under the command area of Thirukoilur barrage shall normally be allowed from 1February-30April (at 55 duties) depending upon the storage position of the reservoir. This will be in addition to the quantities let down for

the command area under the SRP canal and old delta regions till 15 April. A total quantity of not more than 1200 M.cft. will be let down for the 2 crop under Thirukoilur barrage system in three spells.

If any year there is no sufficient storage for irrigating the entire 5000 acres of second crop under the Thirukoilur barrage, the Executive Engineer, PWD, Villupuram district, in consultation with the Executive Engineer, PWD, Thiruvannamalai district, shall fix the extent that can be brought under canal irrigation. The probable dates from which such supply would be made will be duly informed to the beneficiaries downstream not later than the middle January.

The Executive Engineer, PWD, Villupuram district and Sub-Divisional Officers, PWD, Thirukoilur should intimate the Controlling Officer, the quantity that has to be let down from the reservoir for stabilization of the second crop area under the Thirukolilur barrage subjected to the maximum fixed amount.

9. The Sathanur reservoir shall not be normally depleted below level plus 46.00 feet, since this storage available will be required for supplying water to the colony and maintenance of parks and for aquaculture in the reservoir.

Sathanur Right Bank Canal

1. The rules and regulations already approved for the Sathanur reservoir in G.O. Ms. 1074, PWD, dt. 20 July 1976 hold good for the command area under right bank canal system. The current rules form part of these existing rules, as the right bank canal is only an additional source of irrigation from the same reservoir.
2. Release of water for irrigation under SRBC shall normally be allowed from 1 October – 15 February.
3. A minimum impounded storage of 1000 M.cft. should be available in the reservoir on 1st October to open the canal. Earlier opening of the canal can be done in case the storage position of the reservoir is good.

4. The Executive Engineer, Thiruvannamalai, can either postpone the opening date of the canal or restrict the command area to be irrigated after reviewing the storage position of the Sathanur reservoir.
5. For the wet irrigation under the tanks, the supply may be given for filling up the tanks at suitable intervals and the total quantity to be allowed may be limited to 650 M.cft.
6. For the direct dry irrigation the water may be released intermittently at an interval of about 10 days or as per the designed period. The supply may be allowed at a duty of 70 and the total quantity to be allowed may be limited to 1450 M.cft.

Water Deliveries to the Farmers

The authorities and the farmers jointly manage irrigation systems. At the macro level, till the outlets to the villages through the canals and branch canals the distribution of water is the responsibility of the authorities. At the micro level, the field wise distribution of water is managed by the beneficiaries based on their mutual consent and understanding.

Irrigation waters are supplied to the downstream farmers as per the rules. The major outlets are manned and operated by the canal inspector as per the schedules. Water is allocated on rotational basis of seven days in the branch canals as per the norms laid down.

As per the present scenario, there are no measuring devices at the outlets to check the discharge and there are no means to account for the seepage losses. Many such proposals are underway, though, awaiting funds, according to the authorities.

Status of Sathanur Canal System and Tanks in the Sathanur Command Area (SCA)

The main canals--SLBC and SRBC were lined along their entire lengths under the scheme of National Water Management Project (NWMP) funded by the World Bank. Under another scheme, Sathanur-sub-project implementation worth Rs. 16.8 millions, lining works of some branch canals and distributaries were executed. Major portion of the arterial system of sub-branch canals, distributaries, and drains in various villages remains unlined due to the paucity of funds. The unlined canals and tanks are heavily silted and infested with weeds and other vegetation (Chapter 16).

The data pertaining to the irrigation was collected from the offices of the Directorate of Statistics and Economics, Thiruvannamalai and Villupuram. The block-wise intensity of irrigation in SCA is presented in Table—10.1. Intensity of irrigation is calculated as the ratio of gross irrigated area to net irrigated area. The Table—10.2 gives an account of the irrigation potential created and realized in India during the five-year plans. The details of the irrigation patterns in various villages are furnished in Tables—10.3-10.8. Cumulative account of the irrigation patterns in various blocks, command regions and Sathanur project is given in Table 10.9. Table—10.10 presents the cumulative details of the gross cultivated area and gross irrigated area in SCA. The details pertaining to the proposed area under canal irrigation and the actual potential realized for the year 1997-98 are presented in Table—10.11.

Table—10.1 Intensity of irrigation in Sathanur reservoir project

Block	Intensity of irrigation
Thachampattu (SLBC command)	1.32
Thirukoilur (SLBC command)	1.35
Chengam (SLBC command)	1.33
Rishivandiyam (SRBC command)	0.98
Sankarapuram (SRBC command)	1.34
Chengam (SRBC command)	1.76
SLBC command	1.32
SRBC command	1.33
SCA	1.33

Table—10.2 Irrigation potential created and utilized through irrigation schemes during various five year plans in India

Plans	Irrigation potential (Million hectares)	Irrigation utilization (Million hectares)
Pre-first	22.6	22.6
First six	65.2	58.8
Seventh	76.5	68.6
Eighth	89.3	80.5

Table—10.3 Irrigation pattern: Thatchampattu (SLBC command) block (1997-98)

Villages	Area under		
	Canal irrigation (in hectares)	Tank irrigation (in hectares)	Well irrigation (in hectares)
Thatchampattu	51.16.0	9.82.0	214.75.5
Allihondapattu	52.85.0	0.58.5	56.92.0
Kalampoondi	127.63.0	57.61.0	243.53.5
Periyakallipadi	186.64.0	52.11.5	128.08.0
Chinnahallipadi	–	22.50.0	118.45.0
Navampattu	166.58.5	74.42.5	31.02.5
Devanur	–	90.20.0	105.27.0
Velayampakkam	171.19.0	57.16.5	149.21.0
Kallothu	102.37.0	–	70.29.0
Athipadi	148.34.0	–	39.49.5
Palayanur	108.62.0	60.89.0	155.23.5
Kandiankuppam	228.21.0	32.73.0	87.42.5
Thalayampallam	138.67.5	-	192.42.5
Chakkarathamadai	27.55.0	8.73.0	34.54.5
Nariyapattu	227.99.0	36.55.0	227.23.5
Parayampattu	131.69.0	33.12.0	131.38.5
Pavapattu	199.00.5	–	149.47.5
Parithram	NA	N A	NA
Aradapattu	NA	NA	NA

NA - Not Available

Table—10.4 Irrigation pattern: Thirukoilur (SLBC command) block (1997-98)

Villages	Area under		
	Canal irrigation (in hectares)	Tank irrigation (in hectares)	Well irrigation (in hectares)
Melandhal	373.51.5	86.50.5	94.37.0
Kangayanur	202.22.0	–	105.17.0
Pallichandal	91.47.5	15.42.0	121.54.5
Kongamanur	40.45.0	25.87.5	69.28.5
Murukhambadi	10.44.0	43.75.0	178.99.0
Athiandhal	76.33.0	40.07.5	54.12.5
Devaradiarkuppam	119.61.5		69.38.5
Jambai	159.17.5	54.09.5	87.87.5
Chellankuppam	–	–	25.24.5
Sithapatinam	25.43.0	93.05.0	102.06.0
Manalurpet	–	50.85.0	36.16.0

Table—10.5 Irrigation pattern: Chengam (SLBC command)—1997-98

Villages	Area under		
	Canal irrigation (in hectares)	Tank irrigation (in hectares)	Well irrigation (in hectares)
Jambadai	–	–	196.42.0
Olgalpadi	3.00.0	18.40.0	180.38.5
Thenmudiyanur	78.75.0	3.80.0	784.22.5
Edalkanur	100.75.0	36.52.0	185.61.5
Allappanoor	–	24.98.0	91.07.5
Vanapuram	152.95.5	39.65.5	244.50.0
Kunglinatham	21.68.5	28.04.0	73.34.5
Perunthuraipattu	42.28.5	14.86.0	48.70.5
Valavackanur	22.79.0	31.28.0	102.18.0
Kottaiyur	115.03.5	71.26.0	134.69.5
Agarampallipattu	21.54.5	119.30.5	165.30.5
Thenkarimbalur	298.72.0	124.64.0	126.69.0

(Table Contd...)

1	2	3	4
Radhapuram	–	17.84.0	257.38.0
Varagur	16.25.0	200.96.0	546.59.5
Serapapattu	2.10.5	20.41.0	93.84.0
Vakillapattu	20.00.0	17.53.5	53.58.0
Sadakuppam	25.00.0	61.70.5	80.00.0
Unnamalaipalayam	25.00.0	32.31.5	60.00.0
Edakkal	168.26.5	–	–
Mazhuvampattu	52.22.5	43.44.5	39.00.0
Peraiyampattu	30.70.0	69.64.5	36.75.0

Table—10.6 Irrigation pattern: Rishivandiyam (SRBC command) block (1997-98)

Villages	Area under		
	Canal irrigation (in hectares)	Tank irrigation (in hectares)	Well irrigation (in hectares)
Athiyur	–	96.30.0	242.45.0
Mangalam	3.18.0	35.22.0	379.69.0
Adalhanur	0.10.0	156.52.5	–
Erudayampattu	12.19.5	–	100.86.0
Maniaandal	–	–	164.33.0
Vanapuram	136.69.5	–	322.57.0
Odiyanthal	24.68.5	–	31.26.0
Edathanur	–	47.34.0	85.41.0
Kadambur	112.58.5	–	194.88.0
Arambarampattu	42.27.5	21.43.0	304.33.5
Thiruvarangam	–	47.01.0	67.38.5
Sirpanandal	26.33.5	160.03.5	237.35.0
Jambadai	28.46.5	61.39.0	222.42.5
Periyakolliyur	6.02.5	92.04.0	390.74.0
Sirupanaiyur	98.25.0	–	185.33.0
Chinnakolliyur	–	20.07.0	113.71.5
Theluvanthangal	112.47.5	74.26.0	–
Kadavanur	–	260.43.0	336.96.0
Pakkam	–	139.99.0	391.20.5
Vadamamandur	79.46.5	93.19.0	115.53.5
Nagalkudi	51.73.0	–	61.12.0

Table—10.7 Irrigation pattern: Sankarapuram (SRBC command) block (1997-98)

Villages	Area under		
	Canal irrigation (in hectares)	Tank irrigation (in hectares)	Well irrigation (in hectares)
Rayasamudram	–	47.21.0	136.27.5
Arulampadi	51.44.0	21.98.5	145.08.5
Moongilthuraipattu	153.82.5	32.62.0	215.75.0
Porasapattu	–	15.45.5	91.26.5
Olgalapadi	119.87.5	39.99.5	225.00.0
Poravalur	113.88.5	14.00.5	269.97.5
Melsiruvalur	185.65.5	62.86.5	411.78.5
Vadasiruvalur	–	53.26.0	214.23.0
Varagur	2.03.5	27.38.5	60.91.5
Arur	2.69.0	31.10.5	75.35.5
Moohanur	13.67.0	119.55.0	429.94.0
Thimmendhal	–	20.46.0	66.81.5
Chellakakuppam	13.24.5	–	47.50.5
Viriyur	–	41.07.5	119.73.0
Arasampattu	–	16.08.5	55.46.0
S.Kolathur	21.46.5	128.78.5	171.23.5
Vadakeeranur	21.41.5	48.29.0	89.77.5
Vadaponparappi	98.21.5	–	190.73.5
Kidagudayampattu	1.08.5	32.51.0	78.66.5
Sivapuram	16.27.5	–	92.98.5

Table—10.8 Irrigation pattern: Chengam (SRBC command)—1997-98

Villages	Area under		
	Canal irrigation (in hectares)	Tank irrigation (in hectares)	Well irrigation (in hectares)
Rayandapuram	90.56.0	137.21.0	283.35.0
Thiruvadathanur	87.81.0	0.16.0	230.97.0.
Thondamanur	-	119.30.5	165.30.5
Puthurcheekadi	NA	NA	NA

NA - Not Available

Figure—10.1 is the graphical representation of the irrigation patterns in various blocks in SCA. The cumulative account of the patterns in SCA is presented in Figure—10.2. Figure—10.3 illustrates the gross irrigated area in SCA while the efficiency of the canal irrigation in SCA is reflected in Figure—10.4.

IMPACTS

Irrigation in SCA is practised with the conjunctive use of surface water through canals and tanks, and ground water in the form of dug-wells and bore wells. All the villages in the command area are electrified enabling the farmers to install motor pumps in their dug-wells. The Figure—10.1 depicts the extent of area (block-wise) under various categories of irrigation. As is obvious from the figure, ground water-based well irrigation assumes the major share of irrigation in all the blocks with the exception of Thirukoilur (served by SLBC). Indeed SLBC command has better proportion of land under canal irrigation as compared to the SRBC command. Canal irrigation constitutes 12 per cent, 17 per cent and 18 per cent of the gross irrigated area in Rishivandiyam, Sankarapuram and Chengam blocks respectively, in SRBC command areas. More than 50 per cent of the land is irrigated by groundwater in these blocks. Tanks claim 12 per cent 25 per cent of the share in the gross irrigated area.

The patterns of irrigation for SRBC command, SLBC command and cumulatively for the Sathanur project are presented in Figure—10.2. The figure reiterates that ground water is the major source of irrigation in SCA followed by canal irrigation and tank irrigation.

The contribution of canal irrigation is very low in SRBC command (14%) and is not even 25 per cent in the SCA. SRBC command has its 65 per cent of land under ground water based irrigation and SLBC command has 50 per cent of the cultivated land under ground water based irrigation, thus highlighting the role of groundwater in irrigation process in SCA.

Table—10.9 Irrigation pattern in SCA (1997-98)

Blocks	Area under canal irrigation (in hectares)	Area under tank irrigation (in hectares)	Area under well irrigation (in hectares)
Thachampattu (SLBC command)	2068.80.5	570.98.0	4775.34.0
Thirukoilur (SLBC command)	1098.65.0	409.62.0	2452.48.0
Chengam (SLBC command)	1172.06.0	857.29.0	5294.24.0
Rishivandiyam (SRBC command)	734.46.0	1305.23.0	5987.24.0
Sankarapuram (SRBC command)	814.77.5	752.64.0	4755.89.5
Chengam (SRBC command)	228.37.0	306.67.5	1264.67.0
SRBC command	1777.60.5	2364.54.5	12007.80.5
SLBC command	4339.51.5	1837.89.0	12522.15.0
SCA	6117.12.0	4202.43.5	24529.95.5

Table—10.10 Gross cultivated area and gross area under irrigation in SCA (1997-98)

Blocks	Gross sown area (in hectares)	Gross irrigated area (in hectares)
Thachampattu (SLBC command)	5545.47.5	4775.34.0
Thirukoilur (SLBC command)	3930.70.0	2452.48.0
Chengam (SLBC command)	7719.52.0	5294.33.0
Rishivandiyam (SRBC command)	8597.09.0	5987.24.0
Sankarapuram (SRBC command)	6581.23.0	4755.89.5
Chengam (SRBC command)	2563.75.0	1264.67.0
SRBC command	17742.07.0	12007.80.5
SLBC command	17195.69.5	12522.15.0
SCA	34937.76.5	24529.95.5

Table—10.11 Proposed area under canal irrigation and actual area irrigated in SCA (1997-98)

Blocks	Proposed area under canal irrigation (in hectares)	Actual area under canal irrigation (in hectares)
Thachampattu (SLBC command)	3518.78.0	2068.80.5
Thirukoilur (SLBC command)	2207.06.0	1098.65.0
Chengam (SLBC command)	2815.15.0	1172.06.0
Rishivandiyam (SRBC command)	3445.51.5	734.46.0
Sankarapuram (SRBC command)	4235.27.5	814.77.5
Chengam (SRBC command)	737.30.0	228.37.0
SRBC command	3418.09.0	1827.60.5
SLBC command	8540.99.0	4339.51.5
SCA	16959.08.0	6167.12.0

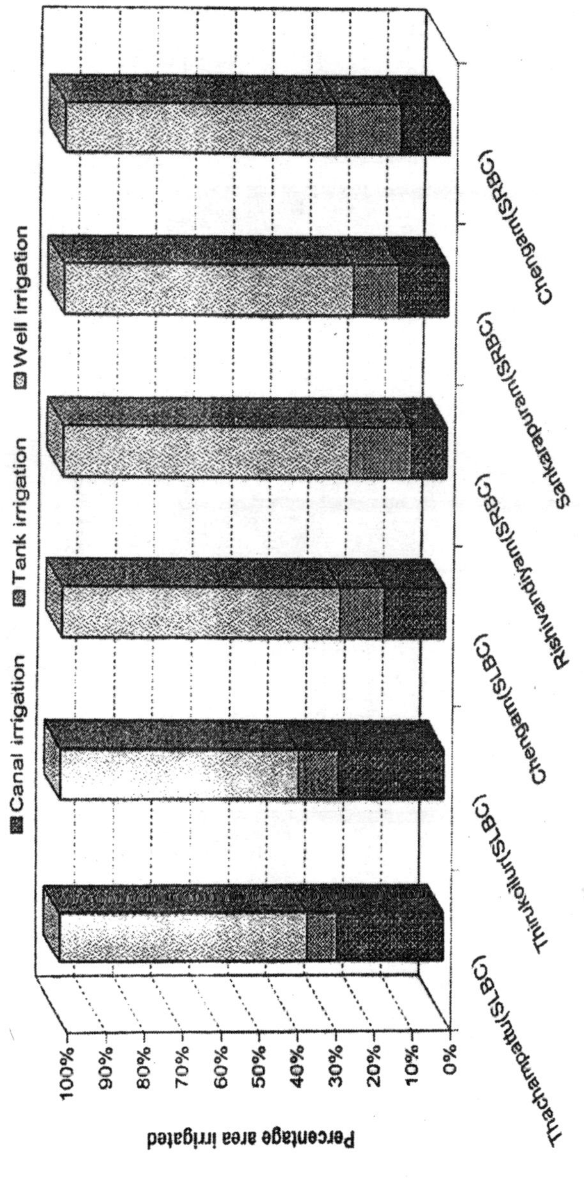

Fig. 10.1: Irrigation pattern in SCA (block wise) for 1997-98

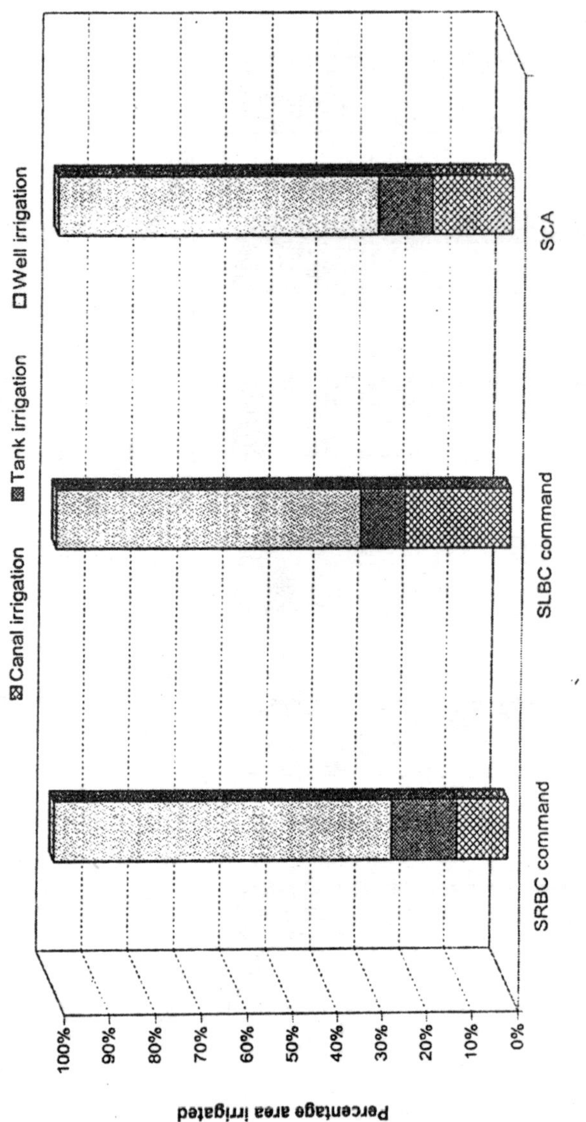

Fig. 10.2: Irrigation pattern in SCA (block wise) for 1997-98

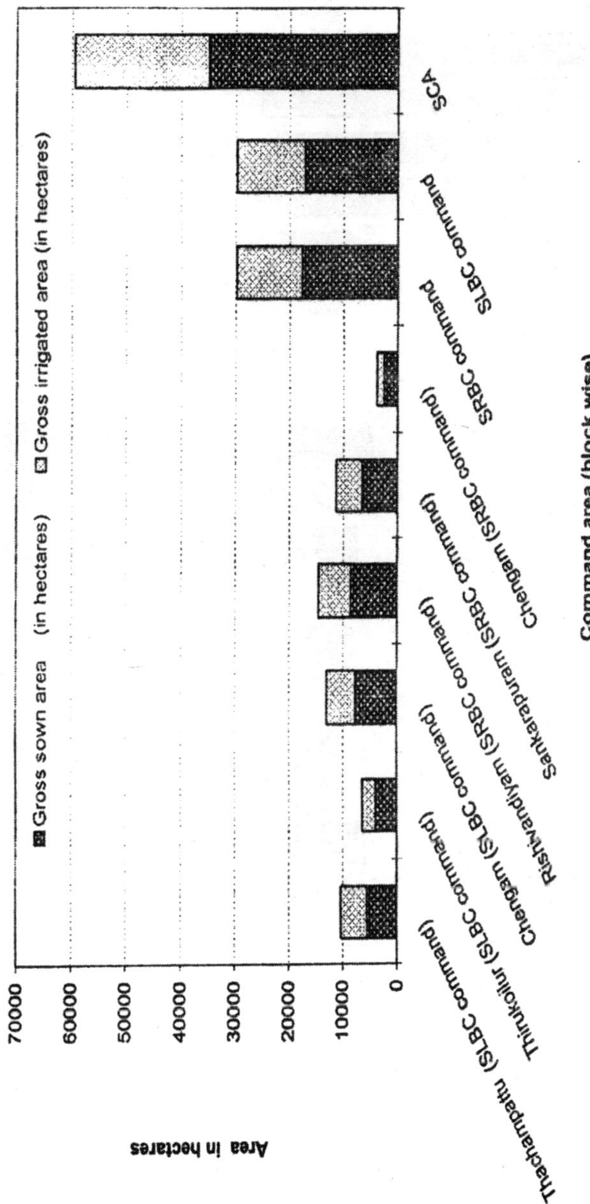

Fig. 10.3: Gross irrigated area to gross cultivated area in SCA

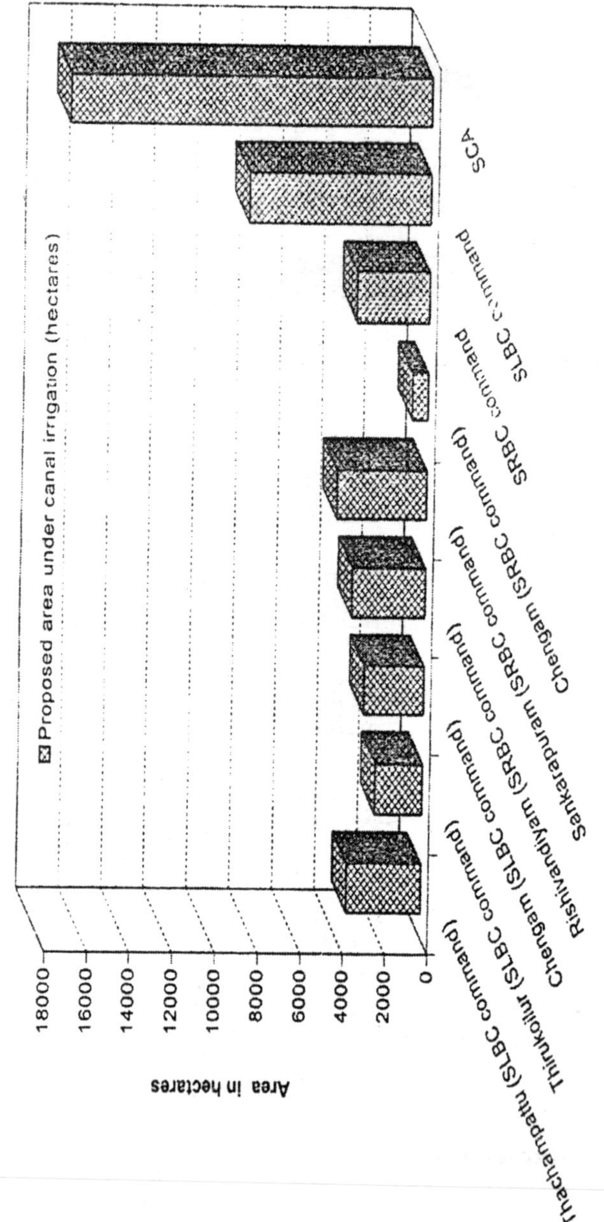

Fig. 10.4: Status of canal irrigation in SCA (1997-98)

The higher utilization of groundwater as compared to surface water for irrigation in the SCA is a manifestation of a nation-wide phenomenon. Since 1950-51 considerable importance has been attached to the provision of canal irrigation nation-wide as the canal irrigated area increased from 8.3 million hectares to 16.9 million hectares during the last four decades. But, still, it is groundwater-based irrigation, particularly tube-well irrigation that has intensified to a very great degree and has made spectacular progress (Dutt and Sundharam, 1999). Groundwater based irrigation in 1990-91 accounted for 51 per cent of the total irrigated area as compared to only 29 per cent in 1950-51. In SCA 58 per cent of the irrigated land is under groundwater irrigation while only 25 per cent under canal irrigation (Figure—10.2). Higher proportion of area under groundwater utilization in SRBC command (65 %) is due to irregular flow of water in these canals, either due to damaged reaches or faulty regulations, which have made the cultivators of the region more dependent on the groundwater which according to them is a more reliable source of water for irrigation than surface water. As inferred from the extensive field survey of the command area backed up by the interactions with the authorities concerned, the SLBC command enjoys the maximum benefits of the canal irrigation: a fact confirmed by the statistics of canal irrigation (Figure 10.1) which depict that 50 per cent of the irrigated area is under canal irrigation in SLBC command as compared to 21 per cent in SRBC command. Under the present scheme of regulation of the water allocation (2.10.2), SLBC command has priority over SRBC command. The groundwater level substantially increases during the flow of water in the canal and the Thiruvannamalai region receives good rains during the monsoons. Thus, availability of surplus water is the reason behind the fact that gross irrigated area is more in SLBC command (75%) while 67.6 per cent of the gross cultivated area is irrigated in SRBC command (Figure—10.3).

In stark contrast, the SRBC command, in the Villupuram district, is mostly a dry land. In fact SRBC was introduced to divert excess water of the reservoir (as per the regulation) to these regions. Many a times when the water is below the prescribed limit in the regulation of water release, water for irrigation is either not released at all (with prior notice) or is released in controlled fashion as per the availability in the reservoir. The groundwater level of these regions is generally low but when the reservoir in full and water is released, the villagers have reported an increase of 10-14 feet and during these times groundwater is exploited to the maximum for irrigation purposes.

Figure—10.3 illustrates the total area irrigated as against the total area cultivated (in percentage). Gross irrigated area was high in all the SLBC blocks with Thachampattu block having the highest (86.1 per cent) followed by Chengam (68.6%). SRBC blocks registered less area under irrigation compared to SLBC with Chengam block dismally low at 49.3 per cent, Sankarapuram block had the highest irrigated area of 72.3 per cent in the SRBC command.

The gross irrigated area from all the sources in the Sathanur project was 70.2 per cent. In the SRBC and SLBC commands the gross irrigated area was 67.6 per cent and 72.8 per cent respectively.

Table—10.1 provides the intensity of irrigation in various blocks in the SCA, individual command areas and cumulatively for Sathanur project. It is evident that both the SLBC and SRBC commands have almost similar intensity of irrigation, 1.32 and 1.33 respectively. This implies that both the command regions are way short of the irrigation facilities for double cropping. Chengam block (SRBC command) has the maximum intensity with 1.76 and Rishivandiyam block has the least with mere 0.98. Thus, the irrigation facilities are just enough for a single crop. Irrigation intensities for all the remaining blocks are in a narrow range of 1.32 to 1.35. Cumulative intensity of irrigation for of the Sathanur project is 1.33. This could also mean the heavy shift towards cultivating water intensive perennial crops like Paddy (*Oryza sativa*), Sugarcane (*Saccharum officinarum*) or banana (*Musa paradisiaca*, which prevents double and triple cropping (Chapter 13).

Figure—10.4 Illustrates the extent of irrigation actually realized through the canal as against the irrigation potential created in the SCA. All three blocks under SLBC command have higher percentages of canal irrigated land ranging from 41 per cent - 58 per cent as compared to the blocks under SRBC irrigation where the areas irrigated range from a mere 19 per cent to 30 per cent. SRBC command registered only 21% of the proposed area under canal irrigation where as the situation was better in SLBC command where 50 per cent of proposed area was brought under canal irrigation. Cumulatively for Sathanur project, the figure stood at a mere 36 per cent, which projects that only one-third of the potential created for canal irrigation in SCA was actually realized (as per the statistics of 1997-98). The gulf between irrigation potential created through major, medium and minor irrigation schemes and actual utilization is also a countrywide phenomenon as furnished in Table—10.2.

The situation is even more dismal in SCA where a meagre 36 per cent of the irrigation potential created through canals is actually being utilized (Figure—10.4). When compared to other projects such as Ukai-kakrapur project across Tapi which has realized the irrigation potential of 88 per cent and the Mahi Kadna project across Mahi river in Gaujrat, India (Purohit et al, 1992), the situation seems to be particularly worrisome.

The inefficiency and unreliability of the canal irrigation can be attributed to the unlined sub-canals and distributaries and to the faulty drainage. Not only the flow of water in these canals gets reduced, they also give rise to the problems of seepage, water logging and weed infestation, in the command area as evident from the various studies accomplished worldwide (Dixon et al, 1987; Singh, 1989; Purohit et al, 1992). In SCA water logging problems were reported mostly in the fields adjacent to the canals and tanks. Even the lined canals in some places were potential source of water logging because of the poor quality of cement used (Chapter 16). The seepage loss shoots up the water table of the area and reduces the surface flow. The tanks of the command area are silted and full of weeds thus leading to reduction in their storage potential and making the adjacent fields water logged. Sadakuppam, Valavachanur, Velayampakkam, Parayampattu, Pavapattu, Palayanur, Kallottu, Pudur, Pakkam, Arundhatiar colony and several other villages had seepage and water logging problem owing to the unlined canals and silted and weed-infested tanks.

Introduction of perennial irrigation springs forth the hazards of water logging, seepage, salinity and alkalization as evident from the case studies of all-major and minor irrigation projects worldwide. Tarbela dam, in Indus basin, Pakistan; Aswan dam on Nile river, Egypt (Dixon et al, 1987); Pravara and Mula dams in Maharashtra, India (Bhave, 1992); Ukai-Kakrapur irrigation project, India (Sahai et al, 1985); Krishnaraj Sagar project, Karnataka, India (Rao et al, 1990); Nizam Sagar command area, India (Venkatratnam et al, 1991) and numerous other projects have been reported to have their command areas degraded leading to loss of agricultural lands.

Introduction of surface irrigation and availability of surplus water in SCA have resulted in the general tendency of the farmers to switch over to paddy (*Oryza sativa*) or sugarcane (*Saccharum officinarum*), which require high water applications. Very intensive irrigation by the farmers also tends to cause alarming rise in water table of the command area. This trend has also been reported by many authors for other major projects viz. Aswan dam, Egypt (Zeid, 1990); Tawa reservoir, India (Sahu, 1992), Ukai- Kakrapur project across Tapi, India (Purohit et al, 1992) Krishanaraj Sagar project, India (Rao et al, 1990).

Interestingly many farmers in the Sathanur command area weren't aware of the phenomenon of water logging and they considered it to be a boon as the paddy crops require standing water. So the water logged fields adjacent to the canals and tanks have also been devoted to paddy cultivation.

The Government of Gujarat, India, in 1991-92 decided not to sanction the planting of perennial crops such as sugarcane in its major reservoirs command areas. (Purohit et al 1992). Master plans to line the canals were formulated and successfully implemented to check the menace of water logging and seepage in the major and medium irrigation project under command area development (CAD) scheme. All these measures towards efficient water use and better water management can be successfully applied to Sathanur irrigation project as well.

There exists no well-defined policy or management framework for irrigation scheduling for effective and efficient water management in the Tamil Nadu state (Pundarikanthan and

Santhi, www.fao.org). The adverse implications of this drawback are more than obvious in Sathanur irrigation project.

Cropping Pattern

There is an accepted practice to include all the cultivable land in command area under irrigation. This has led the beneficiaries to assume that they have the freedom of choice of the crops to be cultivated in their fields. There is a heavy tilt towards perennial, cash crops of paddy, sugarcane and groundnut (*Arachis hypogea*) (Chapter 13), which suggests a cropping pattern driven entirely by commercial considerations with little regard to sustainability of the land for long-term output. The noteworthy fact is that Sathanur irrigation project in SLBC command is meant for dry irrigated crop of groundnut only according to the Agricultural Officer, Thiruvannamalai. But villagers devote their fields to water intensive paddy cultivation (Chapter 13).

Water Allocation and Distribution from the Reservoir

As the Governmental agencies and the cultivators jointly manage the canal irrigation systems, very often clashes are witnessed in the absence of proper communication. At the village level and field level, there is a general complaint that tail-enders suffers the most. They don't get their due share because of the location of village or the farm towards the end, as also due to most of the water having been utilized by the farmers of the head reaches (Chapter 16).

There was a general complaint by the farmers that the water is not enough for their crops, especially towards the end of the main season (April) when it is the time for harvesting the crops. According to the Agricultural Officers and Irrigation Engineers, there is a tendency to over-irrigate their fields by the cultivators downstream. So some times when due to less inflow and less storage even when the water is adequate there is a panic among the farmers that water is inadequate for their crops.

Studies done on the farmers of Nathiyunni command in Tamiravani system, Tamil Nadu reveal that in farmer's perception stagnating water in the paddy fields is like building up stock of water and when it is not possible they have a sense of inadequacy. The relative water supply (RWS), a measure of water adequacy, during the same season was computed to be 110 per cent of the water requirement (Rajan, 1993). The officials vehemently denied the allegations of bribery for releasing water in the canals by the cultivators.

Conflicts Between the Users of the SRBC and SLBC Commands

The two canal systems have bifurcated the SCA not only in terms of supplying the water through the separate canal systems but also by way of pattern of utilization in the command regions.

Originally the project was constructed to serve SLBC command, but it deprived the other traditional and productive delta areas of irrigation water. The rights of the downstream irrigators are recognized in the dam's operating rule, but most of the regulated flow below the dam is diverted into the upper channels. Losses have greatly increased in the wide sandy bed and no surface water has reached the sea for over twenty years.

Continued spills to the extent of 50 per cent, were the reasons used to justify the subsequent construction of the right bank irrigation command, further aggravating the shortages in the delta and producing endless conflict between the two commands.

The inhabitants of SLBC command had the complaint that after the construction of SRBC their privilege of getting water for irrigation was reduced from six months to mere three months. On the other hand residents of SRBC command complained that due to the regulation policy and also due to faulty and mismanaged canals, they never got their due share of water except for first five years or so after the opening of the right bank canal. They maintain that it's the cultivators of the SLBC command who enjoy the maximum benefits of canal irrigation.

Further, there is an apprehension among the farmers downstream that after the construction of one more (Nandan) canal, the share of water allocated to them may further go down. Moreover, additional storage dams on tributaries upstream are increasing the evaporation losses in what was earlier a fully developed basin.

Fragmented management of water resources in south India is exemplified in case of Chittar river and Amravati dam too. Constructing the storage dam without adequately considering downstream users and the storage capacity prevailing in the basin causes great deal of economic losses and social problems to the end users downstream.

Seepage and Conveyance Losses

There are no considerations based on the crop water requirement at various stages. There are no means to measure seepage or conveyance losses and neither is there any framework regarding the nighttime utilization of water. The cultivators tend to irrigate the fields only during daytime and the water remains unutilized in the canals during the nights.

With a view to narrow the gap between the potential created through the major and medium works and its ground level utilization, the government has started Command Area Development (CAD) programme. The basic objective of the CAD is to maximize productivity in the irrigation commands through an integrated approach covering the developmental works of the supply channels and provides equitable and assured distribution of water to individual farm holdings.

Unfortunately SRP was not covered in this CAD programme, as confirmed by the authorities, for a long time. It is only very recently that with the help of World Bank financing and other independent schemes the lining and developmental works of the canals and branch canals has been undertaken and is an ongoing process. There are no schemes to reclaim the damaged and silted tanks as yet.

The Irrigation Engineers and project authorities blame the cultivators for their lack of organization and mutual understanding. The cultivators in the head reaches are not supposed to take water till it reaches the tail-end farmers' fields but the cultivators at head reaches disregard this and utilize the maximum available water thus depriving the others from getting their share. Also the cultivators who own pump sets ought to let the poorer cultivators to use more of surface water but this seldom happens, according to the authorities. There is a pressing need for a proper balance between the infrastructure, water flow and organizational dimension of the canal irrigation.

11

Land Use

INTRODUCTION

'Land use' may be defined as 'a dynamic pattern of human utilization of the land resources for various purposes'. Such pattern is the manifestation of the influence of geo-morphology, availability of natural resources, climate and socio-cultural and techno-economic factors operating in an area. The present study evaluates the current status of land use in the Sathanur Command Area (SCA). The possible impact of Sathanur Reservoir Project (SRP) and other related environmental factors on the evolution of the land use pattern in the area has also been assessed.

STATUS

Officially, the following land-use classification is employed by the Government of Tamil Nadu; it applies to Sathanur as well:

- Forests
- Barren and uncultivable land
- Land for non-agricultural usage
- Cultivable waste
- Permanent pastures and grazing land
- Land under miscellaneous trees and groves

- Current fallow land
- Other fallow land
- Net sown area.

The data pertaining to the land-use was collected from the offices of the Directorate of Statistics and Economics, Thiruvannamalai and Villupuram. The patterns of land use, in various blocks under the SCA, in the individual command areas and SRP overall, are illustrated in the Figures—11.1 to 11.9. The cumulative percentage figures have been arrived at based on the land-use data for the years 1997-98 for the individual villages in the SCA. The data has been furnished in Tables—11.1 to 11.6. The land use pattern of the Thiruvannamalai district is depicted in Figure—11.10 and that of Tamil Nadu (TN) state in Figure11.11. The cumulative block–wise land use data for the command area and the catchment has been presented in the Tables—11.7 and 11.8.

IMPACTS

Forest

The forest cover in the study area is mere 1.1 per cent of the total reported area (TRA) of 84219.3 ha. (Figure.11.11). Thiruvannamalai district, which encompasses most of the Sathanur Left Bank Canal (SLBC), has its 24.28 per cent of the total land under forest (Figure—11.10). SLBC command has 2 per cent of the total land as forest cover (Figure—11.4) as against 0.3 per cent in Sathanur Right Bank Canal (SRBC) command (Figure—11.8). Thirukoilur block (SLBC command) has the highest share of 7.7 per cent, in the command region (Figure—11.2) followed by Sankarapuram block (SRBC command) with 0.9 per cent. Thachampattu block (Figure—11.1) has mere 0.5 per cent of Total Reported Area (TRA). There is no forest cover at all in the blocks of Rishivandiyam (Figure—11.6), Chengam, (SLBC command) (Figure—11.3) and Chengam (SRBC command) blocks (Figure—11.7).

The only villages in the SCA with forest cover are Nariyapattu, Parayampattu, Pavapattu and Manalurpet in SLBC command with there 3.8 per cent, 1.2 per cent and 49.9 per cent of the respective TRA under forest. In SRBC command, Kadambur,

Table—11.1 Land use (in hectares): Thachampattu (SLBC command) block (1997-98)

Villages	Forest	Barren and uncultivable land	Land for non-agricultural uses	Cultivable waste	Permanent pasture and grazing land	Land under miscellaneous trees and groves	Current fallow land	Other fallow land	Net sown area	Total area
Thatchampattu	-	26.48.5	80.91.5	5.56.5	13.01.0	-	14.13.5	-	270.85.0	410.96.0
Allikondapattu	-	1.39.0	22.77.0	0.11.0	-	-	2.27.0	1.93.5	82.13.0	110.60.5
Katampoondi	-	42.27.0	23.64.5	47.01.0	-	-	62.21.0	13.17.5	373.38.5	561.69.5
Periyakallipadi	-	23.60.0	13.89.5	13.21.5	-	2.69.0	85.22.5	24.02.0	367.21.5	529.86.0
Chinnakallipadi	-	23.64.0	14.21.5	23.20.5	-	-	64.30.5	28.64.0	212.89.0	366.89.5
Navampattu	-	18.67.5	16.32.0	33.42.5	-	16.49.5	80.93.0	-	324.68.5	490.53.0
Devanur	-	17.60.0	64.35.0	2.33.0	-	-	8.82.5	-	230.80.5	323.91.0
Velayampakkam	-	2.65.5	108.76.0	-	-	-	13.71.0	-	317.61.0	442.73.5
S. kollathur	-	0.58.5	25.72.5	-	-	-	9.12.0	0.44.0	107.41.5	138.28.5
Athipadi	-	-	141.79.5	41.85.5	-	-	29.62.0	1.64.0	164.57.0	379.48.0
Palayanur	-	24.21.5	17.63.0	8.39.5	-	26.61.5	84.43.5	11.20.5	294.91.0	467.40.5
Kandiankuppam	-	20.70.5	53.86.5	8.66.5	6.28.5	-	10.21.0	1.17.0	292.36.0	393.26.0
Thalayampallam	-	-	12.65.5	55.47.0	-	-	88.94.0	2.29.0	337.21.0	496.53.5
Chakkar thamadai	-	2.72.0	11.30.0	-	-	-	5.61.0	-	70.10.0	89.73.0
Nariyapattu	12.60.5	-	4.00.0	60.84.0	-	-	13.79.0	36.33.0	-	332.59.5
Parayampattu	2.71.5	14.89.5	-	10.23.5	-	-	-	-	194.81.5	222.66.0
Pavapattu	19.13.5	27.95.5	48.42.0	18.45.5	15.90.0	2.86.0	20.00.0	-	293.94.5	446.67.0
Pavithram	-	5.31.5	9.17.0	2.47.5	19.21.5	27.19.0	310.11.0	110.19.5	368.22.5	851.89.5
Aradapattu	-	3.63.0	57.22.5	12.16.0	-	2.08.0	61.32.5	6.33.0	273.83.5	416.58.5

Table—11.2 Land use (In hectares): Thirukoilur (SLBC command) block (1997-98)

Villages	Forest	Barren and uncultivable land	Land for non-agricultural uses	Cultivable waste	Permanent pasture and grazing land	Land under miscellaneous trees and groves	Current fallow land	Other fallow land	Net sown area	Total area
Melandhal	–	26.20.0	21.96.5	15.22.5	–	0.66.5	40.50.0	126.03.0	515.17.0	745.75.5
Kangaiyanur	–	23.33.0	66.12.5	0.73.5	–	–	25.88.5	3.00.5	276.92.0	396.00.0
Pallichandal	–	95.03.5	44.05.0	1.21.0	–	–	15.28.0	–	242.29.0	397.86.5
Kongamanur	–	0.56.0	59.82.5	–	–	–	15.10.0	–	132.32.5	207.81.0
Murukkambadi	–	6.35.5	53.71.5	15.73.5	–	3.01.0	45.56.0	–	350.89.5	475.27.0
Athiandhal	–	0.29.0	44.20.0	0.20.0	–	–	12.00.0	2.76.5	149.87.5	209.33.0
Devaradiarkuppam	–	108.24.5	–	–	–	29.45.0	–	6.92.5	189.00.0	333.62.0
Jambai	–	33.25.0	151.54.5	–	3.69.0	–	57.63.0	0.31.5	448.20.5	694.63.5
Chellankuppam	–	0.35.5	17.60.5	–	–	–	–	12.73.5	117.49.5	148.19.0
Sithapatinam	–	19.11.5	22.0.0	2.22.5	0.98.0	10.43.0	40.48.0	125.18.0	288.39.5	479.00.5
Manalurpet	369.74.0	123.12.0	45.96.0	12.67.0	–	–	7.81.0	23.74.5	157.94.0	740.98.5

Table—11.3 Land use (in hectares): Chengam block (SLBC command)-1997-98

Villages	Forest	Barren and uncultivable land	Land for non-agri-cultural uses	Cultivable waste	Permanent pasture and grazing land	Land under miscellaneous trees and groves	Current fallow land	Other fallow land	Net sown area	Total area
Jambadai	–	37.87.5	26.01.5	15.54.0	–	–	98.04.0	2.46.0	121.12.5	301.05.5
Olgalapadi	–	130.10.5	86.20.0	17.54.0	–	–	120.37.5	17.07.0	241.97.5	613.26.5
Thenmudiyanur	–	10.11.35	118.33.5	15.40.5	–	–	188.00.0	12.35.0	83.77.30	1273.45.5
Edathanur	–	11.09.5	–	10.30.0	–	–	339.08.0	141.16.0	288.09.0	648.56.0
Alappanoor	–	102.15.5	4.16.0	–	–		36.42.5	41.85.0	102.90.0	287.49.0
Vanapuram	–	29.93.5	97.45.5	7.68.0	–	1.75.0	89.41.5	10.00.0	378.88.0	615.11.5
Kunglinatham	–	12.93.5	30.75.5	–	–	–	10.85.0	–	100.90.5	155.48.5
Perunthuraipattu	–	3.07.0	18.80.5	0.87.0	–	–	30.50.5	3.03.5	98.91.0	155.19.5
Valavackanur	–	–	105.49.0	–	–	–	8.00.5	–	131.01.0	244.50.5
Kottaiyur	–	–	150.48.0	–	–	–	20.00.0	60.00.0	292.08.5	522.56.5
Agarampallipattu	–		71.31.5	97.60.5	–	6.50.0	116.37.0	22.45.5	284.96.0	599.20.5
Thenkarimbalur	–	22.00.5	134.75.5	9.32.5	–	–	130.75.0	–	510.81.5	807.65.0
Radhapuram	–	50.20.5	90.35.5	110.25.0	–	10.15.0	84.55.0	200.54.0	761.51.0	1307.56.0
Varagur	–	38.67.5	36.12.0	8.13.5	–	–	77.49.0	–	182.31.0	342.73.0
Serapapattu	–	2.69.5	46.38.0	1.84.0	–	–	27.96.0	0.46.5	163.58.0	242.92.0
Vakillapattu	–	7.71.0	17.52.5	0.68.5	–	–	8.70.5	–	64.32.5	98.55.0
Sadakuppam	–	20.00.0	75.01.0	–	–	–	10.00.0	24.83.0	165.74.5	295.57.5
Unnamalaipalayam	–	20.00.0	17.58.5	–	–	–	30.00.0	–	112.16.0	179.74.5
Edakkal	–	28.23.0	26.67.0	3.33.0	–	–	73.06.5	–	254.52.0	385.80.5
Mazhuvampattu	–	6.81.5	18.15.5	–	–	0.37.0	10.29.0	–	103.03.0	138.66.0
Peraiyampattu		72.57.0	25.02.5	–	–	–	9.50.5	–	113.32.0	220.42.0

Table—11.4 Land use (in hectares): Rishivandiyam (SLBC command) block (1997-98)

Villages	Forest	Barren and uncultivable land	Land for non-agricultural uses	Cltivable waste	Permanent pasture and grazing land	Land under miscellaneous trees and groves	Current fallow land	Other fallow land	Net sown area	Total area
Athiyur	–		132.76.0	3.10.5	–	–	55.40.0	77.92.5	465.63.5	734.82.5
Mangalam	–	4.42.0	58.36.5	–	–	–	51.33.5	–	362.68.0	476.80.0
Adathanur	–	4.72.0	8.56.0	-	–	–	62.61.5	–	156.98.5	232.88.0
Erudayampattu	–		5.83.0	19.50.0	–	–	12.31.0	–	104.46.5	142.10.5
Maniyandal	–	62.58.0	2.36.5	–	–	–	76.67.0	–	198.67.0	340.28.5
Vanapuram	–	125.95.0	8.00.0	–	–	–	131.31.5	–	555.90.0	821.16.5
Odiyanthal	–		43.19.5	2.97.0	–	–	32.63.5	12.51.0	101.23.5	192.54.5
Edathanur	–	26.80.0	7.12.0	4.92.0	–	–	81.06.0	–	235.64.0	355.54.0
Kadambur	6.98.0		175.66.0	6.41.5	–	–	165.45.5	–	379.16.5	733.67.5
Arambarampattu	–	7.15.0		15.15.0	–	–	265.13.0	–	368.04.0	655.47.0
Thiruvarangam	–		0.30.0	–	–	–	022.21.5	105.34.5	155.24.0	283.10.0
Sirpanandal	–	301.06.0	24.12.0	21.46.0	–	–	475.54.0	–	630.16.0	1452.34.0
Jambadai	–	-		7.55.5	–	–	179.70.5	155.02.0	420.15.5	762.43.5
Periyakolliyur	–	165.83.5	10.80.5	34.65.5	–	–	104.12.0	–	622.94.5	938.36.0
Sirupanaiyur	–	142.67.0	6.47.5	8.94.0	–	–	88.75.0	–	392.16.0	638.99.5
Chinnakolliyur	–	55.70.5	5.05.5	1.17.0	–	–	62.96.5	–	130.37.5	255.27.0
Tholuvanthangal	–		2.37.0	31.52.0	–	–	36.91.5	–	231.48.5	302.29.5
Kadavanur	–	220.10.0	31.43.0	0.80.0	–	0.73.5	31.53.0	9.35.5	471.83.0	765.78.0
Pakkam	–	151.19.0	165.60.0	8.21.0	3.89.0	1.70.0	32.17.0	–	563.03.0	925.79.0
Vadamamandur	–	22.15.0		8.12.0	–	–	169.56.0	–	285.19.0	485.02.0
Nagalkudi	–	42.66.0	6.50.0	0.63.5	–	–	20.28.5	–	116.17.0	186.25.0

Table—11.5 Land use (in hectares): Sankarpuram (SRBC command) block (1997-98)

Villages	Forest	Barren and uncultivable land	Land for non-agricultural uses	Cultivable waste	Permanent pasture and grazing land	Land under miscellaneous trees and groves	Current fallow land	Other fallow land	Net sown area	Total area
Rayasamudram	–	1.16.5	2.26.0	0.45.5	–	–	126.05.0	–	257.49.5	387.42.5
Arulampadi	–	29.19.0	5.10.6	1.02.5	–	–	34.55.5	–	181.55.0	251.42.5
Moongilthurai-pattu	66.82.5	187.12.5	14.77.0	–	–	–	54.00.5	–	394.15.0	721.87.5
Porasapattu	–	44.71.0	12.15.0	54.12.5	–	–	67.43.0	–	135.79.5	314.21.0
Olgalapadi	–	58.43.5	8.25.0	2.90.0	–	0.42.0	48.42.5	–	357.86.5	476.29.5
Poravalur	–	60.30.0	81.73.0	2.99.5	–	–	41.91.0	–	408.15.0	595.08.5
Melsiruvalur	–	41.10.0	30.04.0	4.10.0	–	1.05.0	67.93.0	3.33.5	388.20.0	535.75.5
Vadasiruvalur	–	3.94.5	94.60.0	–	0.56.0	2.04.0	133.42.0	–	351.15.5	585.72.0
Varagur	–	–	40.41.5	–	–	–	45.60.0	–	98.19.5	184.21.0
Arur	–	4.17.5	44.73.5	–	–	–	32.98.0	–	112.95.0	194.84.0
Mookanur	–	4.00.5	112.10.5	1.60.0	–	–	6.10.0	88.84.0	595.13.5	807.78.5
Thimmendhal	–	–	39.13.5	–	–	–	4.40.0	–	60.76.5	104.30.0
Chellakakuppam	–	–	18.00.0	–	–	–	6.80.0	–	58.35.0	83.15.0
Viriyur	0.89.0	–	113.80.0	–	–	–	62.64.5	–	228.02.0	405.36.0
Arasampattu	–	2.69.0	35.60.5	–	–	0.20.0	39.35.5	–	185.45.0	263.30.0
S.Kolathur	–	13.91.5	94.94.5	0.26.0	–	–	30.00.0	–	313.53.0	452.65.0
Vadakeeranur	–	1.54.0	8.96.0	0.38.5	–	–	54.61.0	72.39.5	142.60.0	280.49.0
Vadaponparappi	–	35.01.5	2.29.5	2.00.0	–	–	30.23.0	–	194.20.0	263.74.0
Kidagudayampattu	–	9.08.0	13.50.0	8.11.0	–	–	12.60.0	5.96.5	128.00.0	177.25.5
Sivapuram	–	0.68.0	11.00.0	1.90.5	–	–	26.18.0	14.00.0	107.48.0	161.24.5

Table—11.6 Land use: Chengam (SRBC command)—1997-98

Villages	Forest	Barren and uncultivable land	Land for non-agricultural uses	Cultivable waste	Permanent pasture and grazing land	Land under miscellaneous trees and groves	Current fallow land	Other fallow land	Net sown area	Total area
Rayandapuram	–	–	263.15.5	16.46.0	–	–	123.71.0	–	609.97.0	1154.45.5
Thiruvadathanur	–	–	89.50.0	37.54.5	–	–	236.52.5	–	325.83.0	689.40.0
Thondamanur	–	–	102.40.0	–	–	–	20.00.0	202.65.0	329.80.5	836.95.5
Puthurcheekadi	–	–	28.23.5	9.72.5	–	–	117.83.5	–	254.73.0	410.52.5

Table—11.7 Cumulative land use (block-wise) in hectares

Villages	Forest	Barren and uncultivable land	Land for non-agricultural uses	Cultivable waste	Permanent pasture and grazing land	Land under miscellaneous trees and groves	Current fallow land	Other fallow land	Net sown area	Gross sown area
Thachampattu (SLBC command)	34.45.5	247.39.0	660.26.0	328.77.5	35.19.5	48.66.0	593.33.5	120.84.5	4134.89.5	6203.81.0
Thirukoilur (SLBC command)	369.74.0	435.85.5	507.19.0	48.00.0	4.67.0	33.55.5	260.24.5	300.70.0	2868.51.0	4828.46.5
Chengam (SLBC command)	-	64.66.75	1160.47.0	290.08.0	-	18.77.0	1453.39.0	394.65.5	5128.46.0	9092.50.0
Rishivandiyam (SRBC command)	6.98.0	1333.99.0	691.51.0	172.12.5	3.89.0	2.43.5	2157.68.0	360.15.5	6952.16.0	11680.92.5
Sankarapuram (SRBC command)	67.71.5	49.70.70	783.40.0	79.86.0	0.56.0	3.71.0	930.22.5	184.53.5	4699.04.0	7246.11.5
Chengam (SRBC command)	-	-	483.38.0	63.73.0	-	-	680.17.0	343.81.0	1520.33.5	3091.42.5

Table—11.8 Cumulative land use in hectares

Village	Forest	Barren and uncultivable land	Land for non agricultural uses	Cultivable waste	Permanent pasture and grazing land	Land under miscellaneous trees and groves	Current fallow land	Other fallow land	Net sown area	Gross sown area
SLBC command	404.19.5	1329.92.0	2327.92.0	666.85.5	39.86.5	100.98.5	2306.97.5	816.20.0	12131.86.5	20124.77.5
SRBC command	74.69.5	1831.06.0	1958.29.0	315.71.5	4.45.0	6.14.5	3768.07.5	888.50.0	13171.53.5	22018.46.5
SCA	1478.89.0	3160.98.0	4236.21.0	982.57.0	44.31.5	107.13.0	6075.04.5	1704.70.0	25303.40.0	42143.24.0
Thiruvannamalai	153318	21039	90840	11553	3818	8722	52319	18755	270841	631205
Tamil Nadu	2141000	481000	1937000	345000	125000	225000	1028000	1229000	5486000	12997000

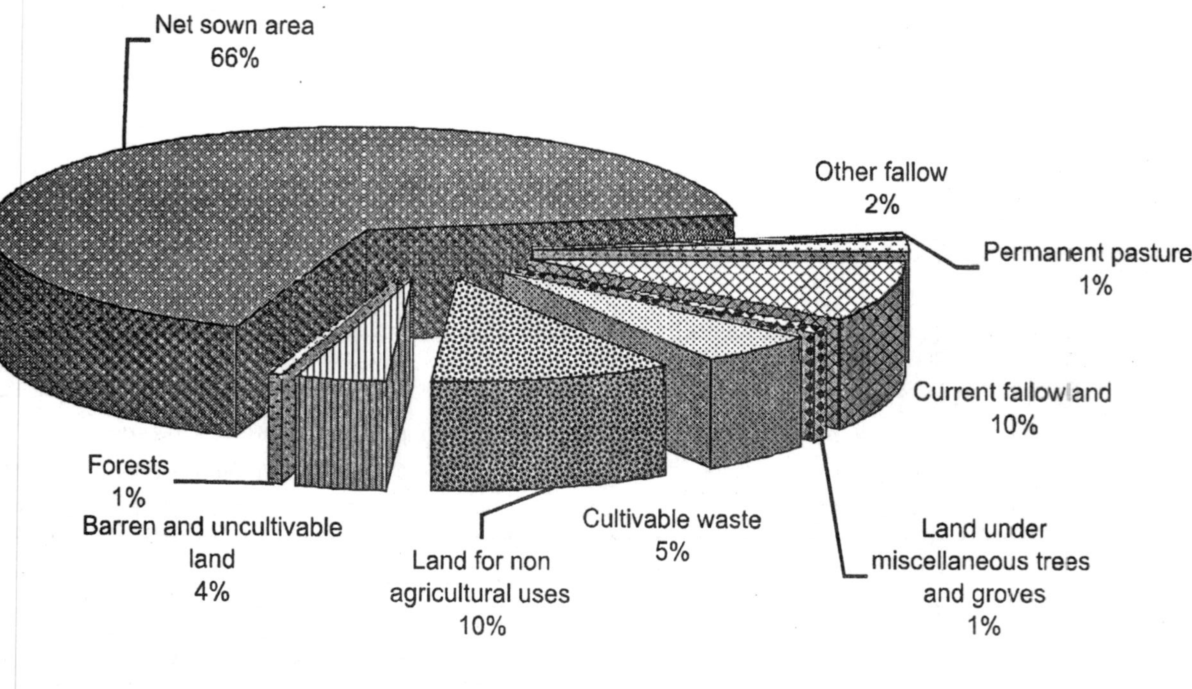

Fig. 11.1: Land-use pattern in Thachampattu block (SLBC command) for 1997-98

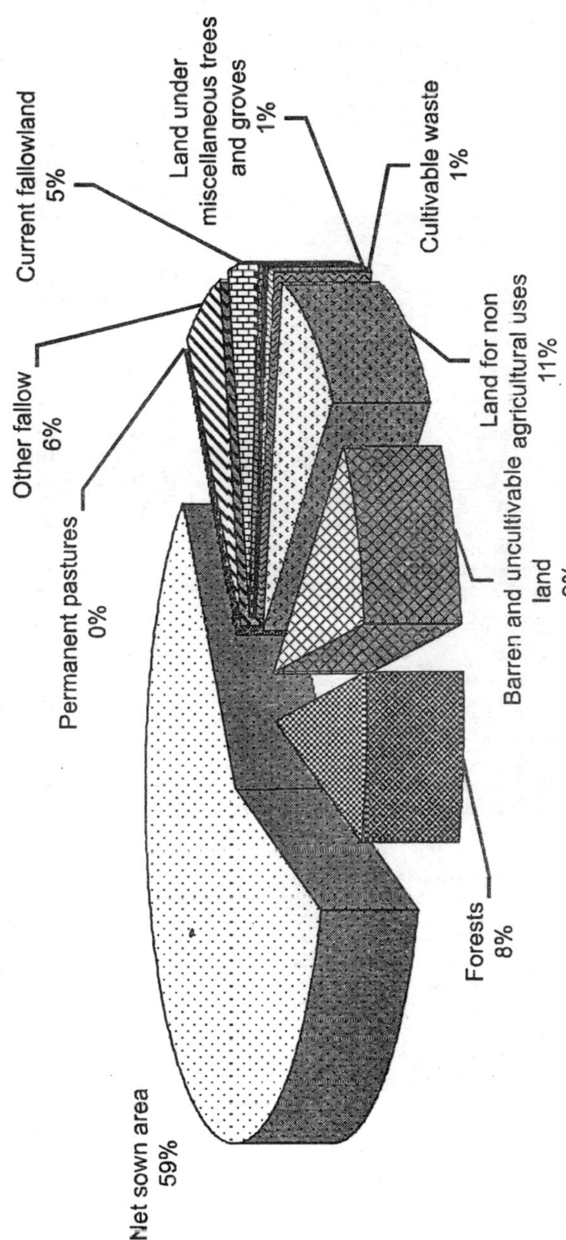

Fig. 11.2: Land-use pattern in Thirukoilur block (SLBC command) for 1997-98

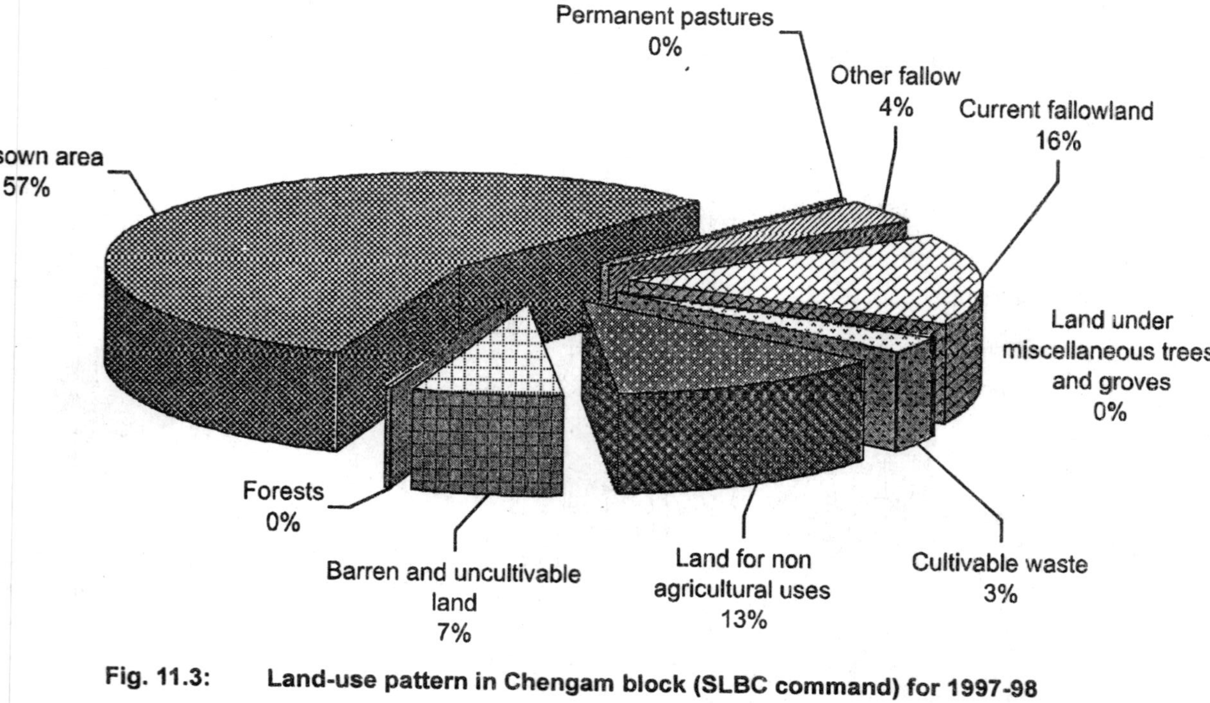

Fig. 11.3: Land-use pattern in Chengam block (SLBC command) for 1997-98

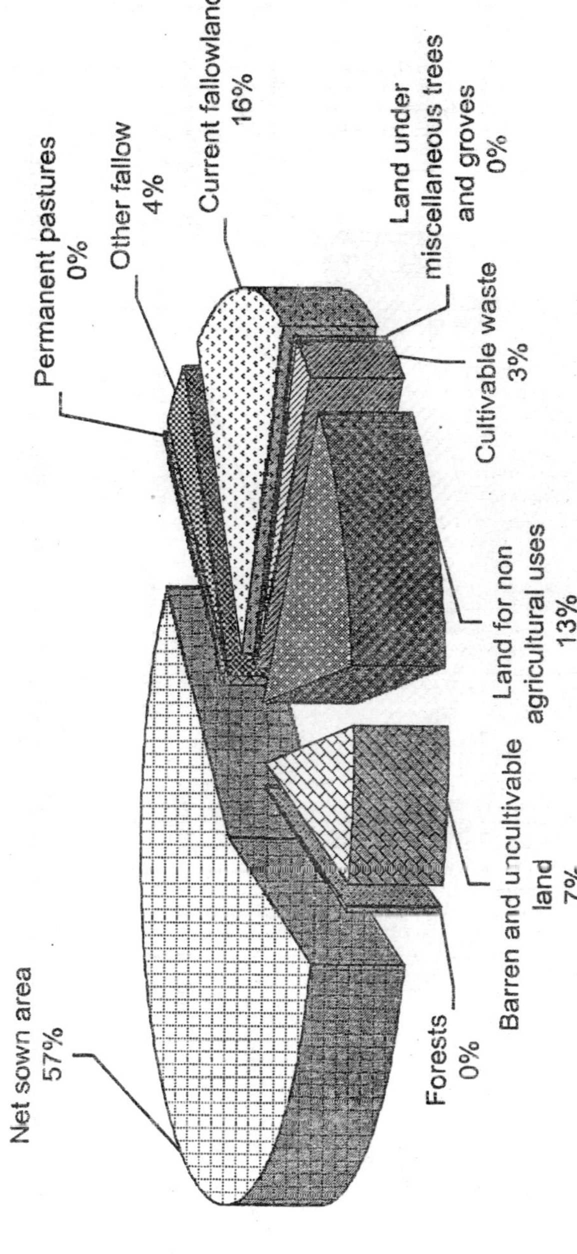

Fig. 11.4: Land-use pattern: SLBC command (1997-98)

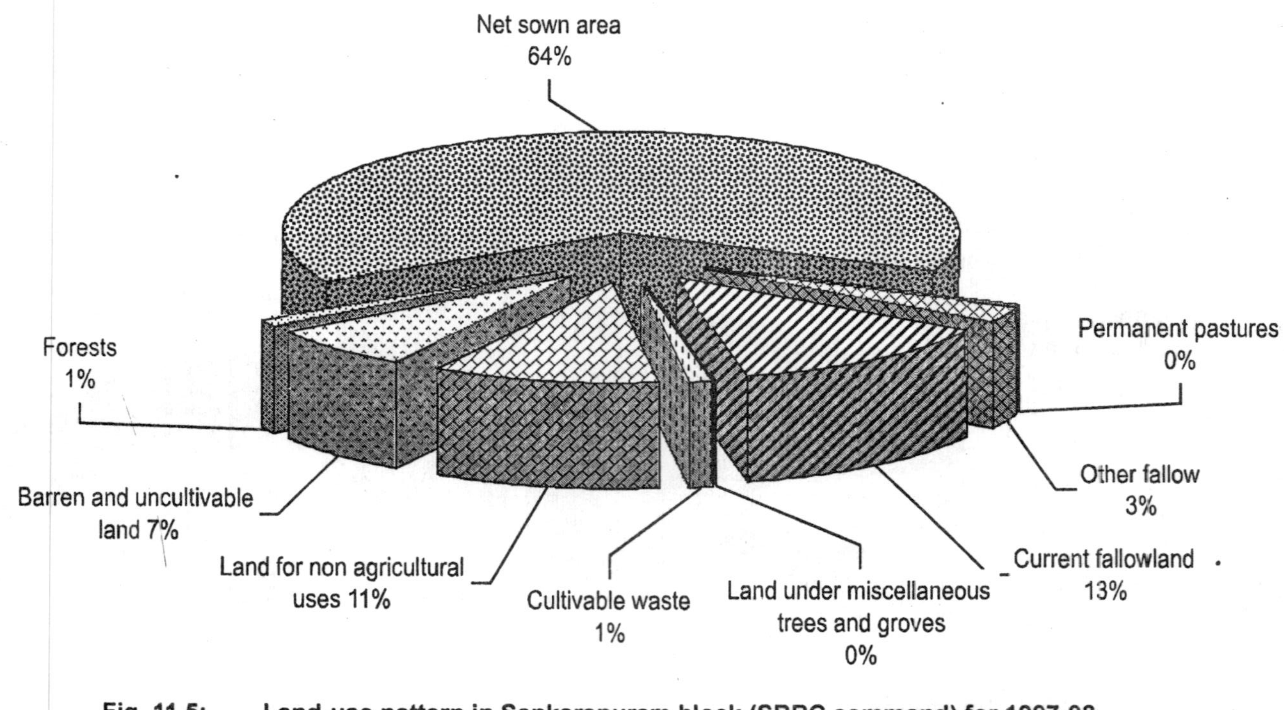

Fig. 11.5: Land-use pattern in Sankarapuram block (SRBC command) for 1997-98

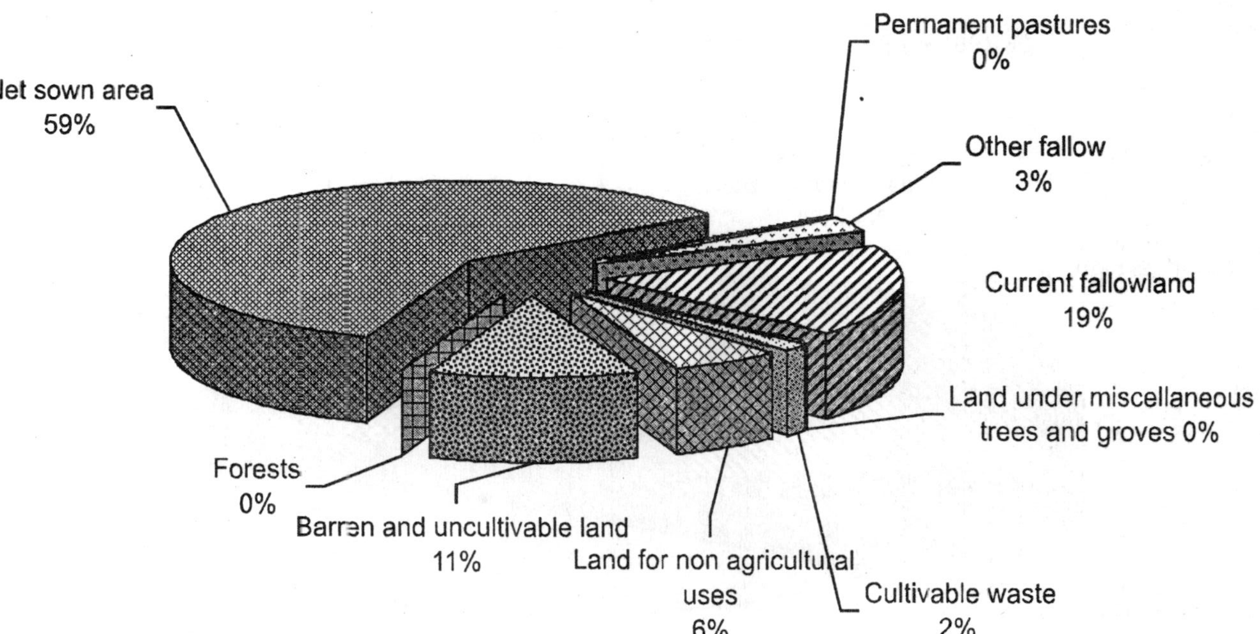

Fig. 11.6: Land-use pattern in Rishivandiyam block (SRBC command) for 1997-98

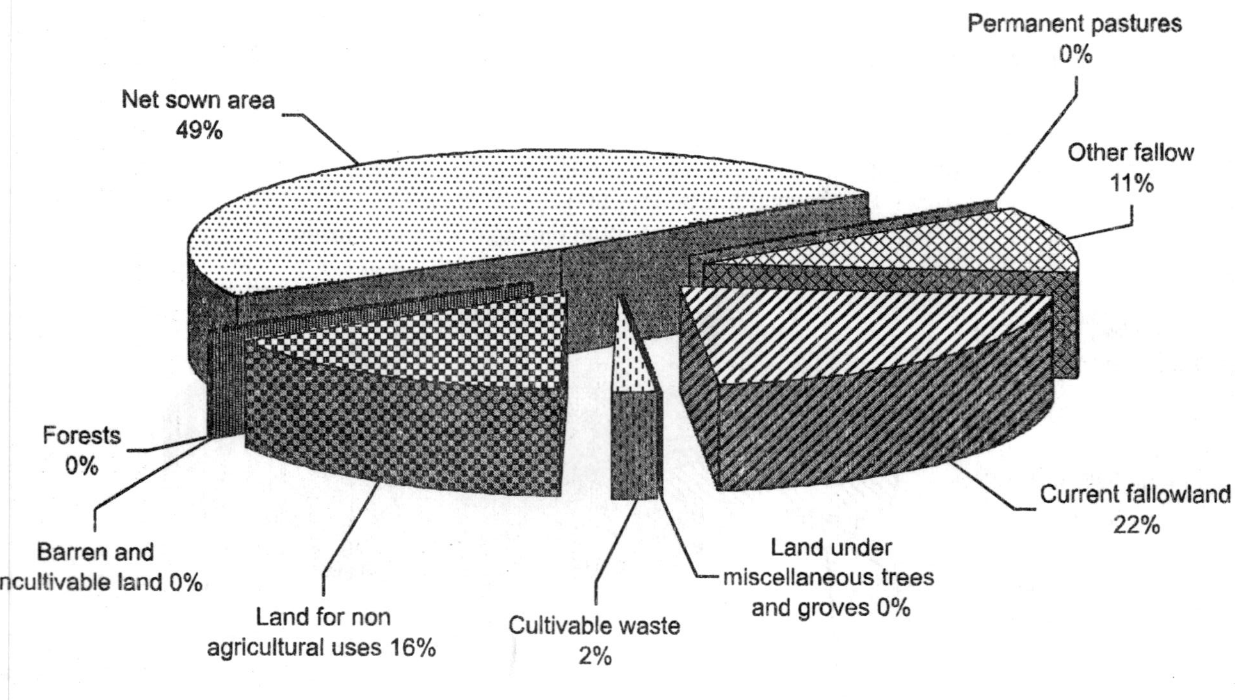

Fig. 11.7: Land-use patter in Chengam block (SLBC command) for 1997-98

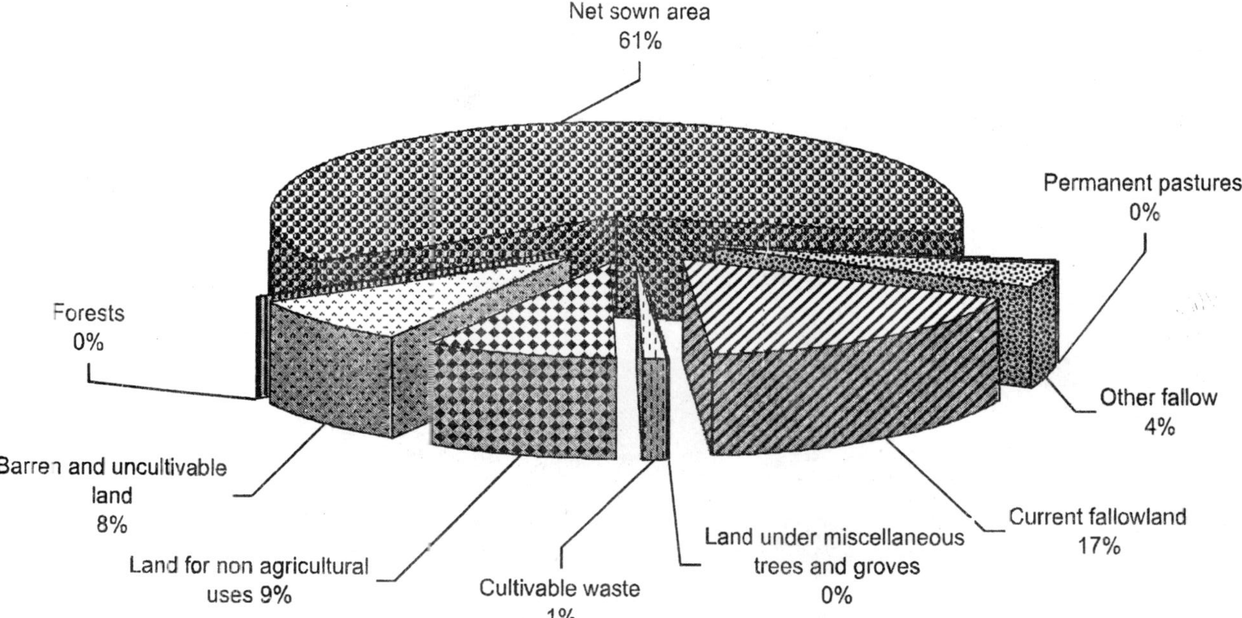

Fig. 11.8: Land-use pattern in SRBC command for 1997-98

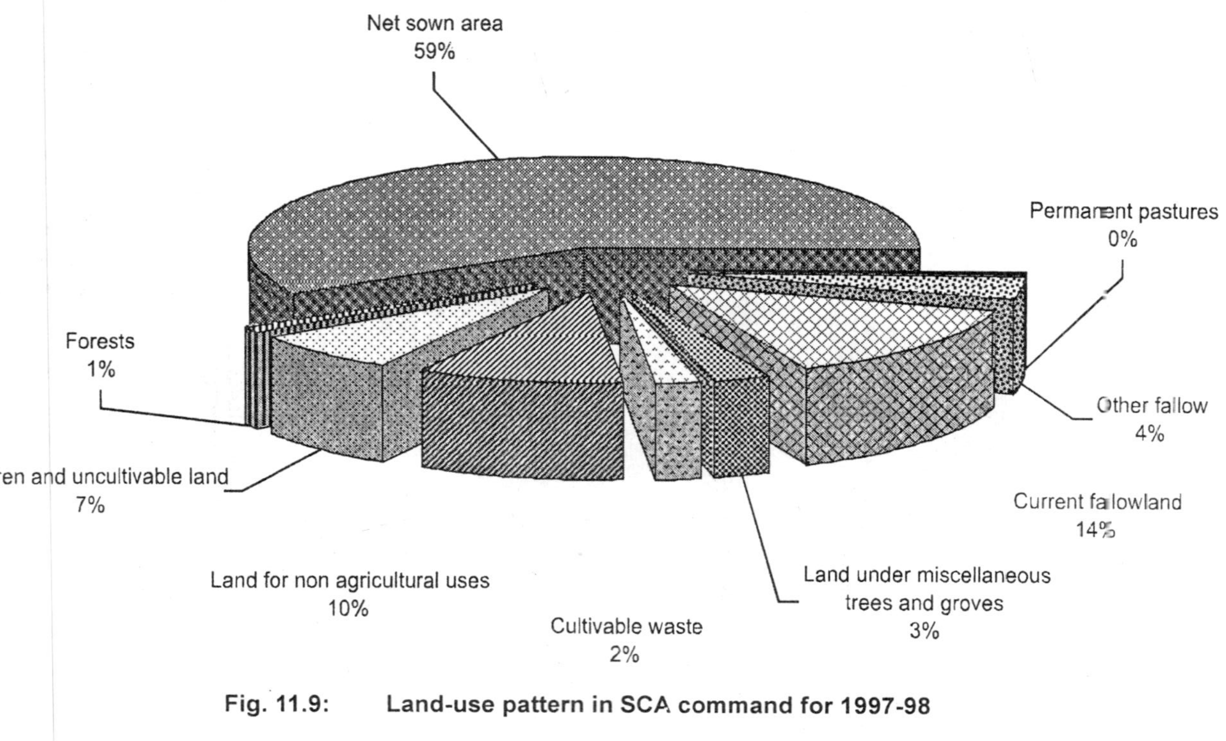

Fig. 11.9: Land-use pattern in SCA command for 1997-98

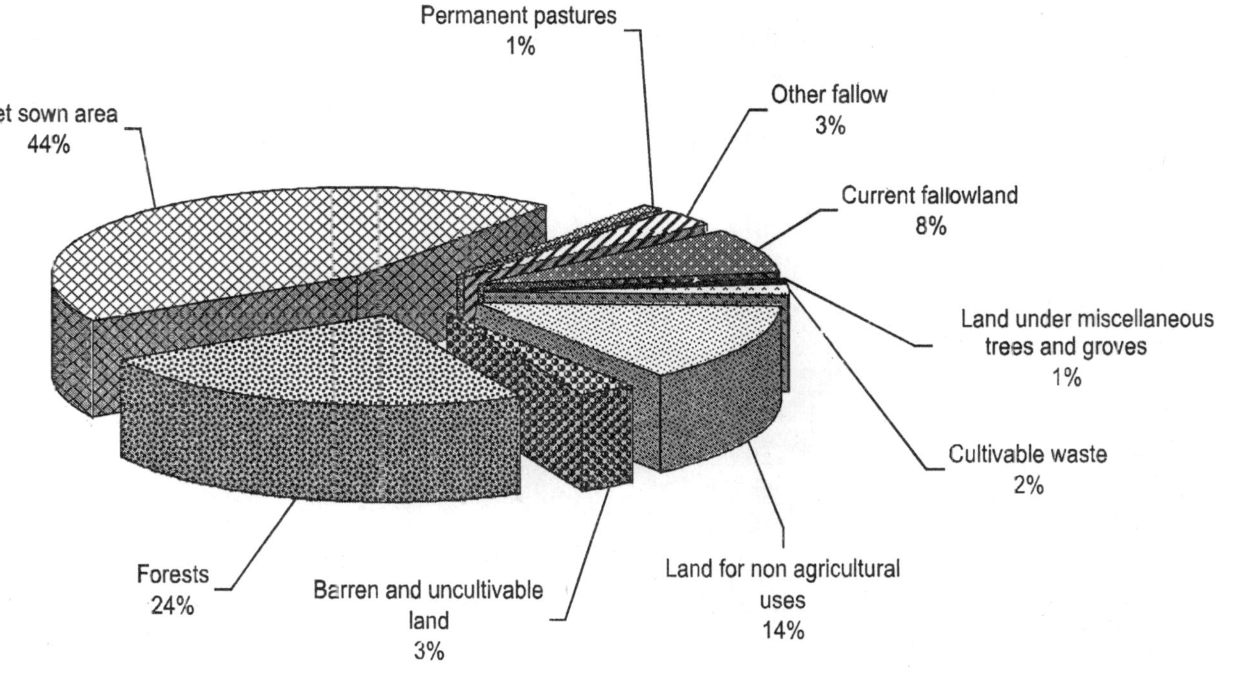

Fig. 11.10: Land-use pattern in Thiruvannamalai for 1997-98

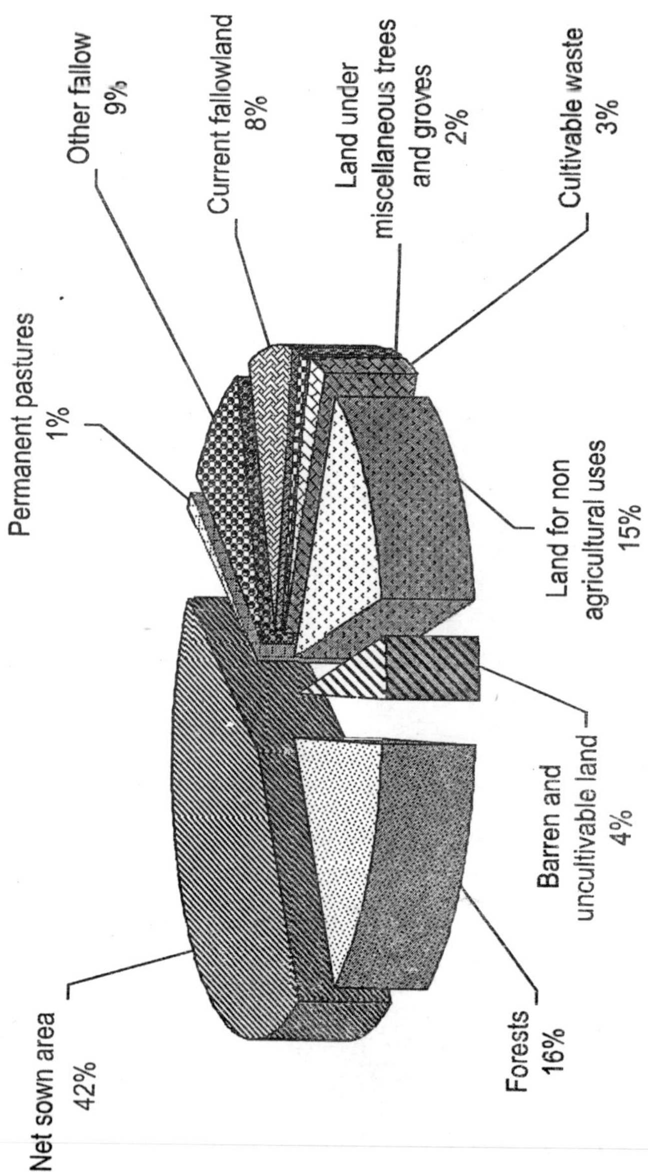

Fig. 11.11: Land-use pattern in Tamilnadu for 1997-98

Moongilthuraipattu and Viriyur are the only villages with forest cover, the forest constituting 1 per cent, 9.3 per cent and 0.2 per cent of their total lands respectively.

In the absence of forest cover the villagers face acute shortage of fuel wood, especially the women who have to either traverse long distances in search of it or have to shell out lump sum to buy. Else they use crop-residues and local weeds as fuel source.

Barren and Uncultivable Land

Barren and uncultivable lands constitute 7.5 per cent of the TRA in SCA (Figure—11.9), which is double of the state's proportion of 3.7% (Figure—11.11). The SLBC command has 6.6 per cent of its land under the category barren and uncultivable (Figure—11.4) while SRBC command's proportion is 8.3 per cent (Figure—11.8). Rishivandiyam block has the highest percentage (11.4%) of barren land (Figure—11.6) in the SCA followed by Thirukoilur which has 9.0 per cent of its land classified as barren and uncultivable (Figure—11.2). Sankarpuram block (Figure—11.5) has 6.9 per cent of its land barren and uncultivable while Chengam block (SLBC command) (Figure—11.3) and Thachampattu block (Figure—12.1) have 7.1 per cent and 4 per cent of their total land as barren and unfit for cultivation.

Amongst the villages, Devariyarkuppam has more than 30 per cent of its land designated as barren and uncultivable. Moongilthuraipattu, Sirpanandal, Kadavanoor and Pallichandal have more than 20 per cent of their total land under the category barren and unfit for cultivation.

SRBC command has the highest proportion of its land classified as barren and unfit for cultivation. This is because the land terrains of Rishivandiyam and Sankarpuram blocks and also of Thirukoilur block are rocky with the soil of kankar and nodules type (Chapter 12). These sorts of lands are not suitable for growing crops. By employing proper measures these lands can be reclaimed for other usages thus easing the pressure on the land.

Land not Available for Agriculture Uses

Land for non-agricultural usage consists of such lands, which are put to other economic use such as human settlements, industrial structures, roads etc. The land not available for agricultural uses form 9.7 per cent of the TRA in the study region as against the state's proportion of 14.9 per cent and Thiruvannamalai district's 14.39 per cent as inferred from Figure—11.10 and 11.11 respectively.

Cultivable Waste

The category 'cultivable waste' includes those lands, which were under agricultural use previously but have been discarded currently, having been laid to waste due to over using or water logging, salinity or alkalinity problems. In SCA 2.3 per cent of the TRA is under the classification of cultivable waste (Figure—11.9), which is almost equal to that of Tamil Nadu state's proportion of 2.7 per cent (Figure—11.11). SLBC command has 3.3 per cent of its land designated as cultivable waste (Figure—12.4) while SRBC command's share is mere 1.4 per cent. Thachampattu block (Figure—11.1) has the maximum proportion of the land i.e. 5.3 per cent of TRA delineated as cultivable waste in the command region followed by Chengam (SLBC command) block with 3 per cent (Figure—11.3). Chengam (SRBC command), Sankarapuram, Rishivandiyam and Thirukoilur blocks have more 2.06 per cent, 1.1 per cent, 1.5 per cent and 1 per cent of their land categorised as cultivable waste, as illustrated in Figures—11.7, 11.5, 11.6 and 11.2 respectively.

The villages Poraspattu, Erudayampattu, Tholuvanthangal, Agarampallipattu, Nariyapattu and Athipadi have more than 10 per cent of their TRA classified under cultivable waste.

SRBC command has less proportion of its land as cultivable waste because of lesser availability of water for irrigation as compared to SLBC command, as discussed elaborately in chapter 10 Lesser water for irrigation means lesser problems by way of water logging, salinity and alkalinity. The lands of SRBC command are high lying, which further negates the possibility of such problems. By employing suitable remedial measures such lands can be reclaimed and put to appropriate usage.

Permanent Pastures and Grazing Land

Permanent pastures and grazing lands constitute a mere 0.1 per cent in the land use pattern of study area (Figure—11.9). The share of T.N. state under this category is also a meagre 0.9 per cent of its total land (Figure—11.11). Thirukoilur, Chengam (SLBC command) Rishivandiyam, Sankarapuram and Chengam (SRBC command) blocks have almost nil pastures and grazing lands as illustrated in Figures—11.2, 11.3, 11.6, 11.5 and 11.7 respectively. Thachampattu, Kandiankuppam, Pavapattu, Jambai, Sithapatinam and Pakkam are the only villages having marginal amount of their total land under fodder cultivation. In the absence of pastures and grazing lands there is almost nil cattle rearing in SCA. Villagers find it profitable to sell the cattle for their flesh rather than maintaining them at a considerably high cost. As a result the cattle population has dwindled to one -fourth in the recent five years as confirmed by the villagers of the command area.

Agricultural Officers and Forest Officers though put the blame squarely on farmers for this loss of forests and pasture lands. According to one official, in the greed of growing more and more crops, the farmers have started encroaching upon the burial grounds also.

Land Under Miscellaneous Trees and Groves

Land under miscellaneous trees and groves constitute 0.24 per cent of the TRA in the study area (Figure—11.9). T.N state and Thiruvannamalai district have 1.73 per cent and 1.38 per cent of their respective lands under the cover of miscellaneous trees and groves (Figures—11.11 and 11.10 respectively).

Fallow Lands

The current fallow and other fallow lands for the year 1997-98 in SCA were reported to be 17.62 per cent (Figure—11.9), which are comparable to state's share of 17.35 per cent (Figure—11.11). The situation is better for Thiruvannamalai district with only 11.25 per cent of the total land left as fallow (Figure—11.10). The situation is dismal in Chengam (SRBC command) block, which has 33.12 per cent of its total land left as fallow followed by Rishivandiyam (21.55%). Chengam (SLBC command) (20.32%) and Sankarapuram (15.37%) blocks. Thachampattu and Thriukoilur blocks have 11.5 per cent and 11.6 per cent fallow lands respectively.

Arambarampattu (40.44% of 655.45 ha.), Jambadi (43.89% of 762.43 ha.), Vadamamandoor (34.95% of 485.02 ha.), Edathanur (52.28% of 642.56 ha.), Thiruvadathanur (34.30% of 689.40 ha.), Thondamanur (48.36% of 836.95 ha.), Rayasamucham (32.53% of 387.42 ha.), Sithapatinm (34.59% of 157.94 ha.) are the villages that had more than 30 per cent of their land under the fallow category.

The SLBC command had 15.5 per cent of the TRA left as fallow while SRBC command had 21.6 per cent of the land left fallow during 1997-98 as depicted in Figures—11.4 and 11.8 respectively. These are the lands that have not been brought under cultivation either due to insufficient irrigation facilities or are left fallow after being used continuously over the period of time. Irrigation facilities are under developed in SRBC command (Chapter 10) as compared to SLBC command. This results in more and more of land being left without cultivation. The absence of proper management of land in relation to fertilizer application, soil-water balance and choice of suitable cropping pattern also results in the land being unfit for cultivating any type of crops and thus being left as fallow. Suitable management steps are needed to check the loss of cultivable land to the fallow land.

A comparative account of the fallow land and the net sown area, in various blocks for the year 1997-98 is depicted in Figure 11.12. The Figure indicates that as the proportion of fallow land rises there is a fall in the net cultivated area. Chengam (SRBC command) had more of fallow land and thus less area under cultivation while Thachampattu block had more of the land under cultivation and less percentage of the TRA left as fallow.

Net Sown Area

The net sown area in the SCA was 60 per cent of the TRA which was higher than the T.N. state's contribution of mere 42.2 per cent (Figure—11.11), and Thiruvannmalai district's 42.9 per cent (Figure—11.10). The SLBC command had 59.8 per cent of the total land as net sown area (Figure—11.8). Thachampattu block's contribution in the command region in the highest with 66.6 per cent of TRA under cultivation as illustrated in Figure—11.9 followed by Sankarapuram block with 64.8 per cent (Figure—11.5). All the other blocks with the exception of Chengam (SRBC

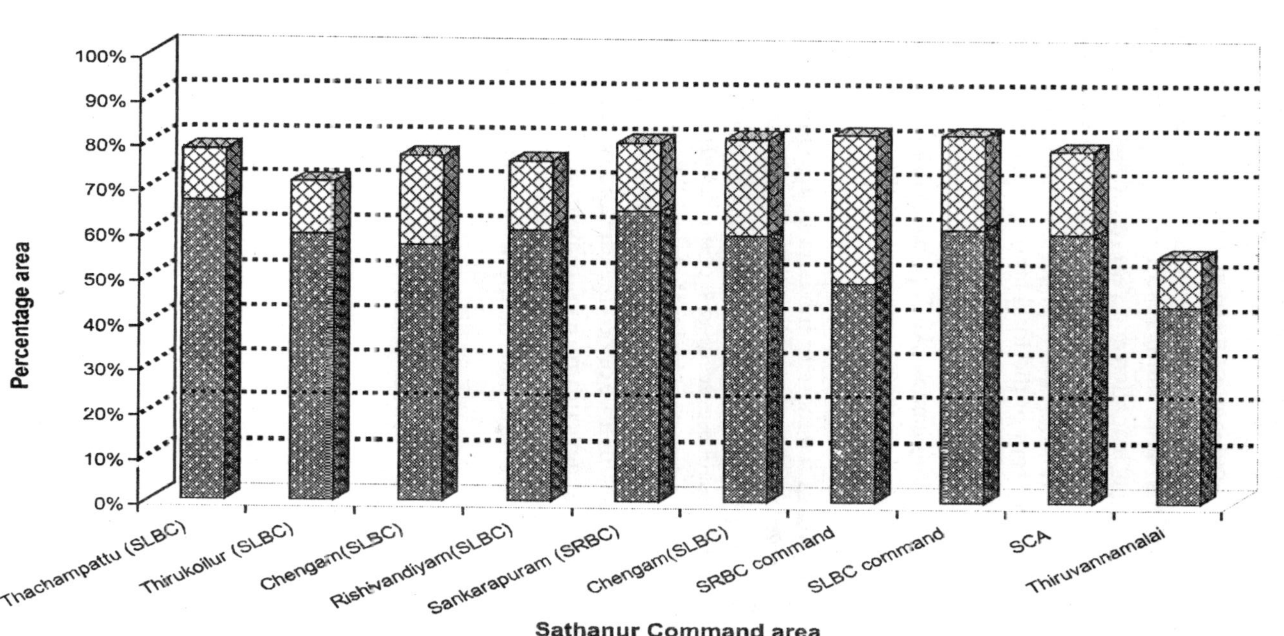

Fig. 11.12: Fallow land Vs net sown area in SCA (1997-98)

command) had more than 50 per cent of their land under cultivation as indicated by the Figures. Chengam block (SRBC command) had 49 per cent of its total area under net cultivation, which, even though lowest in the SCA, was higher when compared to Thiruvannmalai and Tamil Nadu State's figures. With the introduction of canal irrigation and year round availability of water there has been a tendency to bring more and more of land under cultivation by the farmers of SCA. This is apparent from the fact that almost 60 per cent of the total land is under cultivation and another 17 per cent is left as fallow, thus making the total cultivable land to be 79 per cent which is higher than the total cultivable land reported in other basins like Tapi basin (62%), Krishna basin (61%), Sabarmati basin (56%) Yamuna (51.9%) and Brahmani – Baitarni (39.7%) (CPCB, 1994).

As a result of the more of the land being under cultivation, the forest cover and land under miscellaneous trees or groves of grows and the pastures and grazing lands are under tremendous pressure as discussed in preceding section. Poverty of the villagers downstream is an important factor that drives them to desperately bring more and more of land under cultivation in the hope of earning revenues with little or no considerations for any other fact related to planned agriculture, as put by an Agricultural Officer. Illiteracy also adds to their miseries, as most of the cultivators are unaware of the scientific side of the irrigation and agricultural practices.

12

Soil

INTRODUCTION

The properties of soil—physical, chemical, as biological-indicate the potential as well as past history of the utilization of the corresponding land. Ranging from texture to chemical and biological, are the indicators of the land utilization of an area. The physical properties of soil viz. texture and structures directly affect the soil erodibility and sediment transportation and also form a basis for the classification of the land, for various usages. The soil also determines the productivity irrigability and potentiality of the land for suitable cropping pattern.

In this chapter we have presented an assessment of the soil of the Sathanur Command Area (SCA) with special reference to its irrigability and potential for agriculture.

STATUS

A soil survey with respect to was conducted by the Soil Survey and land use Organization, Thanjavur. The survey covered most of the Sathanur Left Bank Command (SLBC) and some part of the Sathanur Right Bank Command (SRBC) command. The sub-divisions Thiruvannamalai, Chengam, Thirukoilur and Kallakurichi, which were earlier, a part of North Arcot and South Arcot districts are not more so after a recent administrtative reorganization of the districts of Tamil Nadu (T.N) state. Thruvannamalai is now a full-fledged district and encompasses most of SLBC and portions of SRBC commands.

The soil of SCA has been classified into Edathanur series and Mudiyanur series.

Edathanur series comprises of moderately deep to deep gravelly soils of very dark grayish-brown, brownish-yellow to dark yellowish-brown surface, with sub soil ranging from dark reddish brown to reddish brown and yellowish red in colour. The texture ranges from loamy sand to sand-loam and gravelly-loam at the surface and sandy-clay-loam to gravelly-clay-loam at the sub surface. The soils are well drained and gravelly are seen at the sub surface. The parent material is laterite over gneissic rock.

Mudiyanur series consists of gray to very dark grayish-brown and deep to very deep soils. The top soil texture ranges from sandy-loam to sandy-clay-loam, and clay-loam to gravelly-clay-loam. Soil is developed from gneissic parent materials. Lime concretions are seen at the sub-surface. Soils are poorly to imperfectly drained and water logging is commonly noticed.

Tables—12.1 and 12.2 present the details of the Edathanur and Mudiyanur soil series. The extent of area under, both the soil series had been indicated in Table—12.3. Land capability classifications based on the soil series of the SCA are represented as Table—12.4. Details regarding the soil distributions in Chengam, Thiruvannamalai, Kallakurichi and Thirukoilur, Tamil Nadu are presented in Tables—12.5-12.8. Information regarding the land capability classification of soils with their limitations in various blocks is presented in Tables—12.9-12.12. Land irrigability classification of soils in various blocks is presented in Table—12.13-12.16. The cropping patterns of the area under types of soil are presented in Tables—12.17-12.20.

Tables—12.21-12.24 are indicated the details pertaining to the suitability of crops under various soil types in various blocks. The productivity ratings of the different soils in the various blocks under SCA are presented in Tables—12.25-12.28.

Table—12.1 Details of Edathanur soil series

Typifying Pedon : Edathanur sandy-clay-loam

Horizon	Depth in cm	Description
AP	0-15	Brown (10 yr.3/2m) very dark greyish Brown, sandy clay loam, medium, weak, sub-angular blocky, few fine faint mottling; dry, hard moist firm and wet and slightly sticky and plastic, common fine vertical and horizontal disroots; effervescence slight; moderately slow permeability with clear smooth boundary.
B1	15.34	(5yr. 4/4 m) Reddish brown clay loam, medium weak, sub-angular blocky; dry hard moist firm; wet sticky and plastic; few fine faint mottling; few, small soft, irregular concretions; common fine discontinuous, horizontal, tubular pores; few, very fine roots; slight effervescence; moderately slow permeability with clear smooth boundary.
B2	34-55	(5yr. 4/3) Dark reddish-brown, gravelly clay loam, medium weak, fine crumb, dry loose, moist, friable wet slightly sticky and non plastic; few small irregular, soft concretions; many fine, discontinuous, vertical tubular pores; common very fine roots; slight effervescence; moderately slow permeability.
Topography		Gently sloping to sloping terrain with 0.3% slope
Drainage and permeability		Well drained with moderately rapid to rapid permeability
Taxonomy		Fine/Coarse Loamy, Iso-hyper-thermic, mixed non calcareous / calcareous typic ustropepts.
Normal solum thickness		Less than 90 cm.

Table—12.2 Details of Mudiyanur soil series

Typifying pedon		Mudiyanur clay loam cultivated
Horizon	**Depth in cm**	**Description**
AP	0-27	(10yr.6/1 d) Grey to very dark grey; (10 yr. 3/1 M) clay loam, coarse, strong sub-angular blocky, dry very hard. Moist, firm wet sticky and platic; very fine, small hard spherical block concentrations; few very fine continuous, vertical tubular pores; common fine roots; violent effervescence moderately slow permeability abrupt smooth boundary.
B2	27-49	(10yr. 4/2m) Dark grayish brown, clay loam coarse, strong, sub-angular, blocky, dry very hard, moist firm, set sticky and plastic; fine to medium, horizontal, discontinuous, tubular pores; few very fine roots; violent effervescence; moderately slow permeability with clear smooth boundary.
B3	49-100	(7.5yr.4/4m) Dark brown, sandy clay loam, fine to medium, weak crumb, moist, friable, wet slightly sticky and non plastic; many, small hard, irregular block concretions; few, fine, discontinuous, horizontal tubular pores, slight effervescence; moderate permeability.
Ranges in characteristics		The texture of the surface soil ranges from sandy loam to clay loam and that of sub surface ranges from sandy clay loam to gravelly clay loam.
Topography		The soils occur on plain terrian and the gradient is 0-1%
Drainage and permeability		The soils are imperfectly drained to poorly drained, water-logging is commonly noticed.
Taxonomy		Fine loamy, iso-hyperthermic, mixed calcareous, typic haplustalfs.
Normal solum thickness		More than 90cm

Table—12.3 Extent of soil series

SI. No.	Soil series	Taxonomy	Area in hectares	Percentage to total area
I	Edathanur	Typic ustropepts	6818.74	52.97
II	Mudiyanur	Typic haplustalfs	6054.86	47.0

Tablc—12.4 Land capability classification

SI. No.	Capability class and capability sub -class	Area in hectares	Percentage of total area
1.	II	3446.07	26.78
2.	IIe	8.74	0.06
3.	IIs	392.44	3.04
4.	IIw	790.70	5.91
5.	III	7281.92	56.58
6.	IIs	647.47	5.03
7.	IIIe	178.60	1.38
8.	IV	157.57	1.22
	Total	12,873.00	100.00

e	*Limiting factors as risks of erosion*
w	*Excess water stagnation which will retard the plant growth*
s	*Soils are either shallower in depth or with saline or alkaline problem.*
II to IV	*Soils suitable for annual or seasonal short-term cultivation with annual or short duration crops.*
II	*Soils can be cropped regularly*
III	*Soils can be cropped regularly, but have narrow range of use and need more careful managements, Moderately good lands with major limitations.*
IV	*Fairly good lands with major limitations and with occasional cultivation. They need very careful managements and have a very narrow range of crops.*

Table—12.5 Distribution of soils in Chengam

Sl. No.	Name of the soil family/ association	Extent in hectares	Percentage of geographical area
1.	Fine loamy, udic ustochrept	156	0.09
2.	Fine loamy,vertic ustchrept	10,684	6.33
3.	Coarse loamy, typic upstochrept	1,101	0.65
4.	Fine loamy, udic haplustalf	7,563	4.19
5.	Fine loamy, vertic haplustaf	583	).35
6.	Fine loamy, udic ustochrept coarse loamy, Typic Ustochrept Association	3,393	2.00
7.	Fine loamy, udic ustochrept, coarse loamy, typic haplustalf association	285	0.17
8.	Coarse loamy, rypic ustothnt Fine loamy, udic ustochrept	14,374	8.51
9.	Fine loamy, typic ustothent fine loamy, udic haplustalf association	55,115	3.02
10.	Fine loamy, udic haplustalf, fine loamy, udic ustochrept association	13,235	7.85
11.	Fine loamy, udic haplustalf, coarse loamy, typic ustorthent association	10,153	6.00
12.	Fine loamy, vertic haplustalf, fine loamy, udic ustochrept association	648	0.69
13.	Fine loamy, typic halplustalf, fine loamy, udic ustochrept association	104	0.69
14.	Fine loamy, vertic haplustalf fine loamy, typic haplustalf association	324	0.19
15.	Fine loamy, vertic ustochrept fkine loamy, typic ustochrept association	958	0.57
16.	Fine loamy, udornthentic chromustrat, fine loamy, vertic ustochrept association	544	0.33
17.	Fine loamy, udic ustochrept coarse loamy, typic ustothent, fine loamy, udich haplustalf, fine loamy, udich haplustalf	129	0.09
18.	Fine loamy, udic rhodustalf, fine loamy, udich haplustalf, coarse loamy, typic ustothent association	790	0.47
19.	Fine loamy, udic haplustalf, coarse loamy, typic ustorthent, fine loamy, udich ustochrept association	1,981	1.17
20.	Fine loamy, udic haplustalf, fine loamy, udic ustochrept. coarse loamy, typic ustothent association	2,5551	1.50
21.	Fine loamy, udic haplustalf coarse loamy, udic ustorthent, fine loamy, udic rhodustalf association	3,717	2.19
22.	Fine loamy, typic haplustalf, fine loamy, udic haplustalf, fine loamy, udic ustochrept	14,440	8.55
	Totai soil area	92,828	55.25

(Table Contd...)

Area under		
Hills	862	0.50
Reserved forests	67.561	39.99
Tanks	6,649	3.94
River	894	0.09
Grand total	1,68,950	100.00

Table—12.6 Distribution of soils in Thiruvannamalai

SI. No.	Name of the soil family/association	Extent in hectares	Percentage of geographical area
1.	Fine loamy, udic rhodustalf	25,650	26.3
2.	Fine loamy, typic rhodustalf	1,370	1.3
3.	Fine loamy, udic haplustalf	2,164	2.1
4.	Fine loamy, typic haplustalf	1,913	2.1
5.	Fine loamy, vertic usotchrept	4,103	4.1
6.	Fine loamy, udic ustochrept	19,840	21.0
7.	Fine loamy, udornthemtic –chromustert	9,384	10.0
8.	Fine loamy, uidc ustochrpt	387	0.3
9.	Fineloamy, udic ustochrept fineloamy, uidc rhodustalf association	10,650	11.0
10.	Fine lomay, udic ustochrept, coarse loamy, typic ustorthent association	3,141	3.1
11.	Fine loamy, uidc ustochrept, fine loamy, udic haplustalf association	121	0.5
12.	Fine loamy, udorthentic chromustert fine loamy, udic ustochrept association	191	0.5
13.	Fine loamy, udic haplustalf, fine loamy, typic haplustalf association	686	0.6
14.	Fine loamy, typic ustochrept, fine loamy, udic rhodustalf, fine loamy, vertic ustochrept association	4,255	4.3
15.	Fine loamy, uidc ustochrept fine loamy, udic rhodustalf, coarse loamy, typic ustorthent	2,278	2.4
	Total soil area	86,133	89.1
Reserved forest		8,976	9.2
House site		1,036	1.1
Hills		547	0.6
Grand total		96,992	100.00

Table—12.7 Distribution of soils in Kallakkaurichi

SI. No.	Name of the soil family/association	Extent in hectares	Percentage of geographical area
1.	Fine loamy, paralithic rhodustalf	23,526.40	15.86
2.	Fine loamy, typic chromustert	19,968.00	13.46
3.	Fine loamy, udic ustochrept	17,280.00	11.65
4.	Fine loamy, vertoc ustochrept	16,640.00	11.22
5.	Fine loamy, typic haplustalf	6,937.00	4.62
6.	Fine loamy, udic rhodustalf	1,510.40	1.02
7.	Coarse loamy, paralithic ustorthent	537.60	0.36
8.	Fine loamy, udic haplustalf	435.020	0.20
9.	Coarse loamy, paralithic ustorthent + coarse loamy, paralithic ustochrept	18,329.60	12.36
10.	Fine loamy, paralithic rhodustalf + fine loamy, paralithic ustochrept	14,329.60	9.86
11.	Coarse loamy, paralithic ustorthent + fine loamy paralithic haplustalf	9,369.60	6.32
12.	Fine loamy, typic chromustert + fine loamy, vertic ustochrept	7,296.00	4.92
13.	Coarse loamy, paralithic ustorthent + fine loamy, paralithic rhodustalf	4,249.00	2.87
14.	Coarse loamy, udic ustochrept + fine loamy, typic haplustalf	4,096.00	2.76
15.	Fine loamy, paralithic rhodustalf + coarse loamy, paralithic ustochrept	3,200.00	2.16
16.	Coarse loamy, paralithic ustochrept + fine loamy, paralithic rodustalf	307.00	0.21
	Grand Total	1,48.300.0	100.00

Table—12.8 Distribution of soils in Thirukoilur

Sl. No.	Name of the soil family/association	Extent in hectares	Percentage of geographical area
1.	Fine loamy, udic ustochrept	5,683.20	12.60
2.	Fine loamy, vertic ustochrept	3,993.60	8.88
3.	Fine loamy, udic haplustalf	486.40	1.00
4.	Coarse loamy, typic ustochrept	76.80	0.11
5.	Fine loamy, udic haplustalf, fine loamy, udic ustochrept	16,665.60	37.07
6.	Fine loamy, typic haplustalf, fine loamy, udic ustochrept	1,894.40	4.21
7.	Fine loamy, typic haplustalf, fine loamy, ludic ustochrept	819.20	1.88
8.	Fine loamy, udic ustochrept, fine loamy, typic haplustalf	588.20	1.89
9.	Fine loamy, udic ustochrept fine loamy, typic haplustalf	486.40	1.0
10	Fine loamy, typic haplustalf, fine loamy, udic haplustalf	204.80	0.20
11.	Fine loamy, udic haplustalf fine loamy, udic ustochrept typic ustipsamment	11,264.00	25.06
12.	Typic ustisamment, fine loamy, udic ustochrept, coarse loamy, fluventic ustochrept	2,150.00	4.78
13.	Fine loamy, typic haplustalf, fine loamy, udic ustochrept, typic ustorthent, coarse loamy	640.00	1.42
	Total	44,853.60	100.00

Table—12.9 Land capability classification of soils with their limitations in Chengam

Sl. No.	Land capability class	Description	Limitation
1.	II Ws	Good cultivable land with moderate limitations that reduce the choice of crops.	Surface hardening heavy texture, drainage problem, sheet erosion.
2.	IIIes	Moderately good cultivable land, severe limitations that reduce the choice of crops.	Run off, slow permeability development of cracks, poor fertility, sub-soil gravelliness, stony surface crusting.

Table—12.10 Land capability classification of soils with their limitations in Thiruvannamalai

Sl. No.	Land capability class	Description	Limitation
1.	II Ws	Good cultivable land with moderate limitations that reduce the choice of crops.	Surface hardening heavy texture, drainage problem, sheet erosion.
2.	IIIes	Moderately good cultivable land, severe limitations that reduce the choice of crops.	Run off, slow permeability development of cracks, poor fertility, sub-soil gravelliness, stony surface crusting.

Table—12.11 Land capability classification of soils with their limitation in Kallakurichi

Sl. No.	Land capability class	Description	Limitation
1.	Iies	Good cultivable land, moderate limitations that reduce the choice of crops.	Surface texture, poor fertility, calcium deficiently, surface crusting

Table—12.12 Land capability classification of soil with their limitations in Thirukoilur

Sl. No.	Land capability class	Description	Limitation
1.	IIes	Good cultivable land, moderate limitations that reduce the choice of crops.	Surface texture, poor fertility, calcium deficiently, surface crusting
2.	IIIes	Moderately good cultivable land. Severe limitations that reduce the choice of crops.	Erosion, soil depth, poor fertility, sub-soil gravelliness, stony surface crusting.

Table—12.13 Land irrigability classification of soils of Chengam with their limitations

Sl. No.	Land irrigability class	Description	Limitation
1.	2s	Moderate soil limitations for sustained use under irrigation	Texture, excessive sub surface drainage
2.	2sd	Severe soil limitations for sustained use under irrigation	Texture impeded drainage

Table—12.14 Land irrigability classification of soils of Thiruvannamalai with their limitations

SI. No.	Land irrigability class	Description	Limitation
1.	2s	Moderate soil limitations for sustained use under irrigation	Texture, excessive sub surface drainage
2.	2sd	Severe soil limitations for sustained use under irrigation	Texture impeded drainage

Table—12.15 Land irrigability classification of soils with their limitations in Kallakkurichi

SI. No	Land irrigability class	Description	Limitation
1.	2ts	Lands that have severe limitations for sustained use under irrigation.	Topography presence of kankar nodules
2.	2ts	Lands that have moderate limitations for sustained use under irrigation.	Topography soil erosion, medium water holding capacity.
3.	3ds	Lands that have severe limitations for sustained use under irrigation.	Poor drainage, High water table.
4.	3ts	Land that have severe limitations for sustained use under irrigation.	Topography coarse, fragments, low water. holding capacity.

Table—12.16 Land irrigability classification of soils with their limitations in Thirukoilur

SI. No.	Land irrigability class	Description	Limitation
1.	2ts	Lands that have severe limitations for sustained use under irrigation	Topography presence of kankar nodules
2.	2ts	Lands that have moderate limitations for sustained use under irrigation	Topography soil erosion, medium water holding capacity.
3.	3ds	Lands that have severe limitations for sustained use under irrigation	Poor drainage, high water table
4.	3ts	Land that have severe limitations for sustained use under irrigation	Topography coarse, fragments, low water holding capacity.

Table—12.17 Crops grown in various soil types in Chengam

Sl. No.	Land under use		Mapping unit	Soil name
	Dry	Irrigated		
1.	Groundnuts millets vegetables	Groundnut, paddy millets, sugarcane	3	Fine loamy, udic rhodustalf
2.	Groundnuts millets vegetables	Groundnut, paddy, millets, sugarcane.	2	Fine loamy, udic haplustalf
3.	Groundnuts millets	Groundnut, paddy chillies	6	Fine loamy, vertic haplustalf
4.	Groundnuts	Paddy, sugarcane millets	12	Fine loamy, vertic haplustalf
5.	Groundnuts millets	Groundnut, paddy, millets, vegetables	1	Fine loamy, vertic ustochrept
6.	Groundnuts	Paddy, ragi (*Eleucine coracana*)	5	Fine loamy, udorthentic, chromustert, coarse loamy.
7	Groundnut	Groundnut, paddy,	4	Coarse loamy, millets typic ustorthent

Table—12.18 Crops grown in various soil types in Thiruvannamalai

Sl. No.	Land under use		Mapping unit	Soil name
	Dry	Irrigated		
1.	Groundnuts millets	Groundnut, paddy millets, sugarcane vegetables	3	Fine loamy, udic rhodustalf
2.	Groundnuts, millets	Groundnut, paddy, millets, sugarcane vegetables	2	Fine loamy, udic haplustalf
3.	Groundnuts, millets	Groundnut, paddy chillies	6	Fine loamy, typic haplustalf
4.	Groundnuts	Groundnut, paddy sugarcane	13	Fine loamy, vertic rhodustalf
5.	Groundnuts	Paddy, ragi, sugarcane	5	Fine loamy, vertic ustochrept
6.	Groundnuts	Groundnut, millets, groundnut	8	Fine loamy, udorthentic, chromustert,.
7.	Groundnuts	Paddy, ragi	10	Typic ustipsmment

Table—12.19 Crops grown in various soil types in Kallakurichi

Sl. No.	Land under use		Mapping unit	Soil name
	Dry	Irrigated		
1.	Gingelly, cumbu Coriander,	Groundnut, tapioca sugarcane, ragi cotton	7	Fine loamy, paralithic rhodustalf
2.	Coriander, ragi cumbu	Paddy, tapioca, sugarcane,	2	Fine loamy, typic chromosturt
3.	Ragi, cumbu	Paddy, cotton tapioca	1	Fine loamy, udic ustochrept
4	Cumbu, ragi	Paddy, tapioca cotton, groundnut, turmeric, ragi	4	Vertic ustochrept
5	Cumbu ,ragi, coriander	Paddy, cotton ragi, tapioca, groundnut	9	Fine loamy, typic haplustalf
6.	Cashew, cumbu groundnut	Paddy, cotton, sugarcane, turmeric (*Curcuma domestica*) ragi, tapioca (*Manihit esculenta*)	5	Fine loamy, udic rhodustalf

Table—12.20 Crops grown in various soil types in Thirukoilur

Sl. No.	Land under use		Mapping unit	Soil name
	Dry	Irrigated		
1.	Groundnut, cumbu, ragi	Groundnut, paddy ragi, casuarina	1	Fine loamy, udic ustochrept
2.	Blackgram, ragi	Paddy, cotton, blackgram, ragi	4	Fine loamy, vertic ustochrept
3.	Groundnut, pulses, cholam, cumbu	Groundnut, sugarcane tapioca, ragi, paddy cumbu	3	Haplustalf
4.	—	Paddy, cotton, ragi	8	Coarse loamy, typic ustothent

Table—12.21 Potentiality of the land under different soils for raising various crops in Chengam

Sl. No.	Suitable crop		Mapping unit	Soil name
	Dry	Irrigatedm		
1.	Groundnut, pulses, millets	Groundnut, pulses, millets, sugarcane, banana, coconut, horticultural crops, chillies, tree crops	3	Fine loamy, udic rhodustalf
2.	Groundnut, pulses, cholam, cumbu	Groundnut, pulses millets, sugarcane, banana, coconut, horticultural crops, chillies, tree crops	2	Fine loamy, udic haplustalf
3.	Groundnut pulses, millets	Groundnut, pulses, millets, chillies, sugarcane, banana, vegetables	6	Fine loamy, typic haplustalf
4.	Pulses, millets	Paddy, cotton, coriander, pulses millets, banana	12	Fine loamy, vertic haplustalf
5.	Groundnut, pulses sunflower, cumbu, Sorghum (*Sorghum bicolor*)	Groundnut, millets, pulses, tapioca, vegetables, horticultural crops, coconut, grapes, trees	1	Fine loamy, udic ustochrept
6.	Pulses, millets	Paddy, cotton, coriander, pulses millets, banana, koraı	5	Fine loamy, vertic ustochrept
7.	Pulses, millets	Paddy, cotton, coriander, pulses millets, banana, korai	8	Fine loamy, udorthentic, chromustert
8.	Settlement pastures, forestry, groundnut, millets	Groundnut, pulses, millets, vegetables	4	Coarse loamy, typic ustorthent

Table—12.22 Potentiality of the land under different soil for raising various crops in Thiruvannamalai

Sl. No.	Suitable crop		Mapping unit	Soil name
	Dry	Irrigated		
1.	Groundnut, pulses, millets	Groundnut, pulses, sugarcane, coconut, horticultural crops, chillies, tree crops	3	Fine loamy, udic rhodustalf
2.	Groundnut, pulses, sunflower, cholam, cumbu	Groundnut, pulses millets, sugarcane, banana, coconut, horticultural crops, chillies, tree crops	2	Fine loamy, udic haplustalf
3.	Groundnut pulses, millets	Groundnut, pulses, millets, chillies, sugarcane, banana, vegetables	6	Fine loamy, typic haplustalf
4.	Groundnut, pulses, millets	Groundnut, pulses, millets, chillies, sugarcane, banana vegetables	13	Fine loamy, typic rhodustalf
5.	Pulses, millets	Paddy, cotton coriander, pulses, millets, banana, korai	5	Fine loamy, vertic ustochrept
6.	Pulses, millets	Paddy, cotton, coriander, pulses millets, banana, korai	8	Fine loamy, ustochrept
7.	Groundnut, casuarina	Coconut, casuarina eucalyptus, palmyra cashew	10	Typic, ustripsamment

Table—12.23 Potentiality of the land under different soils for raising various crops in Kallakurichi

Sl. No.	Suitable Crop		Mapping unit	Soil name
	Dry	Irrigated		
1.	Groundnut, pulses, millets	Groundnut, pulses millets, sugarcane, banana, coconut, other horticultural crops, chillies, grapes, acid lime	7	Fine loamy, paralithic rhodustalf
2.	Pulses	Paddy, cotton, black-gram, green gram, ragi	2	Fine loamy, typic chromostert
3	Groundnut, pulses, millets	Groundnut, millets chilles and vegetables	1	fine loamy, udic ustochrept
4.	Pulses, millets	Paddy, cotton, coriander, pulses, banana, millets	4	Fine loamy, vertic ustochrept
5.	Groundnut, pulses, millets	Groundnut, pulses millets, chillies, sugarcane, banana, vegtables	9	Fine loamy, typic haplustalf
6.	Groundnut, pulses, millets	Groundnut, pulses millets, sugarcane, banana, coconut other horticultural crops, chillies, grapes	5	Fine loamy, udic rhodustalf

Table—12.24 Potentiality of the land under different soils for raising various crops in Thirukoilur

Sl. No.	Suitable crop		Mapping unit	Soil name
	Dry	Irrigated		
1.	Groundnut, pulses, millets	Groundnut, millets, chillies, vegetables	1	Fine loamy, udic ustochrept
2.	Pulses, millets	Paddy, cotton, coriander, pulses, millets, banana	4	Vertic ustochrept
3	Groundnut, sunflower, cholam,	Groundnut, millets sugarcane, banana, pulses, tapioca, horticultural crops, vegetables, grapes, coconut, tree crops	3	Fine loamy, udic haplustalf
4.	Settlement pastures, forestry, groundnut, pulses, millets	Paddy, cotton, coriander, pulses, millets, banana	8	Coarse loamy, typic ustorthent

Table—12.25 Productivity rating of soils in Chengam

Sl. No.	Soil Name	Symbol	Productivity		Improvements needed
			Rating	Grouping	
1.	Fine loamy, udic rhodustalf	Ursf	20-34	Average	Improvement of soil structure and nutrient status
2.	Fine loamy, udic haplustalf	Uhsf	20-34	Average	Improvement of soil structure and nutrient status, application of organic manure
3.	Fine loamy, typic haplustalf	Thaf	8-19	Poor	Improvement of structure and nutrient structure and supplementary irrigation, application of organic manure.
4.	Fine loamy, vertic haplustalf	Vhaf	20-64	Average	Improvement of structure and texture, supplementary irrigation
5.	Fine loamy, udic Ustochrept	Uuop	20-34	Average	Textural improvement by stone removal, measures to control water erosion
6.	Fine loamy, vertic ustochrept	Vuop	20-34	Average	Improvement of soil drainage, incorporation of organic manure
7.	Fine loamy, udorthentic chromustert	Dscv	20-34	Average	Improvement of soil drainage, incorporation of organic manure
8.	Coarse loamy, typic ustorthent	Truct	0-7	Extremely Poor	Improvement of structure of tank silt,application of farm yard manure, water conservation measures, mulching

Table—12.26 Productivity rating of soils in Thiruvannamalai

SI. No.	Soil Name	Symbol	Productivity		Improvements needed
			Rating	Grouping	
1.	Fine loamy, udic rhodustalf	Ursf	20-34	Average	Improvement of soil structure and nutrient status
2.	Fine loamy, udic haplustalf	Uhsf	20-34	Average	Improvement of soil structure and nutrient status, application of organic manure, fertilizer management, soil conservation measures
3.	Fine loamy, typic haplustalf	Thaf	8-19	Poor	Improvement of structure and nutrient status, supplementary irrigation, application of organic manure
4.	Fine loamy, typic haplustalf	Trsf	8-19	Average	Soil structure improvement application of organic manure, fertilizer management
5.	Fine loamy, vertic ustochrept	Vucp	20-34	Average	Improvement of soil drainage, incorporation of organic manure
6.	Fine loamy, udorthentic, Chromustert	Dcsv	20-34	Average	Improvement of soil drainage, incorporation of organic manure
7.	Fine loamy, ustipsumment	Tupt	0-7	Extremely poor	Improvement of structure of tank silt and farm yard manure mulching, supplementary irrigation

Table—12.27 Productivity rating of soils in Kallakurichi

Sl. No.	Soil Name	Productivity		Improvements needed
		Rating	Grouping	
1.	Fine loamy, paralithic rhodustalf	20-34	Average	Application of tank silt/FYM
2.	Fine loamy, typic chromustert	20-34	Average	1. Drainage improvement
				2. Water management
				3. Fertilizer management
				4. Gypsum/fertilizer management
				5. Summer ploughing
3.	Fine loamy, typic haplustalf	20-34	Average	1. More FYM application
				2. Cultivation of selected crops
				3. Advoidance of deep rooted crops
				4. Soil conservation
				5. Micronutrient application
				6. Fertilizer management
4.	Fine loamy, udic rhodustalf	20-34	Average	1. Application of tank silt, FYM
				2. Gypsum application
				3. Deep ploughing
				4. Soil conservation
				5. Fertilizer management

Table—12.28 Productivity rating of soils in Thirukoilur

SI. No.	Soil name	Productivity		Improvements needed
		Rating	Grouping	
1.	Fine loamy, udic ustochrept	20-34	Average	1. Application of FYM
				2. Soil conservation
				3. Fertilizer management
2.	Fine loamy, vertic ustochrept	20-34	Average	1. Drainage improvement
				2. Gypsum application
				3. Application of more FYM
				4. $ZnSo_4$ application
				5. Fertilizer/water management
				6. Summer ploughing
				7. Soil conservation
3.	Fine loamy, udic haplustalf	8-19	Poor	1. Soil conservation
				2. Application of Tank silt/FYM
4.	Coarse loamy, typic ustorthent	0-7	Extremely poor	1. Soil conservation
				2. Addition of tank silt
				3. Cultivation of selected crops
				4. Soil mulching
				5. Soil conservation

IMPACT

Soil Distribution

In Chengam though the soils belong to 3 to 4 orders, mostly they are fine loamy. Out of the total extent of 68,950 ha, soils constitute 55.25 per cent, hills 0.5 per cent, forests 39.99 per cent, water bodies 4.48 per cent and settlements sites 0.09 per cent of the total area.

In Thiruvannamalai too the soil orders are 3 to 4 and the soils are mostly fine loamy. Of the total extent of 96,992 ha, the soils constitute 89.1 per cent, reserve forest 9.2 per cent, settlements 1.1 per cent, and hills 0.6 per cent of the total area.

As regard to Kallkurichi the total extent of the area is 1,48,300 ha, and the soils are rhodustalfs, ustrochrepts, haplustalf and ustorthents, etc. The soil vary from fine loamy to coarse loamy. Similarly in Thirukoilur the soils orders and their textural classification are some as that of Kallakurichi. The total extent of area is 44,854 ha.

Land Capability Classification

The land capability of Chengam and Thiruvannamalai fall under classification II ws and III es. The lands are highly susceptible to water stagnation due to poor permeability of sub soil strata and erosional hazards on the top surface soils. The suitability of the lands for crop raising can be classified as moderately suitable to good cultivable lands.

The lands of Kallakurichi and Thirukkoilur come under the classification II es, III es and III s. They are susceptible to erosion and salinity and alkalinity problems. The lands III s appear only in Kallakurichi indicating that such soils are with serious limitation for sustained use for crop raising. Generally the lands are moderately suitablc to good cultivablc lands.

Land Irrigability

Chengam and Thiruvannamalai possess good irrigable lands due to the latter's fine loamy texture. Their limitations lie in excessive drainage and impeded drainage at other places.

The lands of Kallakurichi and Thirukoilur are with moderate to severe limitations for sustained use under irrigation. The soils are hard, stony and nodules type with high water table conditions. The permeability is low to moderate. Therefore the problems aggravate under irrigation flooding and the soils develop crack with medium retention capacity.

Current Cropping Pattern

Since the general climate, topography, surface, soils availability of ground and surface waters, rainfall etc. are more or less the same and since there is no clear demarcating physical structural or land forms between the divisions, the farmers grow more or less the same type and variety of crops in SCA. Under such environments, the main crops now grown are paddy (*Oryza sativa*), groundnut (*Arachis hypogea*), millets (*Pennisetum typhoides*), pulses, sugarcane (*Saccharum officinarum*), chillies (*Capsicum annum*), cotton (*Gossypium arborium*), vegetables etc. (Chapter 13).

Suitability of Crops Suggested

After a careful study of the properties of water and soil in the SCA and their mutual interactions, the agricultural scientists have suggested the cultivation of groundnut, pulses, millets, sunflower (*Helianthus annus*), maize (Zea *mays*) and bajra as the dry crops. Similarly, groundnut, pulses, millets, sugarcane, banana (*Musa paradisiaca*), coconut (*Cocos nucifera*), horticulture crops and chillies have been recommended as irrigated crops.

Productivity Rating of Soils

Since careful improvements are needed in soil structure and nutritional status, the productivity ratings of the soils were found to have large variation. For fine loamy soils the ratings are between 20 and 34 under average grouping. In poor grouping, it is between 8 and 19 in typic haplustalfs. In extremely poor grouping it ranges between 0 and 8 for coarse typic ustorthents. The above situation is common for all the four areas studied.

13

Agriculture

INTRODUCTION

The objective of Sathanur project is to provide irrigation water to a predominantly agrarian region.

In this chapter we present a gist of our studies on the agricultural practices followed in the Sathanur Command Area (SCA). The study also focuses on the impact of the Sathanur Reservoir Project (SRP) on the cropping pattern in the region.

STATUS

The general pattern of cropping in the command area is tabulated as Table—13.1 The annual rainfall of Thiruvannamalai district and the outflow from the reservoir is presented in Figure—13.1. The annual rainfall of the state of Tamil Nadu (TN) has been depicted as Figure—13.2. The study area has been categorized into various blocks. Tables—13.2-13.8 present the village-wise details of the cropping patterns in various blocks in SCA. The distribution of population—general and vocation wise—in different blocks of SCA are presented in Tables—13.8-13.14. The intensity of cropping (computed as the ratio of gross cropped area to net sown area) and other demographic details are presented in Table—13.15. Figures—13.3 to 13.11 illustrate the cropping patterns of the various blocks, individual command regions and cumulatively for the SRP. Figure—13.12 indicates the percentage area of the gross sown area in command under paddy cultivation during various seasons. Area under food and non-food crops (percentage wise) is represented in Figure—13.13. Groundnut (*Arachis hypogea*) cultivation in relation to total non-food cultivation in the command region is illustrated in Figure—13.14 Figure—13.15 gives an account of the division of working force in the region while Figure—12.16 illustrates the gender wise distribution of agricultural labourers in the command region.

Table—13.1 Cropping pattern of the study area

	Crop	Season of Cultivation
1.	**Wet land**	
(i)	Paddy *(Oryza sativa)*	
	(a) Sornavari	April-June
	(b) Samba	June-December
	(c) Novarai	January-March
(ii)	Sugarcane *(Saccharum officinarum)*	Throughout the year
2.	**Garden land**	
(i)	Millets *(Pennisetum typhoides)*	June-September (Rainfed)
(ii)	Sorghum *(Sorghum bicolor)*	June-September (Rainfed)
(iii)	Groundnut *(Arachis hypogea)*	June-September (Rainfed) October/November to January/February; March, April-June, July (Irrigated)
(iv)	Gingelly *(Sesamum indicum)*	March-May (Irrigated)
(v)	Millets (*Zea mays*)	March-May (Irrigated)
(vi)	Ragi *(Eleucine coracana)*	March-May (Irrigated)
(vii)	Paddy *(Oryza sativa)*	April-June, July (Rainfed)
(viii)	Sugarcane	Throughout the year
3.	**Dry land**	
(i)	Groundnut *(Arachis hypogea)*	June/July to September October
(ii)	Millets *(Pennisetum typhoides)*	June/July to September October
(iii)	Sorghum *(Sorghum bicolor)*	June/July to September October
(iv)	Pulses Horse gram *(Macrotyloma uniflorum)*	June/July to September October
(v)	Ragi *(Eleucine coracana)*	June/July to September October

Table—13.2 Cropping pattern: Thachampattu (SLBC command) block (1997-98)

Villages	Area (in hectares) under							
	Paddy*				Total cereals˙ (other than Paddy	Total pulses^Δ	Total spices^#	Total sugarcane^@
	Sornavari	Samba	Navarai	Total				
Thatchampattu	8.95.0	17.08.5	24.19.5	50.23.0	74.68.5	8.20.0	0.61.5	123.59.0
Alijondapattu	1.99.0	22.31.0	29.67.0	53.97.0	7.45.5	0.30.0	1.18.5	18.07.0
Katampcondi	166.87.0	–	149.27.5	316.14.5	50.91.0	4.07.0	–	87.36.0
Periyakollipadi	39.70.0	43.16.0	57.80.0	140.65.0	35.87.0	–	12.13.0	152.04.0
Chinnakallipadi	–	21.27.5	0.56.0	21.83.5	112.13.0	3.49.5	0.79.5	80.54.0
Navampattu	–	82.83.5	2.99.0	85.82.5	82.67.5	–	1.50.0	88.92.5
Devanur	50.61.5	69.87.5	1.03.0	121.51.5	51.95.0	1.20.0	–	52.57.0
Velayampakkam	–	106 29.5	65.08.5	171.38.0	68.16.5	4.06.0	8.31.5	57.26.5
Kallotu	35.69.5	10.51.0	16.58.0	62.88.5	49.89.0	1.90.5	1.82.5	24.89.5
Athipadi	7.72.5	3.37.5	11.17.5	22.27.5	35.34.5	0.37.0	–	58.87.0
Palayanur	79.92.5	–	46.25.5	126.18.0	88.99.5	1.70.5	0.46.0	101.16.0
Kandiankuppam	83.43.0	56.44.5	–	139.87.5	123.83.5	0.27.0	14.23.5	85.95.3
Thalayampallam	60.94.5	5.26.5	12.30.0	138.51.0	0.10.0	4.20.0	0.49.0	98.44.0
Chakkarathamadai	26.22.0	–	30.07.0	56.29.0	–	–	–	10.75.0
Nariyapattu	44.18.5	–	36.17.0	80.30.5	6.15.0	–	–	100.91.5
Parayampattu	6.18.5	–	14.12.0	20.30.5	17.94.5	0.03.0	0.03.0	109.66.5
Pavapattu	88.78.5	–	99.82.0	188.60.5	42.78.0	0.74.0	0.74.0	80.91.0

(Table Contd...)

Villages	Area (in hectares) under						
	Total fruits and vegetables	Total Food crops	Total groundnuts+	Other non food crops	Total non food crops	Total food+ non food crops	Net sown area
Thatchampattu	7.64.0	264.96.0	82.35.0	9.71.5	92.06.5	357.02.5	410.96.0
Alikondapattu	29.62.0	110.33.0	24.30.0	8.80.5	33.10.5	143.43.5	110.60.5
Katampoondi	1.19.5	459.68.0	97.99.5	0.95.5	98.95.0	558.63.0	561.69.5
Periyakollipadi	1.00.0	341.69.0	71.44.5	7.32.0	78.76.5	420.45.5	529.86.0
Chinnakallipadi	7.97.0	226.76.5	–	40.21.0	40.21.0	97.5	366.89.5
Navampattu	–	258.92.5	107.13.5	0.97.0	108.10.5	367.03.0	490.53.0
Devanur	0.15.0	227.38.5	61.75.5	1.91.5	63.67.0	291.05.5	323.91.0
Velayampakkam	20.95.5	330.14.0	142.99.0	1.32.0	144.31.0	474.450	442.73.5
Kallotu	6.49.0	147.89.0	44.00.0	27.08.0	71.08.0	218.97.0	138.28.5
Athipadi	1.00.0	117.86.0	72.38.5	5.35.5	77.74.0	195.60.6	379.48.0
Palayanur	9.70.0	328.20.0	71.18.0	10.63.0	81.81.0	410.01.0	467.40.5
Kandiankuppam	2.42.0	369.58.5	116.59.0	3.82.0	120.41.0	489.99.5	393.26.0
Thalayampallam	8.20.0	249.04.0	94.24.0	5.86.0	100.10.0	349.14.0	496.53.5
Chakkarathamadai	–	67.04.0	39.31.0	0.04.0	39.35.0	106.39.0	89.73.0
Nariyapattu	3.00.0	190.37.0	74.80.0	10.40.0	85.20.0	275.57.0	332.59.5
Paragampattu	0.03.5	148.01.0	44.58.5	11.92.5	56.51.0	204.52.0	222.66.0
Pavapattu	15.40.0	329.17.5	38.62.5	47.52	86.15.0	415.32.5	446.67.0

Table—13.3 Cropping pattern: Thirukoilur (SLBC command) block (1997-98)

Villages	Area (in hectares) under							
	Paddy*				Total cereals (other than Paddy	Total pulses[Δ]	Total spices[#]	Total sugarcane[@]
	Sornavari	Samba	Navarai	Total				
Melandhal	62.69.0	36.69.0	–	99.38.0	12.63.5	1.17.0	–	118.43.0
Kangayanur	–	60.00.5	60.00.5	120.00.5	129.80.5	1.21.0	0.11.5	85.93.0
Pallichandal	29.05.0	56.82.0	–	85.87.5	121.07.0	4.06.0	–	63.39.5
Kongamanur	3.69.5	5.94.5	38.76.5	48.45.5	38.67.0	0.45.5	0.13.0	55.78.0
Murukkambadi	28.88.0	34.37.0	38.78.0	102.03.0	128.83.5	26.90.0	0.32.0	157.54.5
Athiandhal	47.80.0	41.82.5	–	89.62.5	50.37.0	0.47.5	0.01.0	48.08.5
Devaradiarkuppam	48.15.0	–	22.85.5	71.06.5	123.33.5	0.06.0	0.22.5	67.07.0
Jambai	6.12.0	57.81.0	188.74.5	252.67.5	264.54.0	7.82.5	0.30.0	42.34.0
Chellankuppam	51.64.5	25.14.5	-	76.79.0	14.01.0	3.79.5	–	17.99.5
Sithapatinam		112.33.5	28.28.0	140.61.5	107.74.5	21.32.0	0.30.0	48.70.0
Manalurpet	7.91.5	55.17.0	63.08.5	126.17.0	48.08.5	6.70.5	0.16.5	23.50.0

(Table Contd...)

Villages	Area (in hectares) under						
	Total fruits and vegetable	Total food crops	Total groundnets+	Other non food crops	Total non food crops	Total food+ non food crops	Net sown area
Melandhal	–	23.16.1	281.31.0	5.28.5	286.59.5	518.21.0	471.30.0
Kangayanur	0.06.0	347.12.5	103.03.0	0.58.0	103.61.0	450.73.5	275.60.0
Pallichandal	0.64.0	275.04.0	87.74.5	5.40.0	93.14.5	368.18.5	230.42.0
Kongamanur	0.51.5	144.00.5	27.43.0	2.23.5	29.66.5	173.67.0	137.81.0
Murukkambadi	3.51.0	419.14.0	108.90.5	0.92.0	109.82.5	528.96.5	366.27.0
Athiandhal	0.34.0	188.90.5	47.77.0	1.45.5	49.22.5	2138.13.0	149.81.5
Devaradiarkuppam	1.51.5	263.27.0	81.61.0	0.18.5	81.79.5	345.06.5	237.19.0
Jambai	0.26.5	567.94.5	99.98.5	20.21.5	120.20.0	688.14.5	448.20.5
Chellankuppam	1.89.0	114.48.0	13.93.0	2.46.5	16.39.5	130.87.5	102.93.5
Sithapatinam	0.93.5	319.61.5	63.52.0	4.59.5	68.11.5	387.73.0	272.20.5
Manalurpet	0.55.0	205.17.5	44.76.5	3.43.0	48.19.5	253.37.0	170.59.5

Table—13.4 Cropping pattern: Chengam (SLBC command) block-1997-98

Villages	Area (in hectares) under							
	Paddy*				Total cereals˙ (other than paddy)	Total pulsesΔ	Total spices#	Total sugarcane@
	Sornavari	Samba	Navarai	Total				
Jambadai	–	43.05.5	–	43.05.5	9.88.0	7.46.0	0.56.0	15.43.0
Olgalapadi	33.55.5	14.79.0	29.20.0	77.54.5	–	4.28.0	2.21.0	11.80.5
Thenmudiyanur	48.12.0	61.57.5	234.91.5	344.61.0	1.89.31.0	533.92.0	2.65.5	40.59.5
Edathanur	22.57.0	83.00.0	–	105.57.0	12.95.0	200.00.0	0.06.0	9.42.5
Allappanoor	–	35.63.0	22.50.0	58.13.0	48.52.0	6.34.0	–	4.97.5
Vanapuram	10.29.0	21.68.0	41.37.5	73.34.5	117.49.5	25.01.0	–	145.07.0
Kunglinatham	4.22.5	2.44.5	11.43.5	18.10.5	5.87.5	1.09.0	–	82.21.5
Perunthuraipattu	2.43.0	1.53.0	18.43.5	22.39.5	5.88.5	1.51.5	0.01.0	62.00.0
Valavachanur	6.09.5	–	23.71.0	29.80.5	6.12.5	–	–	98.44.5
Kottauyur	2.02.4	2.09.0	119.22.0	141.55.0	41.55.5	–	0.06.0	115.22.0
Agarampallipattu	22.79.0	8.71.0	61.55.0	93.05.0	121.70.5	16.92.5	–	9.48.0
Thenkarimbalur	60.02.0	30.89.5	162.26.0	253.17.5	139.43.0	1.88.0	0.39.0	197.39.0
Radhapuram	36.65.0	204.30.0	162.57.5	403.52.5	54.05.0	31.85.0	1.88.5	52.53.5
Vargaur	–	–	–	–	–	–	–	–
Serapapathu	10.38.5	17.34.5	27.81.5	55.54.5	48.39.0	4.75.5	–	33.29.5
Verhillapattu	7.59.5	167.45.3	12.01.0	36.35.0	4.82.0	3.05.5	0.10.0	3.32.5
Sadakuppam	6.93.5	25.00.0	25.06.5	57.00.0	13.72.5	3.26.0	-	37.60.0
Unnamalaipalayam	10,00.0	11.78.0	7.35.5	29.13.5	12.45.5	3.79.5	0.71.5	82.01.5
Edakkal	.97.0	8.32.5	62.84.5	72.14.0	19.58.0	11.04.5	0.10.0	94.82.0
Mazumvampattu	15.37.0	.40.0	33.34.5	49.11.5	13.36.0	1.00.5	0.11.0	60.12.0
Peraiyampattu	10.53.0	0.83.0	22.77.5	34.13.5	45.50.0	3.77.0	0.16.0	56.49.0

(Table Contd...)

Villages	Area (in hectares) under						
	Total fruit & vegetables	Total good crops	Total groundnuts[+]	Other non food crops	Total non food crops	Total food + non food crops	Net sown area
Jambadai	19.52.5	95.91.0	21.32.5	23.83.0	45.17.5	141.08.5	301.05.5
Olgalapadi	12.38.5	108.22.5	115.04.0	89.51.5	204.55.5	312.78.0	613.26.5
Thenmudiyanur	24.56.0	1135.65.0	316.76.5	342.46.5	659.23.0	1794.88.0	1273.45.5
Edathanur	7.11.0	335.11.5	84.66.5	58.50.5	243.17.0	578.28.5	648.56.0
Allappanoor	–	117.96.5	49.58.5	–	49.58.5	167.55.0	287.49.0
Vanapuram	–	360.92.0	186.20.5	24.52.5	210.73.0	571.65.0	615.11.5
Kunglinatham	2.22.5	109.51.0	17.19.0	3.04.5	20.23.5	129.74.5	155.48.5
Perunthuraipattu	1.52.5	93.33.0	19.19.5	1.36.5	20.56.0	113.89.0	155.19.5
Valavachanur	16.70.0	159.07.5	17.99.5	9.22.5	27.22.0	178.29.5	244.50.5
Kottauyur	4.02.0	302.40.5	59.97.0	0.97.5	60.94.5	363.35.0	522.56.5
Agarampallipattu	70.0	241.86.0	167.60.5	0.13.5	167.74.0	409.60.0	599.20.5
Thenkarimbalur	7.50.5	599.77.0	108.17.5	10.41.0	118.58.5	718.33.5	807.65.0
Radhapuram	10.35.5	554.20.0	390.15.0	22.43.5	412.58.5	966.21.5	130.75.6
Vargaur	–	–	–	–	–	–	34.27.3
Serapapathu	0.00.5	141.99.0	47.75.0	2.62.0	50.37.0	192.36.0	242.92.0
Verhillapattu	0.14.0	47.79.0	33.83.5	1.28.5	35.12.0	82.91.0	98.55.0
Sadakuppam	5.87.0	117.45.5	61.56.5	–	61.56.5	179.02.0	29.55.7
Unnamalaipalayam	4.32.5	132.44.0	38.19.0	39.46.0	77.65.0	210.09.0	179.74.5
Edakkal	0.20.0	197.88.5	71.85.5	11.27.0	83.12.5	281.01.0	385.80.5
Mazumvampattu	4.44.0	128.15.0	16.17.0	1.16.0	17.33.0	145.48.0	138.66.0
Peraiyampattu	1.74.5	141.80.0	39.72.0	2.60.5	41.16.5	182.96.5	220.42.0

Table—13.5 Cropping pattern: Sankarapuram (SRBC command) block (1997-98)

Villages	Area (in hectares) under							
	Paddy*				Total cereals˙ (other than paddy)	Total pulses^Δ	Total spices#	Total sugarcane®
	Sornavari	Samba	Navarai	Total				
Rayasamudram	16.43.5	140.97.0	11.26.5	168.67.0	6.42.5	0.21.0	-	34.90.0
Arulampadi	27.46.5	113.35.5	-	140.82.0	21.88.0	0.79.5	-	54.29.5
Moongilthuraipattu	-	171.10.5	40.23.5	211.24.0	141.12.5	9.33.5	1.59.0	34.65.5
Porasapattu	7.83.5	40.92.0	11.25.0	60.00.5	32.14.5	9.85.0	1.07.5	19.86.5
Olgalapadi	60.51.5	230.91.5	2.27.5	293.70.5	38.51.0	-	0.74.5	69.62.0
Poravalur	-	101.72.5	59.87.0	161.59.0	225.93.0	9.57.0	1.15.5	38.30.0
Melsiruvalur	25.86.5	108.47.0	274.00.5	608.34.0	38.97.0	0.07.5	5.00.0	32.66.5
Vadasiruvalur	42.52.0	99.98.5	156.49.5	299.00.0	-	2.07.5	0.18.5	105.79.5
Varagur	9.86.0	56.86.0	5.23.5	71.95.5	29.38.5	3.17.5	-	20.62.5
Arur	23.52.5	77.46.5	6.85.0	107.84.0	19.75.0	0.41.5	-	17.57.0
Mookanur	51.06.5	215.005	125.86.0	391.93.0	162.27.0	1.43.5	1.76.5	79.65.5
Thimmendhal	7.14.5	23.005	14.76.6	44.91.5	57.39.5	3.37.0	0.84.0	8.93.0
Chellakakuppam	2.26.5	47.26.5	3.44.5	52.97.5	7.15.0	0.53.0	0.69.0	7.85.0
Viriyur	0.78.5	107.51.5	7.47.5	115.77.5	70.12.5	13.62.0	2.39.0	67.76.5
Arasampattu	13.77.0	47.42.0	27.26.5	88.45.5	116.55.0	29.81.5	2.32.0	33.90.0
S.Kolathur	-	150.11.0	73.35.0	223.46.0	92.57.0	1.27.5	0.17.5	71.26.5
Vadakeeranur	26.68.5	117.23.5	-	143.92.0	10.00.0	0.02.5	0.34.0	16.14.5
Vadaponparappi	15.15.0	168.86.5	2.35.0	186.36.5	39.28.0	2.65.0	0.74.5	51.15.5
Kidagudayampattu	16.43.5	47.61.5	-	64.05.0	39.45.5	1.13.5	0.15.0	32.40.0
Sivapuram	13.71.5	48.98.5	12.36.5	75.06.5	11.26.5	1.32.5	0.17.5	40.79.0

(Table Contd...)

Villages	Area (in hectares) under						
	Total fruits & vegetables	Total food crops	Total groundnuts+	Other non food crops	Total non food crops	Total food + non Food crops	New sown area
Rayasamudram	3.78.5	213.99.0	82.88.5	0.16.0	83.04.5	297.03.5	257.49.5
Arulampadi	0.94.0	218.73.0	16.02.5	1.78.5	17.81.0	236.54.0	181.55.0
Moongilthuraipattu	3.91.5	401.86.0	110.88.0	12.55.5	123.43.5	525.29.5	394.15.0
Porasapattu	0.54.5	123.48.5	27.27.5	1.25.0	28.52.5	152.01.0	135.79.5
Olgalapadi	4.53.5	407.11.5	58.60.0	1.25.0	59.85.0	466.96.5	357.86.5
Poravalur	2.90.5	439.45.0	68.70.5	2.55.0	71.25.5	510.70.5	408.15.0
Melsiruvalur	2.23.0	687.28.0	55.76.0	0.72.5	56.48.0	743.76.0	388.20.0
Vadasiruvalur	16.41.0	423.56.5	81.93.0	10.77.5	92.70.5	516.27.0	351.15.5
Varagur	–	125.68.0	18.51.0	3.02.5	21.53.5	147.21.5	98.19.5
Arur	0.03.5	145.60.5	21.74.0	1.45.5	23.19.5	168.80.8	112.95.0
Mookanur	9.39.5	646.45.0	103.78.0	2.01.0	105.79.0	752.24.0	595.13.5
Thimmendhal	0.01.0	115.460	1.78.0	4.03.0	5.81.5	121.27.5	60.76.5
Chellakakuppam	1.67.0	70.86.5	5.11.0	1.23.0	6.34.0	77.20.5	58.35.0
Viriyur	9.29.0	278.96.5	19.41.5	8.34.5	27.76.0	306.72.5	228.02.5
Arasampattu	1.31.5	272.35.5	18.09.5	1.72.5	19.82.0	292.17.5	185.45.0
S.Kolathur	0.12.5	389.87.0	78.73.0	0.88.0	79.61.0	469.48.0	313.53.0
Vadakeeranur	1.40.0	171.83.0	15.97.0	0.78.0	16.75.0	188.58.0	142.60.0
Vadaponparappi	0.38.5	280.58.0	18.77.5	2.04.0	20.81.5	301.39.5	194.20.0
Kidagudayampattu	0.98.0	138.17.0	30.72.5	0.65.5	31.38.0	169.55.0	128.00.0
Sivapuram	–	128.62.0	9.39.0	–	9.39.0	138.01.0	107.48.0

Table—13.6 Cropping pattern: Rishivandiyam (SRBC command) block (1997-98)

Villages	Area (in hectares) under							
	Paddy*				Total cereals˙ (other than paddy)	Total pulses^Δ	Total spices#	Total sugarcane@
	Sornavari	Samba	Navarai	Total				
Athiyur	242.28.5	–	–	242.28.5	122.89.5	–	–	88.01.0
Mangalam	188.06.5	–	97.04.0	285.10.5	78.35.5	5.38.0	4.39.5	49.20.5
Adalhanur	60.23.5	21.56.0	–	81.79.5	54.01.0	1.47.0	–	13.82.5
Erudayampattu	34.17.0	–	26.33.5	60.50.5	37.44.5	5.93.0	0.53.5	7.48.5
Mamadal	–	107.21.0	–	107.21.0	45.07.0	–	–	20.74.5
Vanapuram	271.13.0	–	73.21.0	344.34.0	169.54.5	–	–	112.07.0
Odiyantahal	34.76.5	–	–	34.76.5	70.19.0	–	–	20.78.0
Edathanur	–	69.33.0	5.92.5	75.25.5	228.89.5	–	–	44.76.5
Kadambur	–	115.35.5	20.95.0	136.30.5	30.00.0	–	0.08.0	195.92.0
Arambarampattu	–	135.17.5	27.32.0	162.49.5	174.39.0	–	–	188.42.0
Thiruvaranham	–	82.52.0	3.52.0	86.04.0	153.96.5	3.00.0	–	25.79.0
Sirpanandal	–	260.88.0	31.09.0	291.97.0	236.54.0	3.00.0	0.50.0	138.71.5
Jambadi	–	150.23.0	–	150.23.0	231.39.5	10.00.0	–	30.43.5
Periyakoilyur	–	251.44.0	67.67.5	319.11.5	221.21.5	–	–	163.44.5
Sirupanaiyur	–	178.64.5	3.43.0	182.07.5	122.56.0	–	–	51.74.0
Chinnakolliyur	–	38.70.5	–	38.70.5	13.12.0	1.12.5	–	63.14.0
Tholavanthangal	3.26.0	140.56.5	4.11.0	147.93.5	97.48.5	11.38.5	1.68.0	63.57.0
Kodavanur	290.52.0	125.56.0	–	416.08.0	55.90.0	1.56.5	0.94.0	154.85.0
Pakkam	10.70.5	316.08.0	34.50.0	361.28.5	61.78.5	3.35.0	0.33.5	192.60.0
Vadamamandur	–	164.45.5	33.14.0	197.59.5	61.27.0	0.10.5	–	101.23.0
Nagalkudi	–	49.91.5	45.10.5	95.02.0	48.42.5	–	–	17.83.0

(Table Contd...)

Villages	Area (in hectares) under						
	Total food & vegetables	Total food crops	Total groundnuts⁺	Other non food crops	Total non food crops	Total food + non food crops	Net sown area
Athiyur	7.64.0	264.96.0	2.20.5	3.15.0	5.35.5	467.84.5	465.63.5
Mangalam	29.62.0	110.33.0	99.86.5	0.17.5	100.04.0	523.75.0	362.68.0
Adalhanur	1.19.5	459.68.0	49.14.0	–	49.14.0	208.70.5	156.98.5
Erudayampattu	1.00.0	341.69.0	39.35.5	–	39.35.5	162.17.5	104.46.5
Mamadal	7.97.0	226.76.5	33.06.5	–	33.06.5	206.09.0	198.67.0
Vanapuram	–	258.92.5	2.85.5	1.55.0	4.40.5	630.36.0	555.90.0
Odiyantahal	0.15.0	227.38.5	0.40.0	–	0.40.0	126.13.5	101.23.5
Edalthanur	20.95.5	330.14.0	12.73.0	0.10.5	12.83.5	361.75.0	238.64.0
Kadambur	6.49.0	147.89.0	28.68.0	2.28.5	30.96.5	393.27.0	379.16.5
Arambarampattu	1.00.0	117.86.0	14.71.0	1.65.0	16.36.0	541.66.5	368.04.0
Thiruvaranham	9.70.0	328.20.0	2.46.0	1.09.5	3.55.5	272.35.0	155.24.0
Sirpanandal	2.42.0	369.58.5	34.75.5	1.99.5	36.75.0	707.97.5	630.16.0
Jambadi	8.20.0	249.04.0	129.20.5	2.57.0	131.77.5	553.83.5	420.15.5
Periyakoilyur	–	67.04.0	27.72.0	4.05.0	31.77.0	736.05.5	622.94.5
Sirupanaiyur	3.00.0	190.37.0	46.98.0	2.78.5	49.76.5	406.14.0	392.16.0
Chinnakolliyur	0.03.5	148.01.0	14.05.5	5.47.0	19.52.5	135.61.5	130.37.5
Tholavanthangal	15.40.0	329.17.5	36.58.0	2.63.0	39.21.0	362.65.5	231.48.5
Kodavanur	–	–	10.82.5	2.08.0	12.90.5	644.04.0	471.83.0
Pakkam	–	–	–	0.19.0	0.19.0	633.03.0	563.03.0
Vadamamandur	–	–	2.17.5	–	2.17.5	362.37.5	288.19.0

Table—13.7 Cropping pattern: Chengam (SRBC command) block-1997-98

Villages	Area (in hectares) under							
	Paddy*				Total cereals• (other than paddy)	Total pulses$^{\Delta}$	Total spices#	Total sugarcane@
	Sornavari	Samba	Navarai	Total				
Rayandupuram	–	–	–	–	–	–	–	–
Thiruvadathanur	–	–	–	–	–	–	–	–
Thondamanur	7.52.5	95.60.0	95.38.0	198.50.5	144.08.5	29.23.5	–	39.08.0
Pathucheekadi	–	121.80.5	–	121.80.5	77.44.0	5.19.5	0.14.0	19.04.5

(Continued...)

Villages	Area (in hectares) under						
	Total fruits & vegetables	Total food crops	Total groundnuts+	Other non food crops	Total non food crops	Total food + non food crops	Net sown area
Rayandupuram	–	–	–	–	–	1036.22.5	1154.45.5
Thiruvadathanur	–	–	–	–	–	544.81.0	68.94.0
Thondamanur	15.31.0	426.21.5	146.19.0	–	146.27.0	572.48.5	836.95.5
Pathucheekadi	18.54.5	242270	105.03.0	–	83.12.5	410.23.0	410.52.5

Table—13.8 Cropping pattern: (Cumulative)-1997-98

Blocks and Sathanur command	Paddy*						
	Sornavari	Samba	Navarai	Total	Millets[•]	Pulses[Δ]	Spices[#]
Chengam (SLBCcommand)	328.77.0	590.12.5	1078.38.5	1997.28.0	910.61.0	860.95.5	9.01.5
Thirukoilur	205.50.5	487.37.5	356.43.0	1049.31.0	957.28.5	47.54.5	0.29.0
Thachampattu	701.22.0	438.51.5	657.04.5	1796.78.0	848.88.0	30.27.5	45.32.0
Rishivandiyam	1135.13.5	2207.62.5	473.35.0	3816.11.0	2314.46.0	56.31.0	8.46.5
Sankarapuram	361.03.5	2314.68.5	834.35.5	3511.07.5	1160.18.0	91.31.5	19.34.0
Chengam (SRBCcommand)	7.52.5	550.94.0	278.52.0	836.98.5	104.85.0	576.50.0	2.82.5
SRBC command	1503.69.5	5073.25.0	1586.22.5	8164.17.0	3579.49.0	724.12.5	30.63.0
SLBC command	1235.49.5	1516.01.5	2091.86.0	4843.37.0	2716.77.5	938.77.5	54.62.5
SCA	2739.19.0	6589.26.5	3678.08.5	13007.54.0	6296.26.5	1662.90.0	85.25.5

(Continued...)

	Sugarcane@	Fruits + vegetables	Total groundnuts+	Other non food crops	Total food crops	Total non food crops
Chengam (SLBCcommand)	1212.25.0	123.33.5	1862.95.0	743.12.5	5113.44.5	2606.07.5
Thirukoilur	730.21.0	5.15.5	1011.17.0	129.73.5	2789.79.5	1140.90.5
Thachampattu	1331.91.5	114.77.5	1183.68.5	193.84.5	4167.94.5	1377.53.0
Rishivandiyam	1744.57.0	37.63.5	587.76.0	31.78.0	7977.55.0	619.54.0
Sankarapuram	838.14.5	59.87.0	844.08.0	57.22.5	5679.92.5	901.30.5
Chengam (SRBCcommand)	130.37.0	66.77.0	653.80.0	191.65.0	1718.30.0	845.45.0
SRBC command	2713.08.5	164.27.5	2085.64.0	280.65.5	16375.77.5	2366.29.5
SLBC command	3274.37.5	243.26.5	4057.80.5	1066.70.5	1201.18.5	5124.51.0
SCA	5987.46.0	407.54.0	6143.44.5	1347.36.0	27446.96.0	7490.80.5

* *Oryza sativa.*

• Millets (*Pennisetum typhiodes*), ragi *(Eleucine coracana)*,maize(*Zea mays*).

Δ Blackgram (*Phaseolus mungo*), green gram (*P. aureus*), kidney bean (*P.valgaris*), pigeon pea (*Cajanurs cajan*), horse gram(*Macrotyloma uniflorum*) .

Turmeric (*Curcuma domestica*), black pepper (*Piper nigrum*), Cumin (*Cuminum cuminium*), ginger (*Zingiber officinale*), coriander (*Coriandrum sativum*).

@ *Saccharum officinarum.*

\+ Arachis hypogea.

Table—13.9 Division of labour: Thachampatu (SLBC command) block (1991 census)

Villages	Population	Workers	Cultivators	Agricultural labourers		
				Male	Female	Total
Thachampattu	1720	849	296	221	258	479
Allikondapattu	1172	554	122	192	181	373
Katampoondi	2767	1060	393	346	201	547
Periyakallipadi	2016	1191	626	161	364	525
Chinnakallipadi	1747	841	156	337	280	617
Navampattu	1737	546	310	113	91	204
Devanur	1688	654	552	16	40	56
Velayampakkam	1384	824	571	47	163	210
Kallottu	674	378	307	47	16	31
Athipadi	400	229	0	102	101	203
Palayanur	3606	1701	563	282	673	955
Kandiankuppam	1309	600	324	62	193	255
Thalayampallam	2949	1314	303	450	432	882
Nariyapattu	1058	518	126	139	185	324
Perayampattu	1796	703	245	224	164	388
Pavithram	3230	721	344	140	30	170
Aradapattu	2124	847	202	332	215	547

Table—13.10 Division of labour: Thirukoilur (SLBC command) block (1991 census)

Villages	Population	Workers	Cultivators	Agricultural labourers		
				Male	Female	Total
Melandhal	1626	529	276	166	64	230
Kangaiyanoor	3579	1434	400	399	405	804
Pallichandal	845	243	181	38	2	40
Konganamur	1996	763	305	165	228	393
Murukkambadi	389	124	73	44	0	44
Athiyandal	1035	451	162	133	126	259
Devariyarkuppam	999	489	321	80	60	140
Jambai	1487	776	149	289	322	611
Chellankuppam	1949	858	446	105	137	242
Sithapattinam	571	196	99	66	22	88
Manilurpet	553	205	64	109	27	136

Table—13.11 Division of labour: Chengam (SLBC command) block (1991 census)

Villages	Population	Workers	Cultivators	Agricultural labourers		
				Male	Female	Total
Iambadai	–	–	–	–	–	–
Olgalapadi	–	–	–	–	–	–
Thenmudiyanur	4307	2135	633	572	775	1347
Eduthanur	2410	969	525	139	193	332
Allappanour	1042	645	241	71	297	368
Vanauram	4095	1764	276	558	612	1170
Kunglinatham	460	229	133	46	36	82
Perunthuraipattu	1676	546	313	93	72	165
Valavarhanur	339	193	121	26	45	71
Kottaiyur	2124	1042	557	133	301	434
Agarampallipattu	1796	727	204	252	242	494
Thenkarimbalur	2475	1316	399	387	463	850
Radhapuram	–	–	–	–	–	–
Serapapttu	647	374	201	68	97	165
Varkillapattu	1321	636	499	25	88	113
Sadakuppam	1293	733	464	102	129	231
Unnamalaipatayam	710	255	145	70	34	104
Edakkal	790	400	134	66	180	246
Muzhavampattu	732	337	98	136	92	228
Periyampathu	1647	450	259	115	35	150

Table—13.12 Division of labour: Sankarapuram (SRBC command) block (1991 census)

Villages	Population	Workers	Cultivators	Agricultural labourers		
				Male	Female	Total
Rayasamudram	433	208	81	47	60	107
Arulampadi	715	319	156	51	93	144
Moongilthuraipathu	1219	452	276	51	91	142
Porasapattu	1642	806	408	173	158	331
Olgalapadi	1283	473	174	169	108	277
Poravalur	1958	794	279	67	207	274
Melsiruvalur	2623	1382	784	179	279	458
Vadasiruvalur	3455	1646	983	191	439	630
Varagur	5402	2754	1565	413	624	1037
Arur	1188	481	278	16	112	128
Mookanur	324	195	33	74	81	155
Thimmendhal	2411	1346	555	295	393	688
Chellankakuppam	777	344	183	43	97	140
Viriyur	1492	663	110	302	199	501
Arasampattu	1824	923	301	248	355	603
S. Kolathur	1151	499	296	77	64	141
Vadakiranur	–	–	–	–	–	–
Vadaponparappi	2303	1066	432	285	302	587
Kidagudayampattu	1491	513	220	143	106	249

Table—13.13 Division of labour: Rishivandiyam (SRBC command) block (1991 census)

Villages	Population	Workers	Cultivators	Agricultural labourers		
				Male	Female	Total
Athiyur	1664	695	101	264	184	448
Mangalam	1025	360	48	48	41	89
Adathanur	3261	995	209	392	198	590
Erudayampattu	3366	1442	667	169	533	702
Maniyandal	1589	657	380	98	154	252
Vanapuram	3809	1801	661	501	522	1023
Odiyanthal	468	289	156	53	72	125
Edathanur	2845	1386	349	368	519	887
Kadambur	1971	880	128	341	292	633
Arambarampathu	863	423	70	124	110	234
Thiruvarangam	1503	626	193	171	201	372
Sirpanandal	4435	2504	657	739	939	1678
Jambadai	5255	1992	865	396	109	505
Periyakolliyur	229	138	37	29	59	88
Sirpanaiyur	2743	1438	276	520	527	1047
Chinnakolliyur	2741	1445	257	509	569	1078
Tholuranthangal	1403	605	150	208	204	412
Kadavanur	1998	912	185	386	309	695
Pakkam	572	296	172	49	61	110
Vadamamandur	1181	370	133	194	12	206
Nagalkudi	475	284	61	67	138	205

Table—13.14 Division of labour: Chengam (SRBC command) block (1991 census)

Villages	Population	Workers	Cultivators	Agricultural labourers		
				Male	Female	Total
Rayandapuram	2765	1242	693	281	225	506
Thiruvadathanur	1941	513	465	31	12	43
Thondamanur	1116	384	358	1	3	4
Puthurchekadi	1404	442	364	25	13	38

Table—13.15 Division of labour (comulative) – 1997-98

	Total population	Total workers	Total cultivators	Others	Total agricultural labourers	Labourers		Total area (in ha.)
						Male	Female	
Chengam (SLBC command)	27864	12751	5202	1410	6550	2859	3691	9092.50.0
Thirukoilur (SLBC command)	15029	6068	2476	605	2987	1594	1393	4828.46.5
Thachampattu (SLBC command)	34179	14549	6025	999	7114	3309	3805	6203.81.0
Rishivandiyam (SRBC command)	43396	19538	5845	2314	11379	5626	5753	11680.92.5
Sankarapuram (SRBC command)	31691	14864	7114	1158	6592	2824	3768	7246.11.5
Chengam (SRBC command)	7226	2581	1880	110	591	338	253	3091.42.5
SRBC command	82313	36983	14839	35821	18562	8788	9774	22018.46.5
SLBC command	77072	33368	13703	30141	16651	7762	8889	20124.77.5
SCA	159385	70351	28542	6596	35213	16550	18663	42143.24.0

IMPACTS

Cropping Pattern

The general cropping pattern followed in Thiruvannamalai district vis a vis SCA is presented in Table—13.1. Paddy (*Oryza sativa*), sugarcane (*Saccharum officinarum*) and groundnut (*Arachis hypogea*) are the major crops cultivated in various seasons, as indicated in the Table.

Rainfall Pattern in the Study Area

The annual precipitation of the Thiruvannamalai district for the years 1962 to 1998 along with the outflow from the reservoir is depicted in Figure—13.1. The average precipitation is 970.70mm, and standard deviation is 274.8. The annual precipitation for T.N. state is depicted in Figure—13.2. Average rainfall for T.N state is 897.5mm and standard deviation is 144.6. The coefficient of variation for the rainfall in Thiruvannamalai is 28.30 per cent while that for T.N. state is 16.1 per cent. As inferred form Figure—13.1 the rainfall and outflow from the reservoir illustrate wide ranging fluctuations. The outflow from the reservoir was very low from 1982-90 and again from 1993-95.

Cropping Pattern Specifically in SCA

Cropping pattern is the proportion of area under various crops at a point of time. The cropping pattern in SCA is essentially governed by socio-economic factors rather than the ecological considerations. This is obvious from the fact that food crops constitute a major produce of the agricultural land (78.55%) in the study area as illustrated in Figure—13.13. The Sathanur Right Bank Canal (SRBC) command has its 86.66 per cent of the gross cultivated area under food crops and Sathanur Left Bank Canal (SLBC) command has 70.19 per cent of its gross sown area under food crops. The figures are higher than for the state of Tamil Nadu (68.8%). Rishivandiyam block has the highest percentage of area under food crops (92.8%) followed by Sankarapuram (86.3%) and Thachampattu (75.2%) blocks. All the blocks have more than 65 per cent of their land under food crops. Major produce comprises of paddy (*Oryza sativa*), millets (*Pennisetum typhoides*) pulses, spices, sugarcane (*Saccharum officinarum*) and fruits and vegetables.

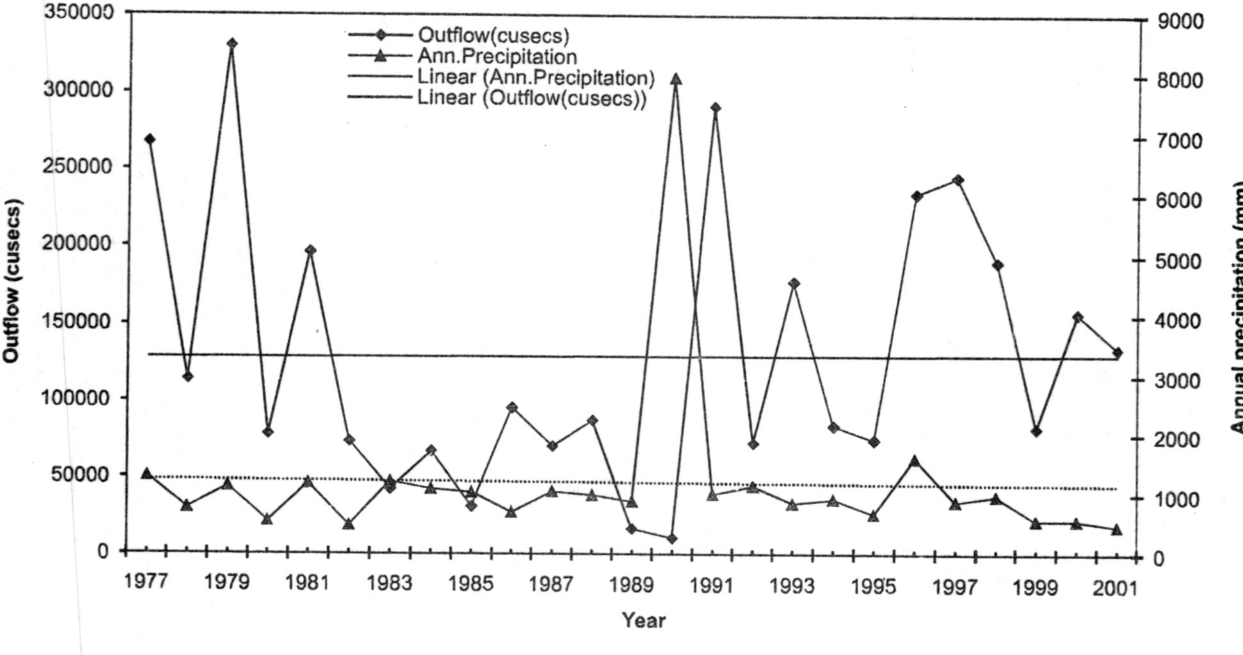

Fig. 13.1: Annual precipitation in SCA Vs annual outflow from the Sathanur dam

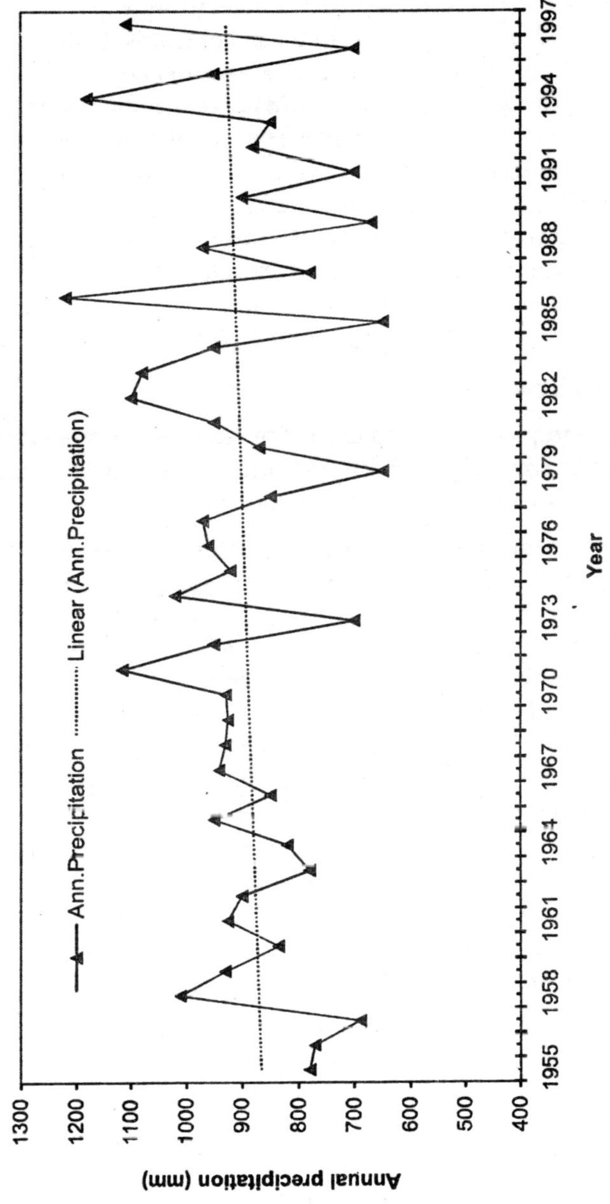

Fig. 13.2: Annual precipitation in Tamil Nadu state, India

Among the non-food crops groundnut (*Arachis hypogea*), cotton (*Gossypium arborium*), and oil seeds like gingelly (*Sesamum indicum*) and castor (*Riccinus communis*) form the major chunk. Groundnut alone shares 82 per cent of the cultivated area under non-food crops in SCA as against the state's share of 44.78 per cent. SRBC command has its 88.13 per cent of land under groundnut cultivation and SLBC command has 79.18 per cent of the land under groundnut cultivation. The prevalence of individual crop is largely influenced by the degree of the availability of water resources and the feasibility in terms of revenue obtained. An attempt has been made to correlate various environmental factors with the spatial distribution of the individual crops in the SCA.

Paddy *(Oryza Sativa)*

Being a tropical monsoon crop, rice requires a temperature of 21°C during sowing and about 37°C during harvesting. It also require high rainfall and assured irrigation facilities.

Paddy occupies about 37.23 per cent of the total cropped area in SCA (Figure—13.11) as against 33.6 per cent for the state of Tamil Nadu (Figure—13.17). Paddy cultivation shows great variations at the block levels. Sankarapuram block (Figure—13.7) is the major paddy growing region with its 53.34 per cent of the cropped area under paddy followed by Rishivandiyam with 44.36 per cent (Figure—13.8) The SRBC command has 46.01 per cent of its gross sown area under paddy cultivation (Figure—13.10) whereas SLBC command has just 28.16 per cent of its gross sown area under paddy (Figure—13.6). Thachampattu (Figure—13.3) and Chengam (Figure—13.5) have 33 per cent of their gross cultivated area under paddy while Thirukoilur and Chengam (SRBC command) blocks have 27 per cent and 25 per cent of their respective cultivated land under paddy (Figures—13.4 and 13.9).

The paddy growing seasons are Samba (June–December), Navarai (January-March) and Sornavari (April- June). Paddy cultivation during the samba season has the maximum hectarage with 50.65 per cent of the total annual paddy grown during the season followed by Navarai (28.27%) and Sornavari (21.05%), (Figure—13.12). SRBC command has maximum land under paddy cultivation during the Samba season (62.14%). Sankarapuram and

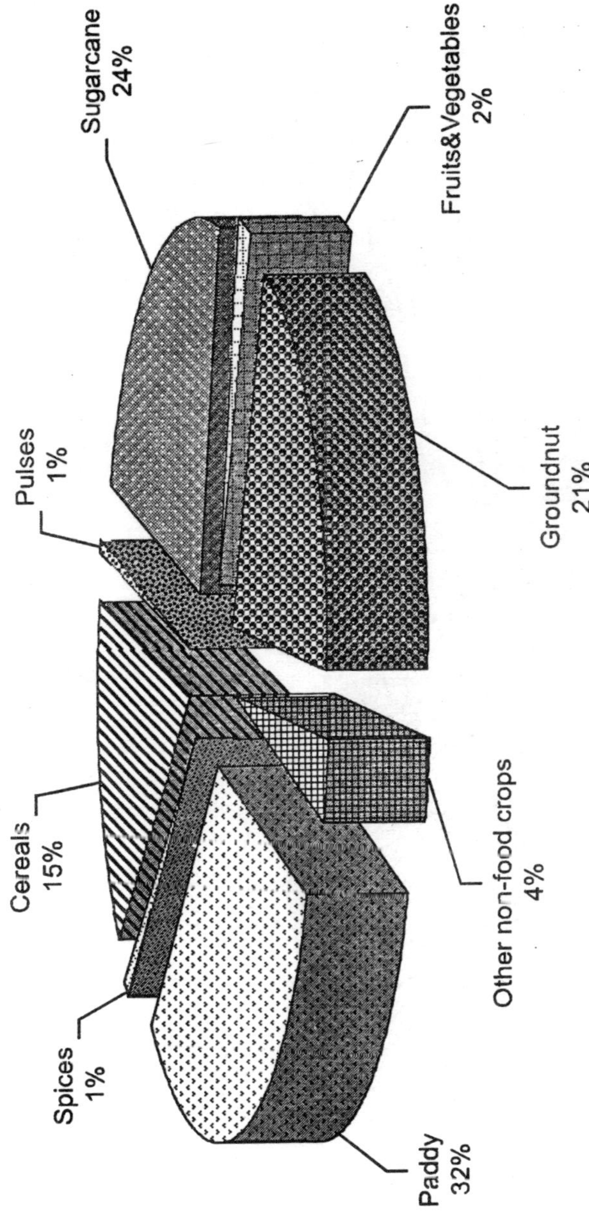

Fig. 13.3: Cropping pattern (1997-98): Thachampattu block (SLBC command)

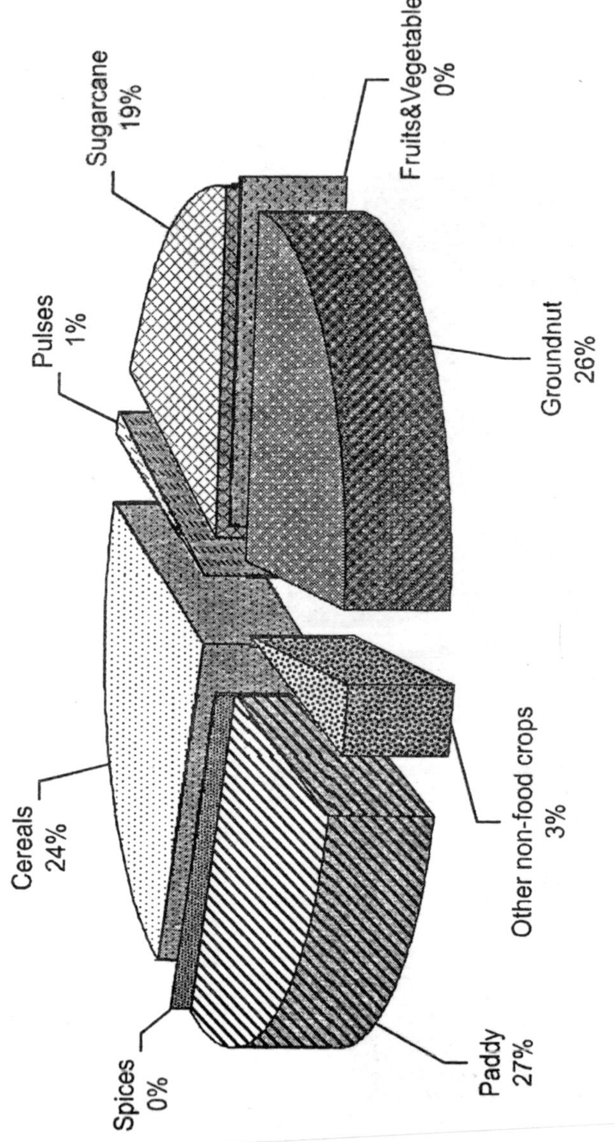

Fig. 13.4: Cropping pattern (1997-98): Thirukoilur block (SLBC command)

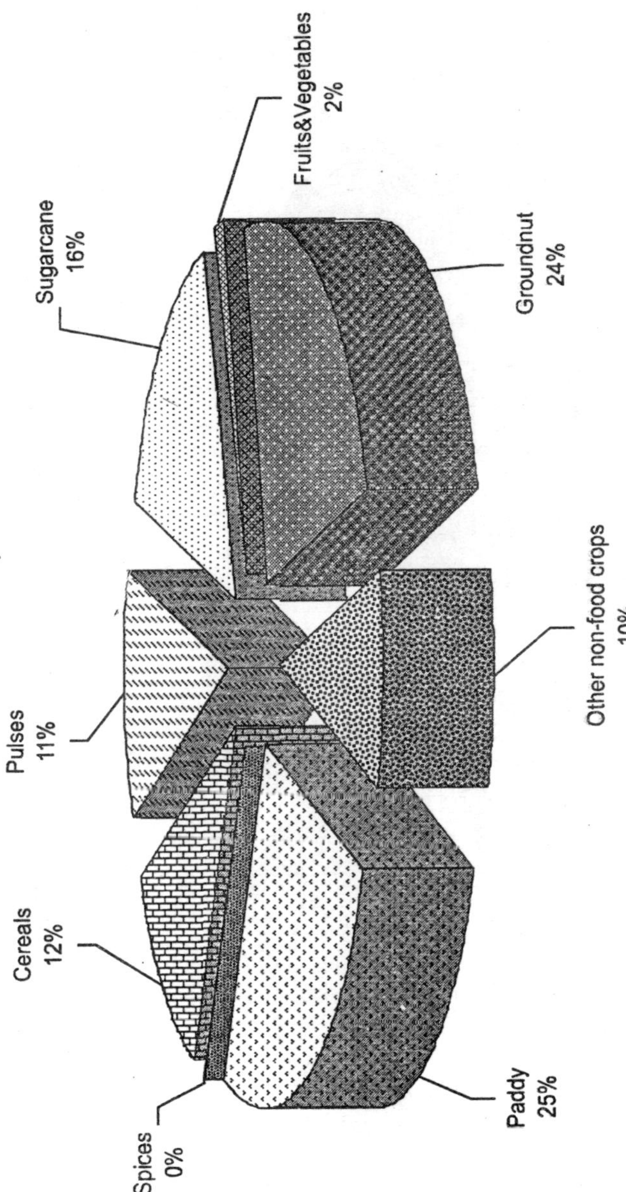

Fig. 13.5: Cropping pattern (1997-98): Chengam block (SLBC command)

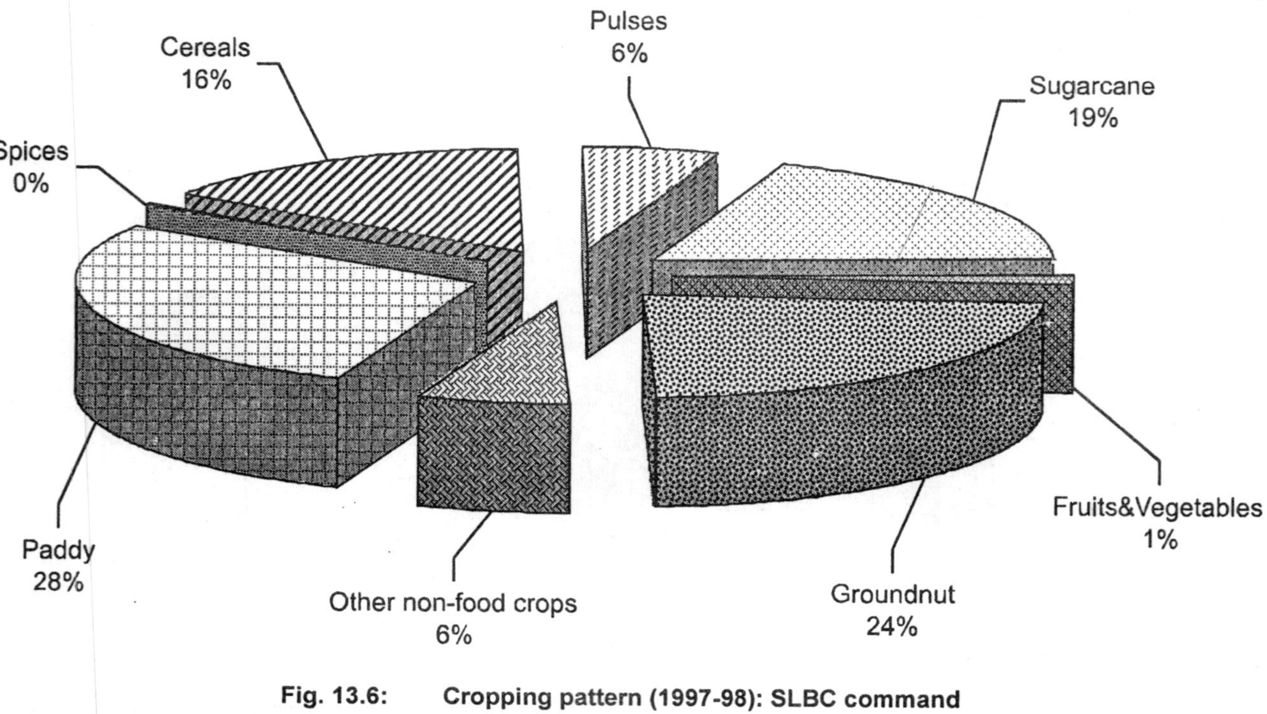

Fig. 13.6: Cropping pattern (1997-98): SLBC command

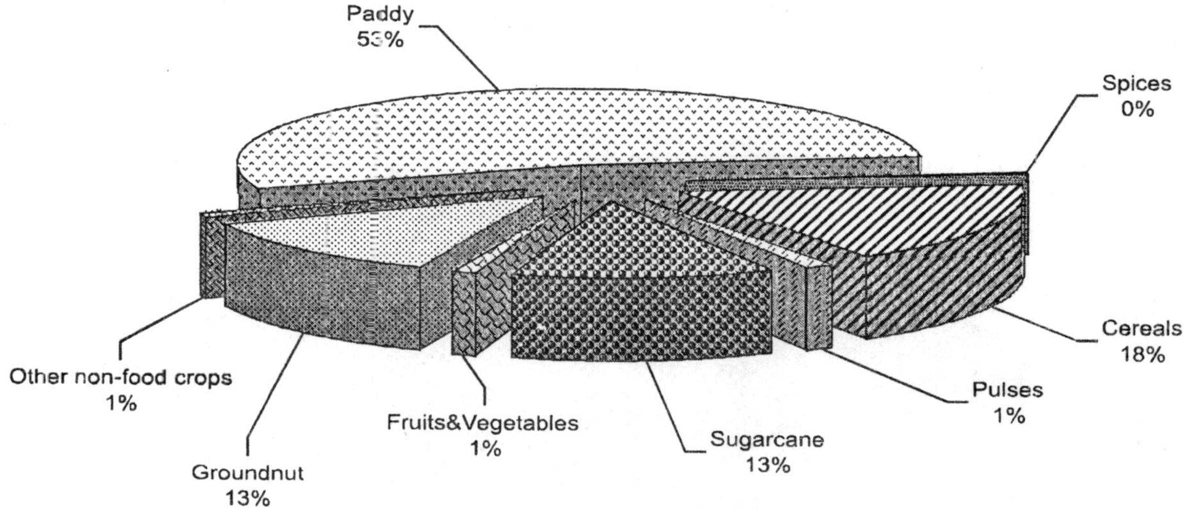

Fig. 13.7: Cropping pattern (1997-98): Sankarapuram block (SRBC command)

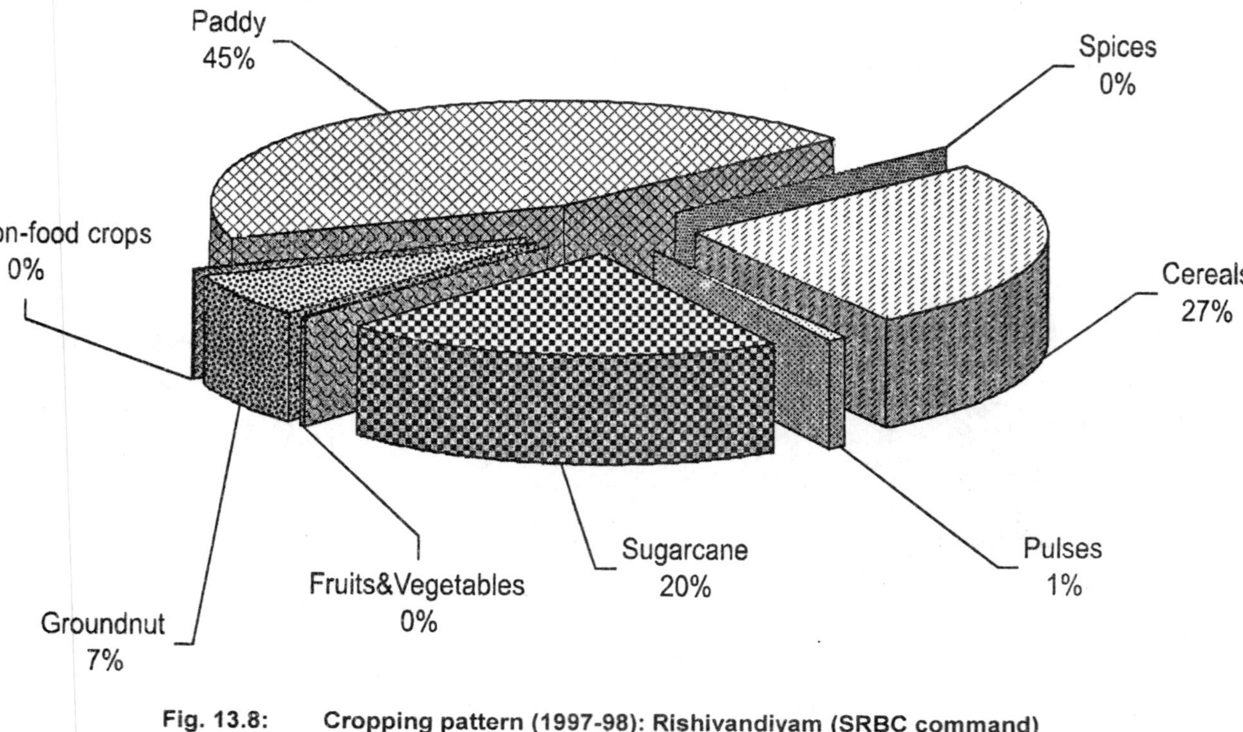

Fig. 13.8: Cropping pattern (1997-98): Rishivandiyam (SRBC command)

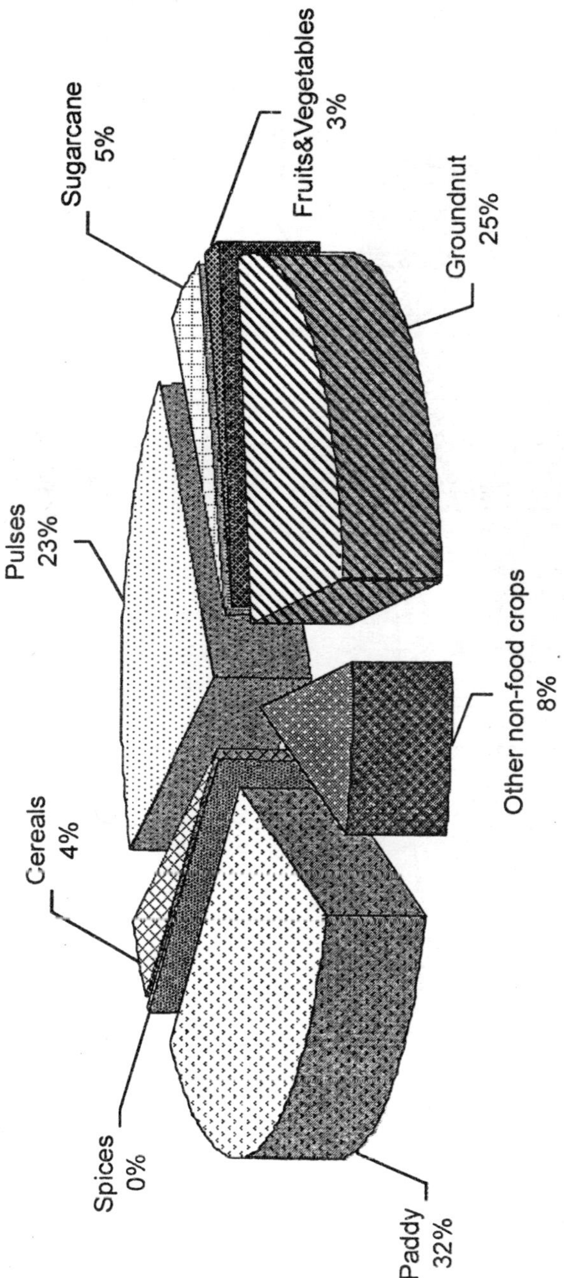

Fig. 13.9: Cropping pattern (1997-98): Chengam (SRBC command)

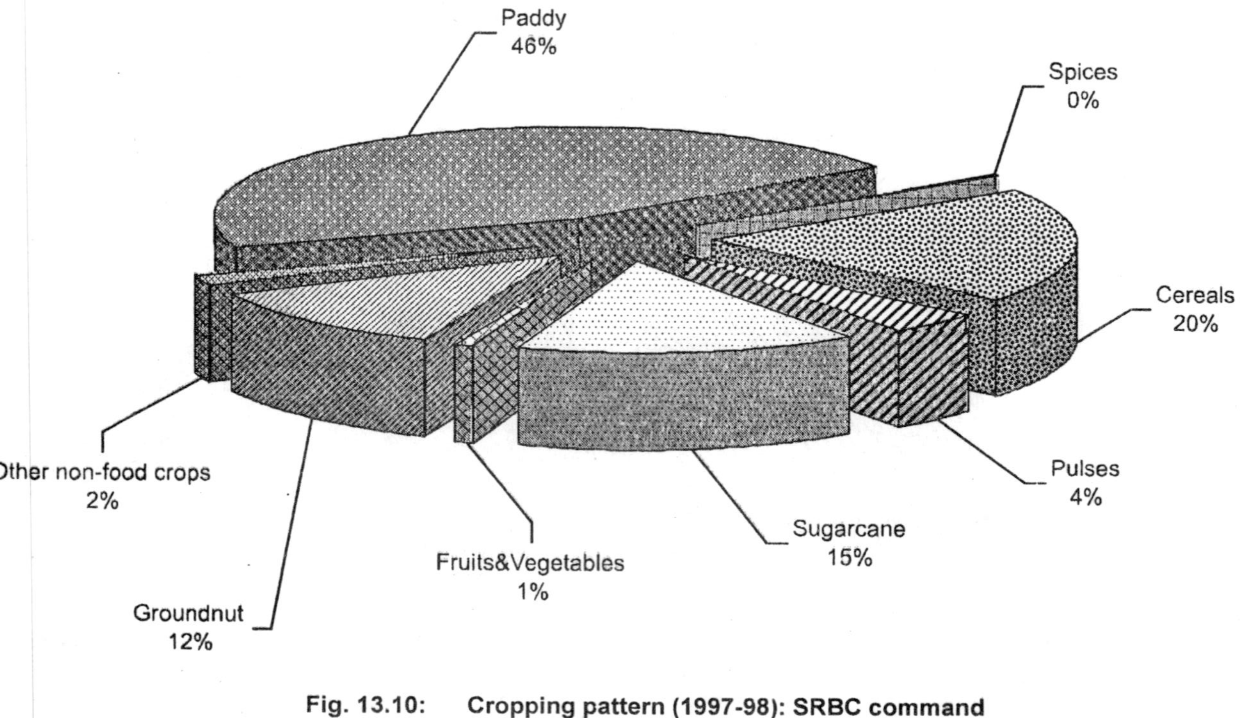

Fig. 13.10: Cropping pattern (1997-98): SRBC command

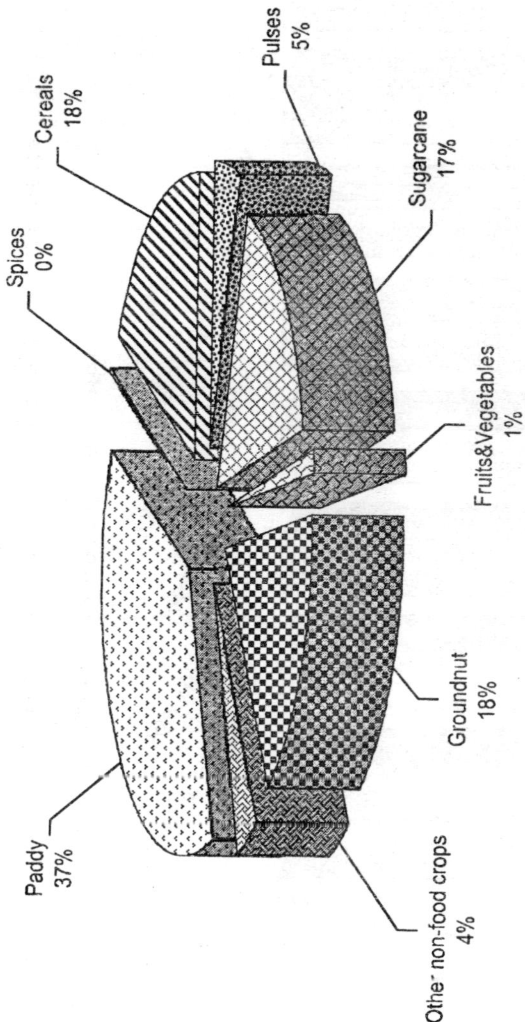

Fig. 13.11: Cropping pattern (1997-98): SCA command

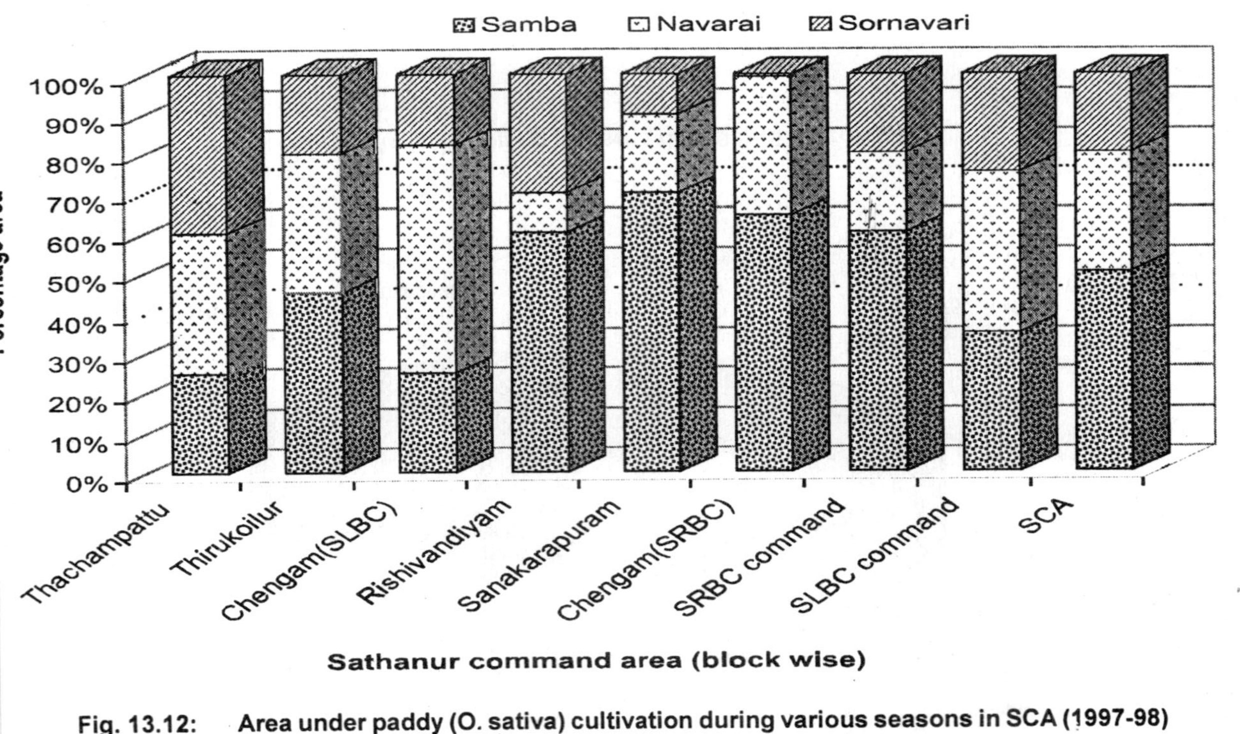

Fig. 13.12: Area under paddy (O. sativa) cultivation during various seasons in SCA (1997-98)

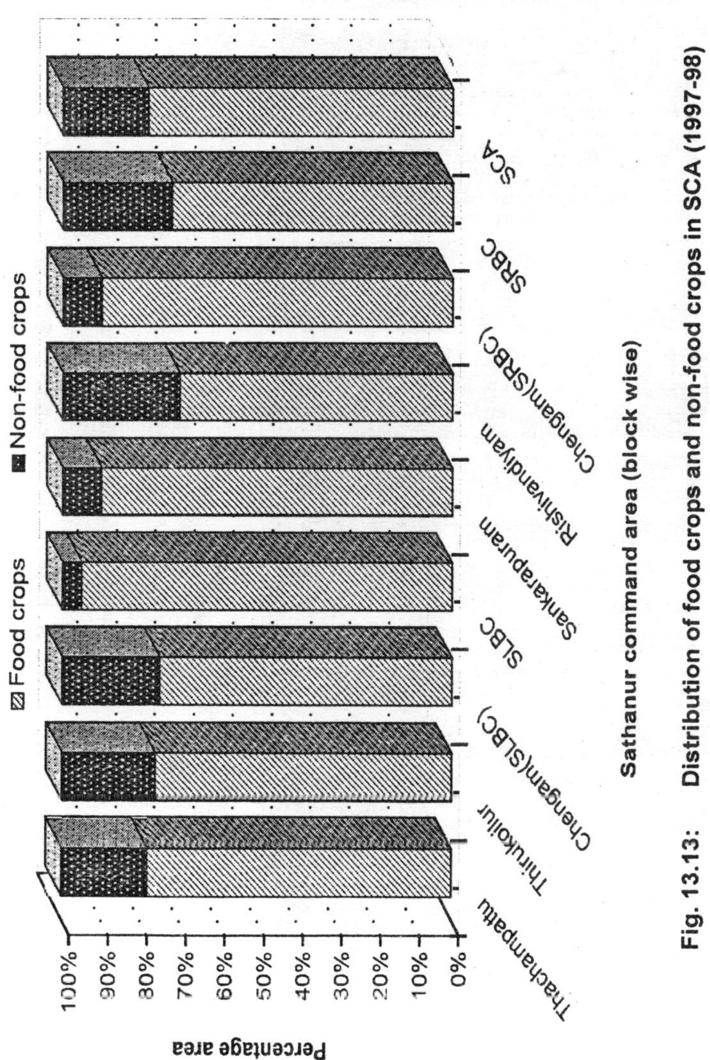

Fig. 13.13: Distribution of food crops and non-food crops in SCA (1997-98)

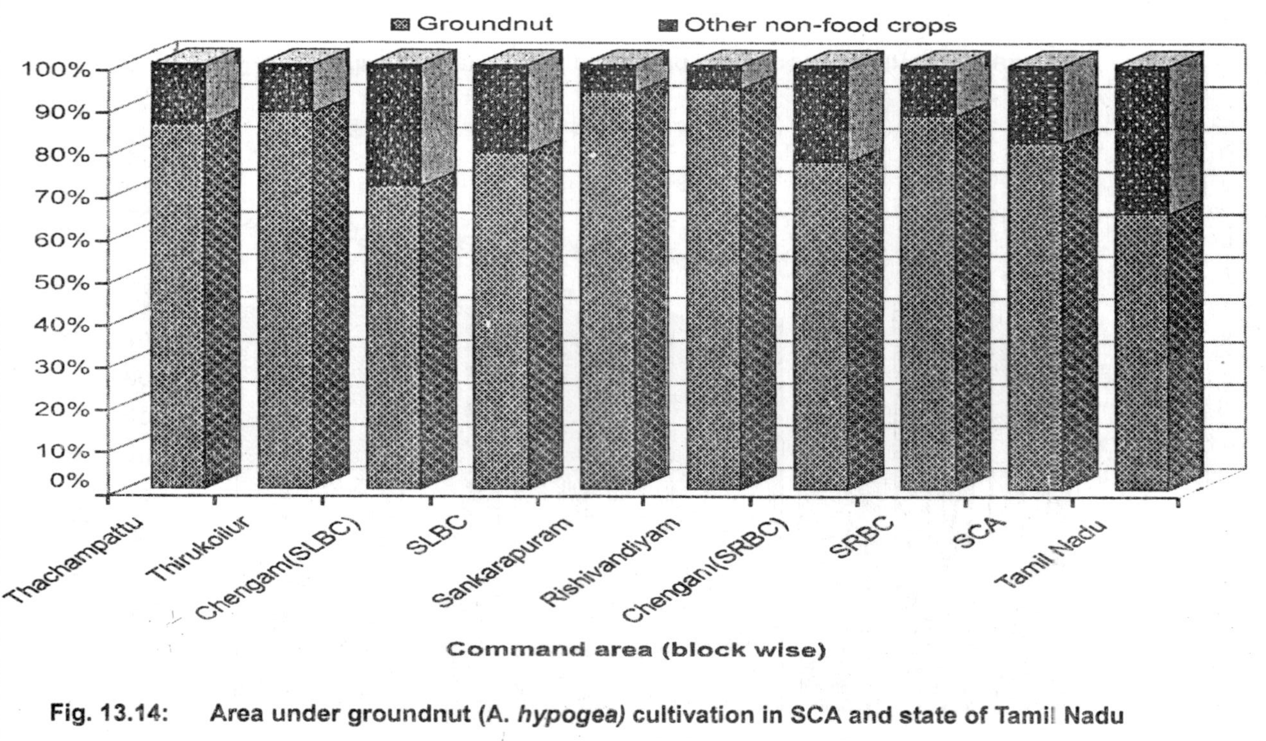

Fig. 13.14: Area under groundnut (A. *hypogea*) cultivation in SCA and state of Tamil Nadu

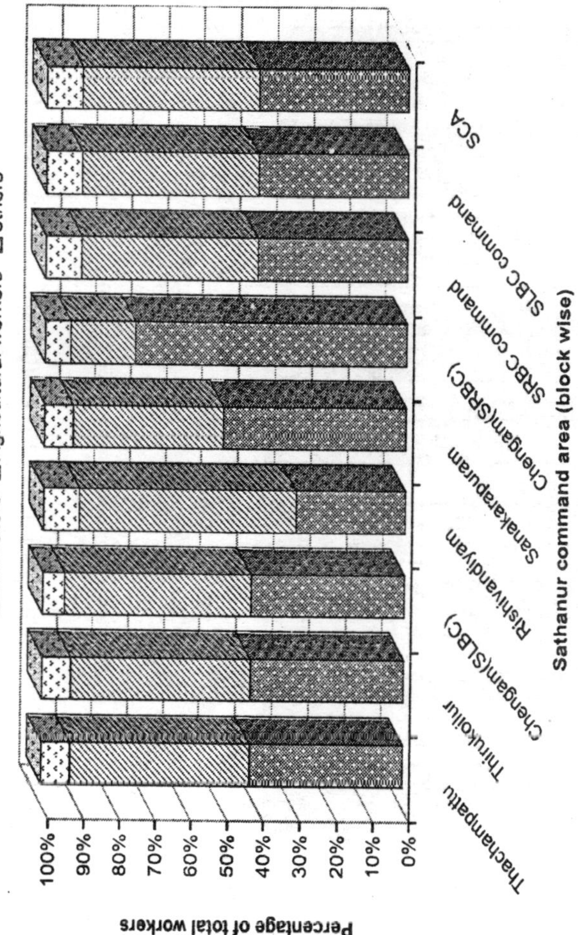

Fig. 13.15: Classification of workers in SCA (1991 census)

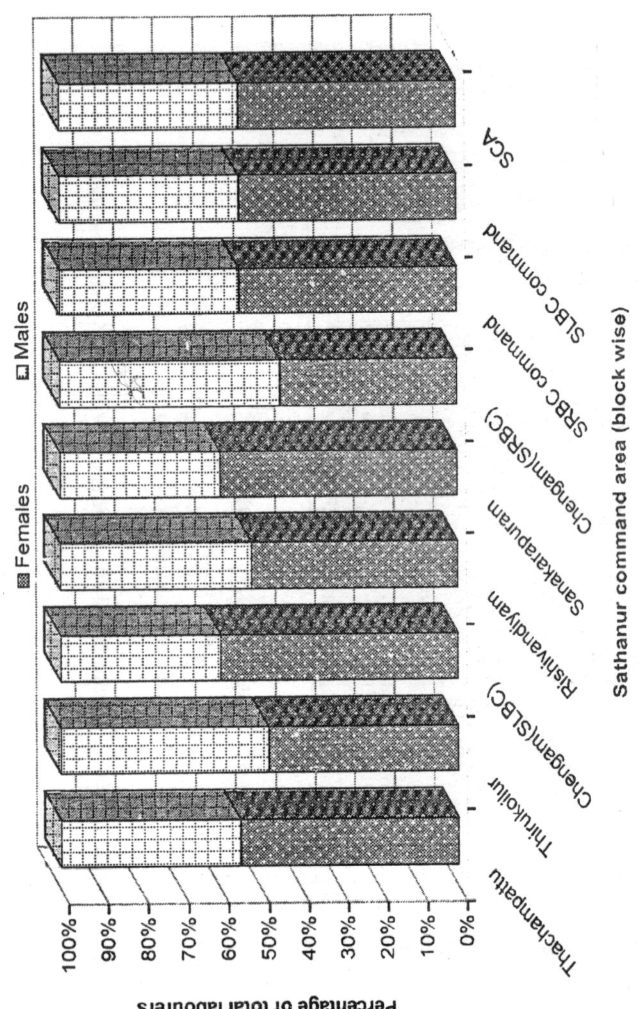

Fig. 13.16: Genderwise distribution of agricultural labourers in SCA (1991 census)

Chengam (SRBC command) blocks have more than 65 per cent of the total land under paddy during the same season followed by Rishivandiyam (57.85%). SLBC command cultivates paddy mostly during Navarai season (43.19%) followed by Samba (31.3%). All the three left command blocks follow more or less equal distribution in terms of area under paddy cultivation during the three seasons with the exception Chengam block which has maximum hectarage, under paddy during the Navari season (53.99%) as indicated in Figure—13.12.

Sugarcane *(Saccharum officinarum)*

Sugarcane a premier cash crop, occupies 17.13 per cent of the total cropped area in the SCA (Figure—13.11) as against the state's share of mere 4.02 per cent (Figure—13.17). SLBC command has its 19.04 per cent of the cropped area under sugarcane plantation (Figure—13.6) whereas SRBC command has 15.29 per cent of the total cropped area under sugarcane plantation (Figure—13.10).

Thachampattu block has the maximum coverage of 24.10 per cent followed by Rishivandiyam block (20.29%) as evident from Figures—13.3 and 13.8 respectively. Thirukoilur (Figure—13.4) and Chengam (SLBC command) (Figure—13.5) blocks have 19 per cent and 16 per cent of their respective cultivated area under sugarcane crop. Sankarapuram (Figure—13.7) block has 13 per cent of its gross cultivated land under sugarcane while Chengam block (SRBC command), 5 per cent, under sugarcane plantation as illustrated in Figure 13.9.

Cereals

Other than paddy, the major cereals grown in the study area are millets (*Pennisetum typhoides*), maize (*Zea mays*) and ragi (*Eleucine coracana*). These occupy 18.02 per cent of the total cultivated land, maximum after paddy, in the command area (Figure—13.11). The T.N state's share in cereals cultivation is 12.4 per cent (Figure—13.17). These are the main crops of the dry lands where they are grown either as rainfed crops or dry summer crops requiring minimum irrigation facilities. SRBC command has 20.17 per cent of its cultivated land under cereals (Figure—13.10) while

SLBC command has 15.79 per cent of its' land under the same (Figure—13.6). Rishivandiyam, Thirukoilur and Sankarapuram blocks have more than 15 per cent of their cultivated land under cereals as evident from Figures—13.8, 13.4 and 13.7 respectively.

Pulses

The major pulses grown in the SCA are black gram (*Phaseolus mungo*), green gram (*P.aureus*), kidney bean (*P.valgaris*), pigeon pea (*Cajanus cajan*), etc Pulses are grown both in rabi and kharif seasons occupying about 4.75 per cent of the total cropped area in the study region (Figure—13.11) which is just half the state's share of 9.10 per cent (Figure—13.17). Chengam block (both SLBC and SRBC commands) has the maximum coverage of 22.48 per cent and 11.15 per cent respectively (Figures—13.5 and 13.9) while the other blocks have less that 2 per cent of their gross cultivated area under pulses cultivation as evident from their cropping pattern figures.

Condiments and Spices

This series of crops includes chillies (*Capsicum annum*), turmeric (*Curcuma domestica*), ginger (*Zingiber officinale*) and methi (*Trigonella foenun-graecum*) among other spices. It forms a very meager share of 2.4 per cent in the total cultivated area. Turmeric as a traditional cash crop is the major crop grown under spices.

Fruits and Vegetables

Banana (*Musa paradisiaca*), Brinjal (*Solanum melongena*), tomatoes and yam (*Dioscorea sativa*) are the major fruits and vegetables cultivated in the region. The area constituting fruits and vegetables is dismally low with 1.16 per cent when compared to the state's share of 6.90 per cent. Thachampattu and Chengam blocks contribute the maximum with almost 2 per cent of their cultivated land under fruits and vegetables.

Groundnut *(Arachis hypogea)*

Groundnuts are the major oilseed crops grown in the command region with 17.58 per cent of the cropped area being occupied by them. The states contribution stands at 13.96 per cent SLBC command region has the maximum coverage of 23.59 per cent while SRBC command has only 11.75 per cent of its cultivated area under groundnuts.

Chengam block is the highest producer with 24.13 per cent and 25.5 per cent of the total cropped area under SLBC and SRBC commands respectively (Figures—13.6 and 13.9) followed by Thirukoilur block (Figure—13.4) with 25.72 per cent. Groundnut crops require a temperature of 20°C to 25°C and five to eight months to grow fully. About 750 to 850 mm rainfall may be considered necessary, though it is grown in areas receiving rainfall below 500 mm also. It is grown both as irrigated and rainfed crop.

Groundnut alone forms 82 per cent of the non food crops in the command (Figure—13.11) which is appreciably higher compared to the state share of 44.78 per cent (Figure—13.17). SRBC has 88.13 per cent of its non-food crops grown as groundnut (Figure—13.10). Rishivandiyam and Sankarapuram blocks have 94.87 per cent and 93.65 per cent of their respected total non-food acreage as groundnuts.

Other non-food Crops

Oil seeds like castor, Gingelly, sunflower (*Helianthus annus*) and other miscellaneous crops like cotton, coconut (*Cocos nucifera*) flowers etc constitute a mere 3.85 per cent of the total cropped area in SCA (Figure—13.11) which is considerable lower compared to the state's figures of 55.32 per cent (Figure—13.17). Chengam block (both SLBC and SRBC irrigated) has the highest coverage of 9.65 per cent and 7.47 per cent respectively under non-food crops; cotton being the major crop (Figures—13.5 and 13.9).

As evident from the cropping pattern of the SCA, the area is dominated by four major types of crops-paddy, cereals (millets and ragi), sugarcane and groundnuts. Paddy is the dominant crop with 37 per cent and 46 per cent of the total cultivated areas under SLBC and SRBC commands respectively. With the year round availability of water, after the introduction of canal irrigation, there has been a heavy shift towards the cash crops viz. groundnut and sugarcane. Sugarcane which was practically unknown to the study area in 1950s, has currently emerged as the major revenue earning crop. As a result many sugar mills, to cater to the needs of the sugarcane cultivators, have sprung up in the vicinity. One of them Kallakurichi Co-operative Sugar Mill is located in the study area itself in Moongilthuraipattu village. Many agro-based cooperatives

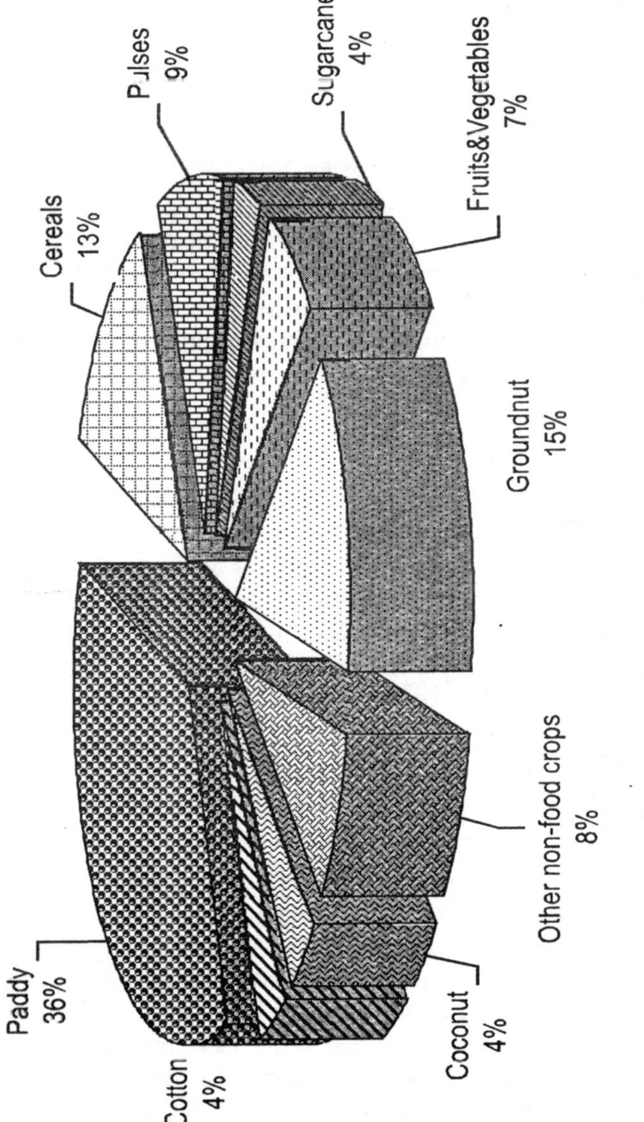

Fig. 13.17: Cropping pattern in the state of Tamil Nadu

dealing in fertilizers, pesticides, farm implements and equipments have also mushroomed. Apart from these, miscellaneous advantages like better road, transportation and marketing facilities have also been reported in the SCA after the introduction of canal irrigation. All these play a significant role in the regional development at large. Such environmental gain due to the introduction of agriculture have been leading to the intensification of agriculture have been by and large reported for many irrigation projects (Purohit et al, 1992).

Cereals (other than paddy) especially millets and ragi are the next major crops after paddy to be cultivated in the SCA, more so in SRBC command where the irrigation facilities are limited and the region is essentially a dry land.

In spite of the 71.37 per cent of the gross cropped area under irrigation, the intensity of cultivation in the SCA has remained 1.38 i.e. it has not yet reached the double cropping status. This discrepancy can be attributed to the fact that major portion of the cultivated land is under perennial crops like sugarcane, paddy and banana. During the months May-September no irrigation facilities are available in the region. During the course of field survey of the command region, it was discovered that cultivators tend to keep the sugarcane crops without harvesting for two-three years, in the absence of the issue of cutting orders from the sugar mills. As a result the cultivators have been suffering major losses for past some five years or so (Chapter 17.2).

The agricultural authorities claim that not only the SCA but the state as a whole has attained the status of self-sufficiency and is the largest producer of food grains. The literature and statistics speak otherwise. In the absence of a long term growth policy and non viability of water-intensive crops like rice and sugarcane, the area under rice shrank. Due to the absence of any perennial river, unresolved disputes of water sharing and erratic rainfall, T.N has slipped from being the sixth largest food grains to tenth (Viswanathan, 2000).

Thus the thrust, in the SCA should be to adopt a policy to switch to less water demanding crops like fruits and vegetables, oilseeds, cotton, pulses which are better suited to its climate and soil (Chapter 12) as per the soil survey report. There is almost nil to 5 per cent of the total cultivated land under crops of above said categories.

With the state having no long-term policy on agriculture and hefty reliance on populism –based sops viz 4.8 kg rice for a rupee; free mid day meals for school children, free electricity for the farmers, suffering a subsidy burden of Rs 2100 millions and promises to increase the price of sugarcane have proved to be extremely harmful for the health of the Tamil Nadu state's economy (Vishwanthan, 2000).

Instead of concentrating the command area with water demanding crops, the farmers should be educated on better alternatives available and provide them the support services needed in the form of awareness regarding their agro-climatic region, soil suitability and regular testing, provision of hybrid viable seeds, plant breeding and most of all introduction of sprinkler, drip and other modern methods of irrigation instead of traditional flooding. Socio-economic condition of the villagers downstream also affects the agricultural practices. Poverty and illiteracy drives the villagers to grow either paddy for their own requirements or groundnut and sugarcane, the cash crops, in the hope of getting good remunerations in the market.

Intensity of Cropping

The intensity of cropping in the study area is 1.38 as against the state's 1.17, SLBC command as the intensity of cultivation of 1.41 while SRBC command has 1.34. Chengam block (SRBC command) has the highest intensity of cultivation (1.68) followed by Chengam (SLBC command) with 1.5. Rishivandiyam block has the lowest of 1.23

Demographic Pressure

Population density per hectare of the sown area is 6 in the SCA. Thachampattu block (SLBC) has the highest population density of 8 people per hectare (Table—13.16). Chengam (SRBC command) has the lowest with 5 people per hectare.

Cultivators constitute 40.57 per cent of the working class in the region whereas 50.05 per cent are landless agricultural labourers (Figure—13.15). Rishivandiyam block has the highest percentage of landless labourers (58.24%). About 53 per cent of the agricultural labourers are females in the command area. With the exception of Thirukoilur and Chengam (SRBC command) blocks, more than 50 per cent of the agricultural labourers in all the blocks are females (Figure—13.16). Number of cultivators per hectare of the sown area in the SCA is 1 and in all the blocks the number is less than two. The male labourers of the commands migrate to the nearby towns and cities to earn their livelihood, thus resulting in the shortage of the labourers in the SCA as confirmed by many agriculturists. These migrants also pose a threat to the region by way of health as stated elaborately in chapter 15.

Table—13.16 Intensity of cultivation and demographic details

Blocks and Sathanur command	Intensity of cultivation	Population density	Population supported per unit net sown area	Cultivators per unit net sown area	Agricultural labourers per unit net sown area
Thachampattu (SLBC command)	1.34	5.50	8.26	1.45	1.72
Thirukoilur (SLBC command)	1.37	3.11	5.24	0.86	1.04
Chengam (SLBC command)	1.50	3.06	5.43	1.01	1.27
Rishivandiyam (SRBC command)	1.23	3.71	6.24	0.84	1.63
Sankarapuram (SRBC command)	1.40	4.37	6.74	1.51	1.40
Chengam (SRBC command)	1.68	2.33	4.75	1.23	0.38
SRBC command	1.34	3.73	6.25	1.12	1.40
SLBC Command	1.41	3.82	6.35	1.12	1.37
SCA	1.38	3.78	6.29	1.12	1.39
Tamil Nadu	1.17	-	-	-	-

14

Ground Water Quality

INTRODUCTION

All through the human history, ground water has catered to the ever-increasing demands of domestic, industrial and irrigation sectors world-wide. Its role in providing water for drinking and irrigation in the regions such as the southern peninsular India has been particularly crucial due to the absence of perennial rivers in these regions. The quality of ground water in an area is essentially a function of initial composition of water, precipitation, land use and the natural geology of the area. Activities, natural and anthropogenic, affect the regional ground water quantity and quality to a great extent. In this context we have conducted an assessment of the ground water quality of Sathanur command area (SCA). As we have elaborated earlier (Chapter 10) groundwater continues to play a dominant role in providing irrigation water to the Sathanur command inspite of the surface water resources generated as a result of the commissioning of Sathanur Reservoir Project (SRP). Ground water is also the principal source of drinking water in the region. The study also focuses on the impact of SRP on the groundwater quality of the SCA.

STATUS

Groundwater samples were collected from the tube wells in the villages in Sathanur Left Bank Canal (SLBC) and Sathanur Right Bank Canal (SRBC) commands. The samples were analyzed for various physico-chemical parameters using the methods detailed in Chapter 9 (Reservoir water quality). The results are presented as Tables 14.1 and 14.2. The World Health Organization (WHO) and Central Pollution Control Board (CPCB) specifications of heavy metals in drinking water are presented in Table—14.3. Tables—14.4 and 14.5 present the physico-chemical characteristics of the groundwater of some villages in SLBC and SRBC commands respectively during the period 1989-90.

The results are graphically depicted as Figures—14.1 to 14.28 for SRBC and SLBC commands.

IMPACTS

pH

pH values in the groundwater of the villages of SLBC command varied from 6.8 to 7.9. Palayanur recorded the least pH of 6.8 while the samples of Kandiankuppam and Kallottu had pH of 7.9. Ground water of Thenmudiyanur, Vanapuram and Melandhal had pH of more than 7.5 while Palayanur, Thenkarimbalur and Jambai recorded acidic pH in their groundwater. (Figure—14.1). In SRBC command the pH values in the groundwater were recorded as low as 6.6 in Porasapattu and as high as 7.7 in Melsiruvallur. Erudayampattu village recorded a pH of 6.9 in its groundwater (Figure—14.2.).

Electrical Conductivity (EC)

Highest values of EC in the groundwater were recorded in Palayanur ground water samples of SLBC command, 2035 and 2040 μ mhos cm^{-1} respectively (Figure—14.3). The EC concentrations in the samples drawn from Allikondapattu, Thachampattu, Periyakallipadi, Thenmudiyanur and Melandhal were greater than 1500 μ mhos cm^{-1}.

In SRBC command, the groundwater samples of S.Kolathur, Erudayampattu, Kadavanur and Pakkam recorded the EC concentration of more than 2000 μ mhos cm^{-1}. Vadaponparappi and Kidagudayampattu recorded EC values of more than 1500 μ mhos cm^{-1} in their groundwater (Figure 14.4). Natives of Pakkam and Kadavanur did complain of a unpalatable taste and pungent odor in the drinking water.

Table—14.1 Physico-chemical characteristics of ground water in SLBC command (1999 field survey)

	Palaya-nur		Allikonda-pattu		Thacham-pattu		Kandian-kuppam	Periya-kallipadi	Devariyar-kuppam	Kallottu	Velayam-pakkam	Sada-kuppam	
	S1	S2	S1	S2	S1	S2						S1	S2
pH	6.8	6.9	7.1	7.3	7.1	7.3	7.9	7.2	7.3	7.9	7.4	7.2	7.4
E.C (μ mho cm^{-1})	2035	2040	1980	1380	1630	1420	1000	1750	1370	900	1370	1250	760
Alkalinity (mg HCO_3 l^{-1})	244	280	239	340	204	217	245	285	290	264	258	357	287
Alkalinity (mg CO_3 l^{-1})	24	–	32	–	–	–	–	–	–	–	8	26	32
Total hardness (mg CaO_3 l^{-1})	448	478	324	308	424	440	468	384	272	248	404	336	152
Calcium hardness (mg $CaCO_3$ l^{-1})	328	392	300	248	316	356	372	348	256	180	300	89.7	96
Calcium (mg l^{-1})	131.4	157	120.2	99.3	122.6	142.6	149.07	139.5	102.5	72.1	120.2	22.4	38.4
Chloride (mg l^{-1})	238.8	256	224.8	183	167.1	173.1	169.1	194	200.9	99.5	187.1	177.9	121.9
Sulphate (mg l^{-1})	132	120	84	71.3	44	101	183	47	96	145	114	184	44
Nitrogen (Nitrate) (mg l^{-1})	11.7	12.3	4.4	3.9	2.7	4.1	6.3	3.4	0.5	4	14.7	15.1	0.6
Phosphate (mg l^{-1})	2.8	3.1	1.8	.9	1.1	1.3	1.9	1.7	1.3	0.9	1.9	1.08	0.9
Sodium (ppm)	29.8	30.4	28	31.4	30.6	–	–	47.5	27.1	–	28	24.6	28.7
Potassium (ppm)	4.1	4.4	2	2.2	4.3	–	–	21.6	12.9	–	2.6	3.8	4.1
Copper (ppb)	12	11	8	8	6	–	–	8	19	–	8	4	7
Chromium (ppb)	3	13	3	4	3	–	–	4	3	–	8	2	2
Zinc (ppb)	2506	2550	160	132	84	–	–	350	111	–	75	64	72
Cadmium (ppb)	14	16	5	10	2	–	–	2	1	–	< 1	< 1	< 1

(Continued...)

	Valavachanur	Unnamalai palayam	Agaram pallipattu	Thenmudiyanur			Vanapuram				Melan-dhal		Thenkarim balur	Jambai	Kangai-yanur	Edath-anur
				S1	S2	S3	S1	S2	S3	S4	S1	S2				
pH	7.0	7.3	7.3	7.4	7.3	7.7	7.3	7.6	7	7.3	7.5	7.0	6.9	6.9	7.2	7.1
E.C (µ mho cm^{-1})	880	950	860	1580	1050	1190	950	1250	750	820	1790	1520	970	1020	940	1530
Alkalinity (mg HCO_3 l^{-1})	270	286	282	500	350	380	375	430	375	425	450	525	350	475	450	266
Alkalinity (mg CO_3 l^{-1})	44	40	36	–	12	18	28	14	22	12	–	–	–	12	–	–
Total hardness (mg CaO_3 l^{-1})	252	244	134	360	196	204	260	340	220	272	88	332	232	248	208	564
Calcium hardness (mg $CaCO_3$ l^{-1})	200	156	184	180	116	180	160	100	176	192	68	264	188	228	104	276
Calcium (mg l^{-1})	80.2	62.5	73.7	72.1	46.5	72.1	64.1	40.1	70.5	76.9	27.3	105.8	75.4	91.4	41.6	110.6
Chloride (mg l^{-1})	97.5	101.9	99.5	203.9	159.9	187.9	185.9	179.8	171.9	197.9	297.9	233.9	197.9	195.9	189.9	252.7
Sulphate (mg l^{-1})	89	38	43.1	109.3	96.2	118	82	76	64	61.2	131.8	127.4	49	74.7	41.8	53.9
Nitrogen (Nitrate) (mg l^{-1})	26	28	12	4.7	3.2	1.9	5.9	12.7	3.2	3.8	8.1	17.4	8.7	5.2	4.7	7.8
Phosphate (mg l^{-1})	0.9	0.8	0.9	0.8	0.9	0.7	0.6	1.2	0.9	0.9	1.1	1.3	0.6	1.6	0.9	2.7
Sodium (ppm)	–	27.1	21	–	–	–	–	–	–	–	–	–	–	–	–	23.5
Potassium (ppm)	–	1.8	1.4	–	–	–	–	–	–	–	–	–	–	–	–	2.1
Copper (ppb)	–	8	6	–	–	–	–	–	–	–	–	–	–	–	–	16
Chromium (ppb)	–	2	1	–	–	–	–	–	–	–	–	–	–	–	–	52
Zinc (ppb)	–	56	249	–	–	–	–	–	–	–	–	–	–	–	–	68
Cadmium (ppb)	–	1	1	–	–	–	–	–	–	–	–	–	–	–	–	1

S1, S2 – Sample   – Not analyzed

Table—14.2 Physico chemical characteristics of ground water in SRBC command (1999 field survey)

	Melsiru-vallur		Moongilthu-raipattu	Vadapon-parappi	S.Kolathur	Porasa-pathu	Erudayam-pattu	Vadama-mandur	Arambaram-pattu	Jambodai
	S1	S2								
pH	7.7	7.3	7	7.2	7	6.6	6.9	7.1	7.3	7.1
E.C (μ mho cm^{-1})	880	1350	1200	1800	2070	910	2080	750	1010	950
Alkalinity (mg HCO_3 l^{-1})	800	315	475	475	196	205	255	285	260	280
Alkalinity (mg CO_3 l^{-1})	–	24	–	–	–	–	–	22	–	–
Total hardness (mg CaO_3 l^{-1})	212	320	396	520	780	320	560	272	352	340
Calcium hardness (mg $CaCO_3$ l^{-1})	196	296	276	360	396	252	332	204	324	212
Calcium (mg l^{-1})	78.5	118.6	110.6	144.2	158.6	100.9	133.04	81.7	134.6	84.9
Chloride (mg l^{-1})	183.9	192.1	125.9	259	352.2	105.4	435.8	77.6	101.4	105.4
Sulphate (mg l^{-1})	73.2	84	131	156	21.4	27.2	99	24	44.9	30:2
Nitrogen (Nitrate) (mg l^{-1})	3.1	5.9	11.9	9.8	10.3	7.9	11.8	5.3	2.8	6
Phosphate (mg l^{-1})	0.8	0.5	1.9	2.1	1.4	1.3	0.9	1.4	0.9	2.1
Sodium (ppm)					29.8	16.6	34.1	12.6	22.3	16
Potassium (ppm)					4.9	2.9	.7	8.4	5.5	2.3
Copper (ppb)					15	19	39	17	9	13
Chromium (ppb)					4	70	8	8	6	13
Zinc (ppb)					151	48	145	73	122	102
Cadmium (ppb)					< 1	1	2	1	1	1

(Continued...)

	Kadavanur		Pakkam	Athiyur	Manar-palayam	Periya-kolliyur	Arur	Kidagudu-gampattu	Varagur
	S1	S2							
pH	7.1	7.2	7.2	7.1	7.4	7.1	7.2	7	7.4
E.C (µ mho cm^{-1})	2300	2030	2020	1160	1400	1290	1120	1750	530
Alkalinity (mg HCO_3 l^{-1})	310	291	298	272	315	280	295	231	205
Alkalinity (mg CO_3 l^{-1})	–	–	20	–	12	–	12	20	–
Total hardness (mg CaO_3 l^{-1})	592	540	608	404	420	372	468	596	240
Calcium hardness (mg $CaCO_3$ l^{-1})	420	344	396	264	360	248	300	212	212
Calcium (mg l^{-1})	168.3	137.8	158.6	105.7	144.3	99.4	120.2	84.9	84.9
Chloride (mg l^{-1})	340.3	403.9	352.2	334	327	159.2	121.3	338.3	57.7
Sulphate (mg l^{-1})	254	216	122.6	98.3	89	62	76	118	48
Nitrogen (Nitrate) (mg l^{-1})	13.7	10.2	19	7.5	2.7	9	1.5	3.4	1.3
Phosphate (mg l^{-1})	1.1	1.2	1.3	0.9	1.2	2.3	2.4	0.6	0.9
Sodium (ppm)	31.2	–	31.7	22.6	33.9	25	22	38	–
Potassium (ppm)	13.2	–	6.6	2.5	15.5	1.6	3.3	3.3	–
Copper (ppb)	16	–	18	13	19	12	14	9	–
Chromium (ppb)	6	–	5	100	6	7	2	2	–
Zinc (ppb)	2184	–	160	170	335	101	1921	86	–
Cadmium (ppb)	30	–	< 1	< 1	2	1	16	2	–

S, S2 – Sample s – Not analyzed

Table—14.3 WHO and CPCB drinking water standards

	WHO (1984)		CPCB	
	HDL*	MPL°	HD L*	MPL°
Copper (ppb)	50	1500	–	1000
Chromium (ppb)	75	200	75	200
Zinc (ppb)	–	–	–	1000
Cadmium (ppb)	5	–	–	10

* Highest desirable limit

° *Maximum permissible limit*

Table—14.4 Physico-chemical characteristics of ground water in SLBC command (1989-90)

S. No.	Village	EC μ mho cm^{-1}	pH	Sulphate mg l^{-1}	Chloride mg l^{-1}	Nitrate mg l^{-1}	Hardness mg l^{-1}
1.	Edathanur	590	8.2	15.8	65.67	.42	164
2.	Agarampallipattu	616	8.2	16.8	66.03	2.9	160
3.	Vanapuram	1055	8.1	39.8	138.45	3.5	221
4.	Palaiyanur	1256	8.2	50.4	222.2	2.8	287
5.	Jambai	645	8.1	35.64	85.9	3.2.	169
6.	Thachampattu	1246	8.18	39.84.	210.8	12.6	367

Table-14.5 Physico-chemical characteristics of ground water in SRBC command (1989-90)

Sl. No.	Village	EC μ mho cm^{-1}	pH	Sulphate mg l^{-1}	Chloride mg l^{-1}	Nitrate mg l^{-1}	Hardness mg l^{-1}
1.	Arumparamapattu	1034	8.2	38.8	14.48	2.1	250
2.	Kaduvanur	638	8.3	15.36	68.1	0.7	217
3.	Athiyur	498	8.04	7.68	24.8	0.84	-
4.	Varagur	774	8.2	8.64	19.8	1.26	137
5.	S.Kolathur	642	8.3	22.5	60.35	0.84	288

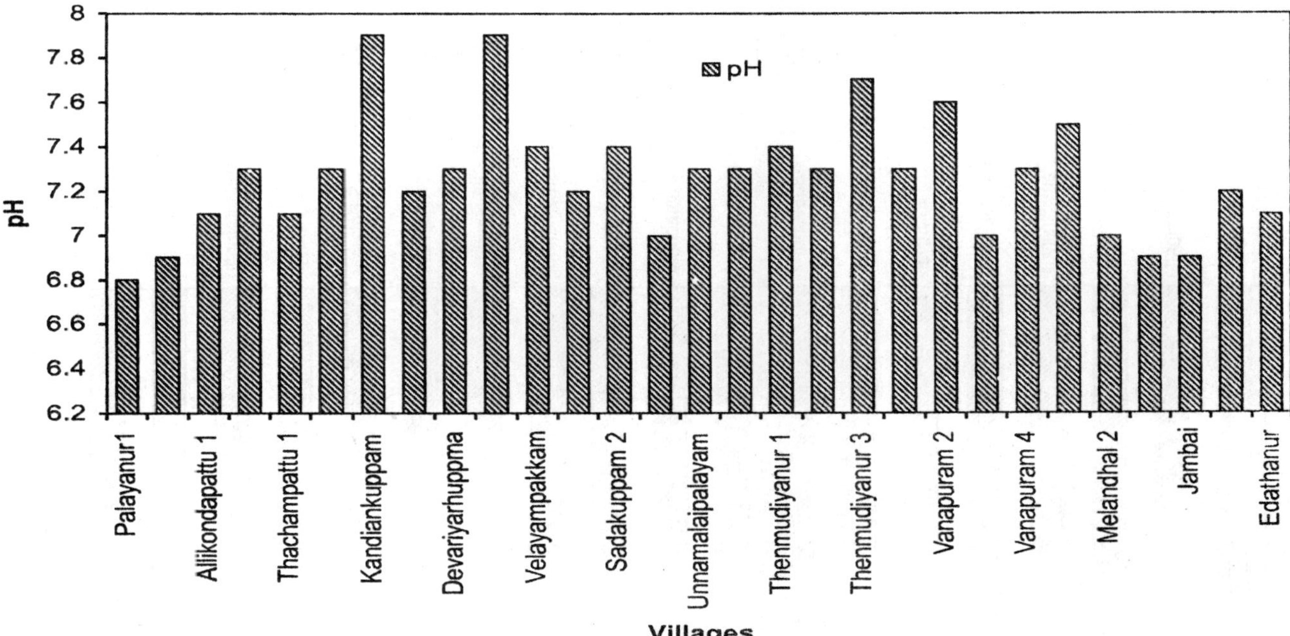

Fig. 14.1: pH values in the ground water samples in SLBC command

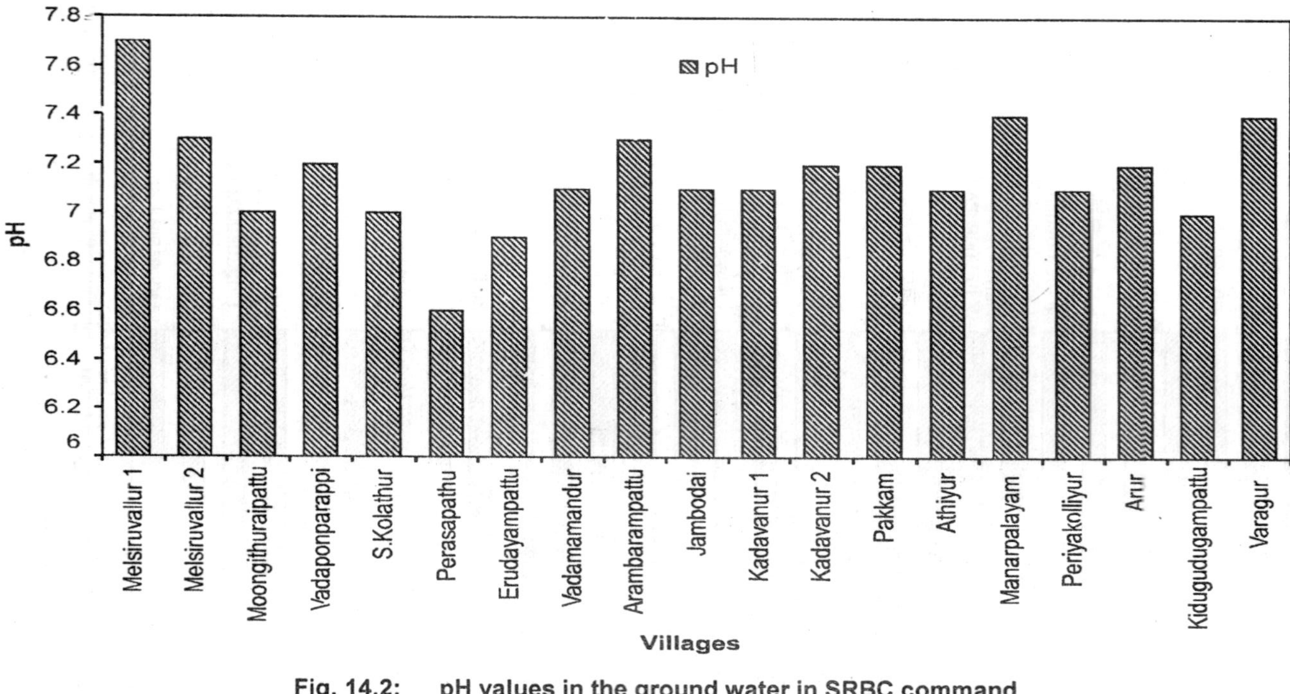

Fig. 14.2: pH values in the ground water in SRBC command

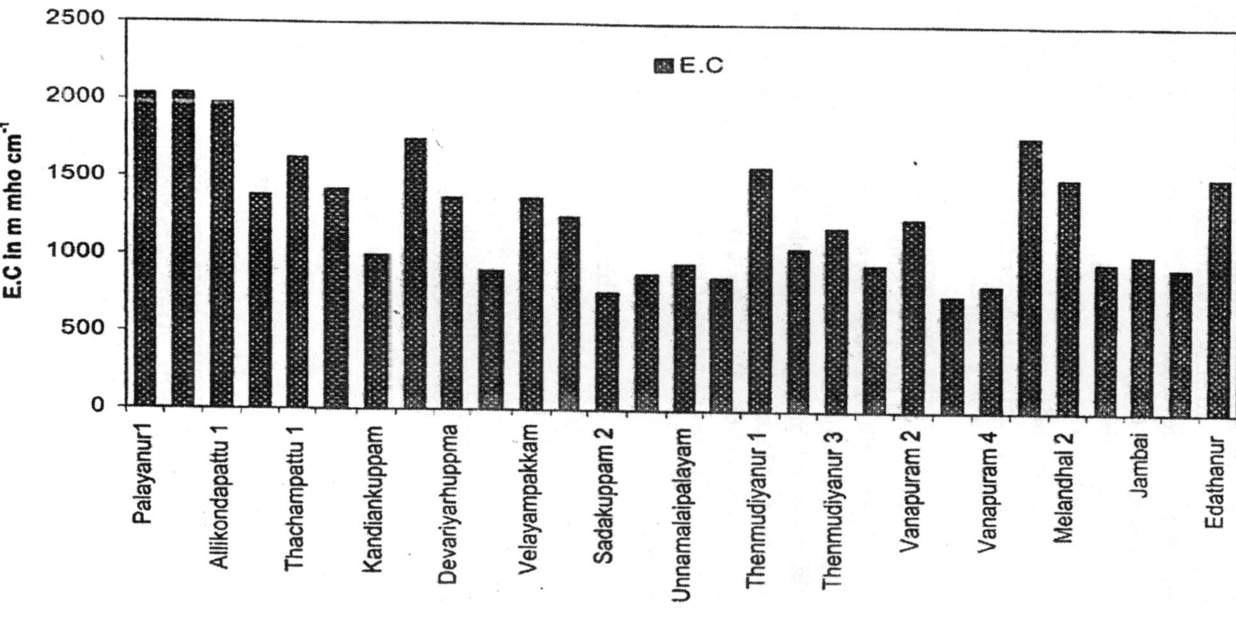

Fig. 14.3: pH values in the ground water samples in SLBC command

Alkalinity

Alkalinity encountered in the ground water samples of SCA was of bicarbonate nature. In SLBC command the alkalinity values in the groundwater sample ranged from 204 mg l^{-1} (Thachampattu) to 525 mg l^{-1} in Melandhal (Figure—14.5). All the villages recorded the alkalinity values more than 200 mg l^{-1} in their groundwater samples thus making the water unfit for drinking. The villages, Sadakuppam, Thenmudiyanur, Vanapuram, Melandhal, Thenkarimbalur and Kangaiyanur recorded a value of more than 300 mg l^{-1} in their groundwater samples.

Alkalinity values in the groundwater samples of SRBC command varied from 196 mg l^{-1} (S.Kolathur) to 800 mg l^{-1} in Melsiruvallur (Figure—14.6). Except for S.Kolathur all the other villages recorded alkalinity values in the groundwater well above the permissible limit of drinking water (Table—9.23). Groundwater in Melsiruvallur, Maongilthuraipattu, Kadavanur, and Manarpalayam recorded the alkalinity values of more than 300 mg l^{-1} in their groundwater samples (Figure—14.6).

Total Hardness

The total hardness values in the groundwater of SLBC command ranged from 88 mg l^{-1} (Melandhal) to 564 mg l^{-1} in Edathanur (Figure—14.7). Concentration of more than 300 mg l^{-1} was observed in the groundwater sample of Palayanur. Allikondapattu, Thachampattu, Kandiankuppam, Periyakallipadi, Velayampakkam, Sadakuppam, Thenmudiyanur, Vanapuram, Melandhal and Edathanur. The water in these villages is thus unfit for drinking as per the BIS criteria for drinking water (Table—9.23).

The hardness values in SRBC command varied from 212 mg l^{-1} (Melsiruvallur) to 780 mg l^{-1} in S. Kolathur (Figure—14.8). Values of more than 300 mg l^{-1} were observed in the water samples of Melsiruvallur, Moongilthuraipattu, Vadaponparappi, S.Kolathur, Porasapattu, Erudayampattu, Arambarampattu, Jambodai, Kadavanur, Pakkam, Athiyur, Manarpalayam and Arur (Figure—14.8). As per the permissible limits provided by the Indian standards of drinking water (Table—9.23) the water of these villages is unfit for drinking.

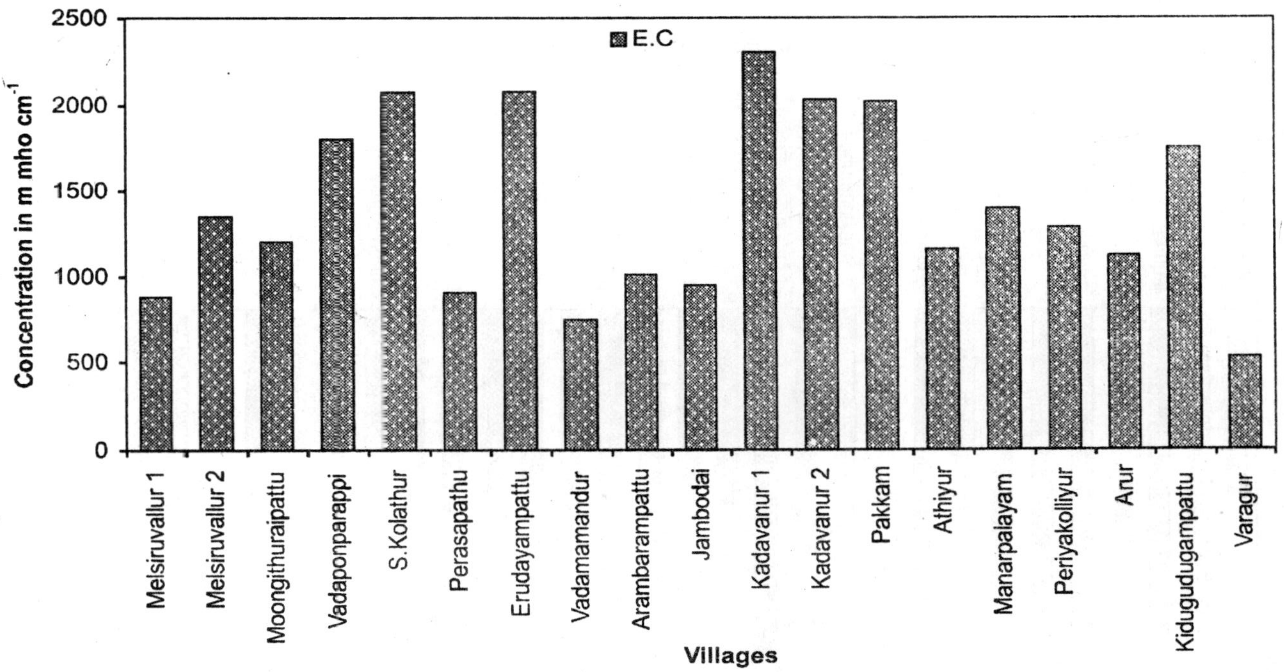

Fig. 14.4: E.C. values in the ground water in SRBC command

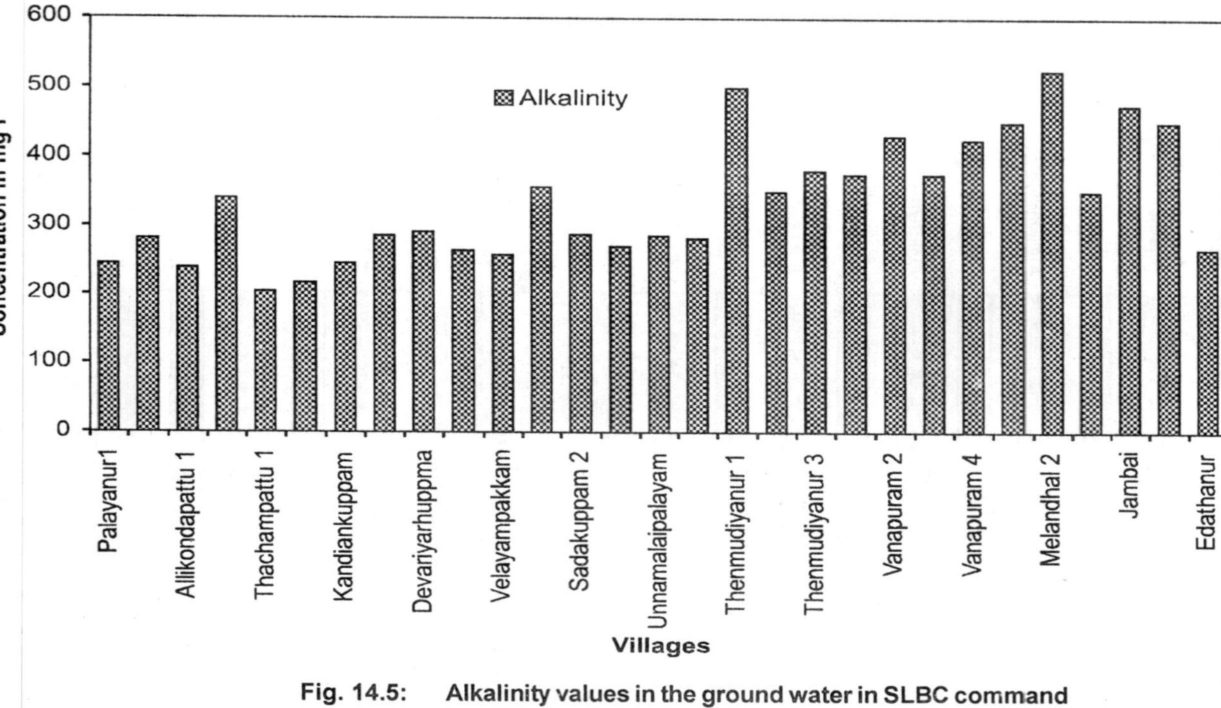

Fig. 14.5: Alkalinity values in the ground water in SLBC command

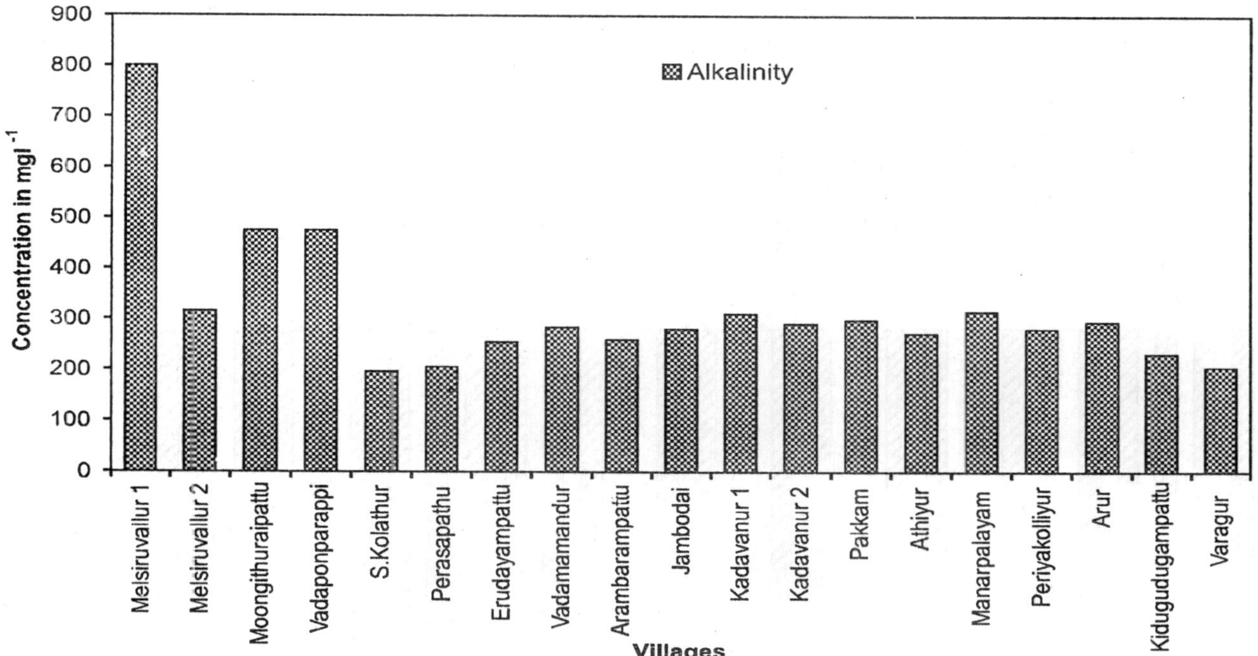

Fig. 14.6: Alkalinity values in the ground water in SRBC command

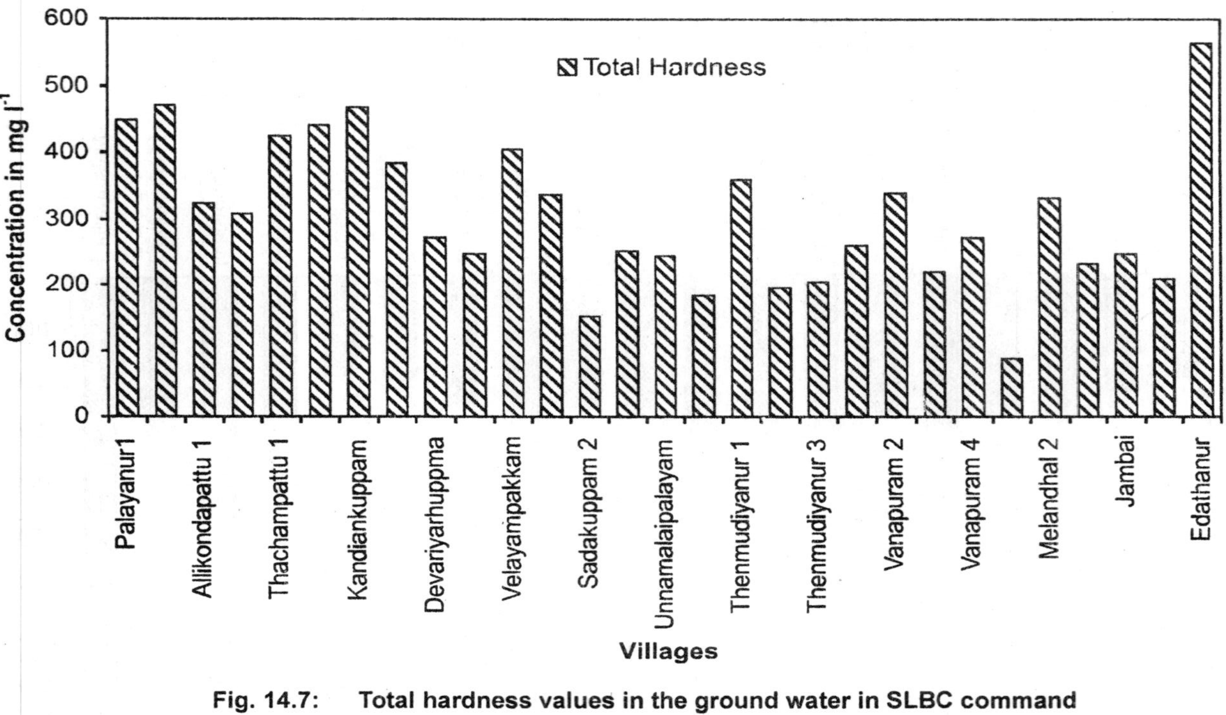

Fig. 14.7: Total hardness values in the ground water in SLBC command

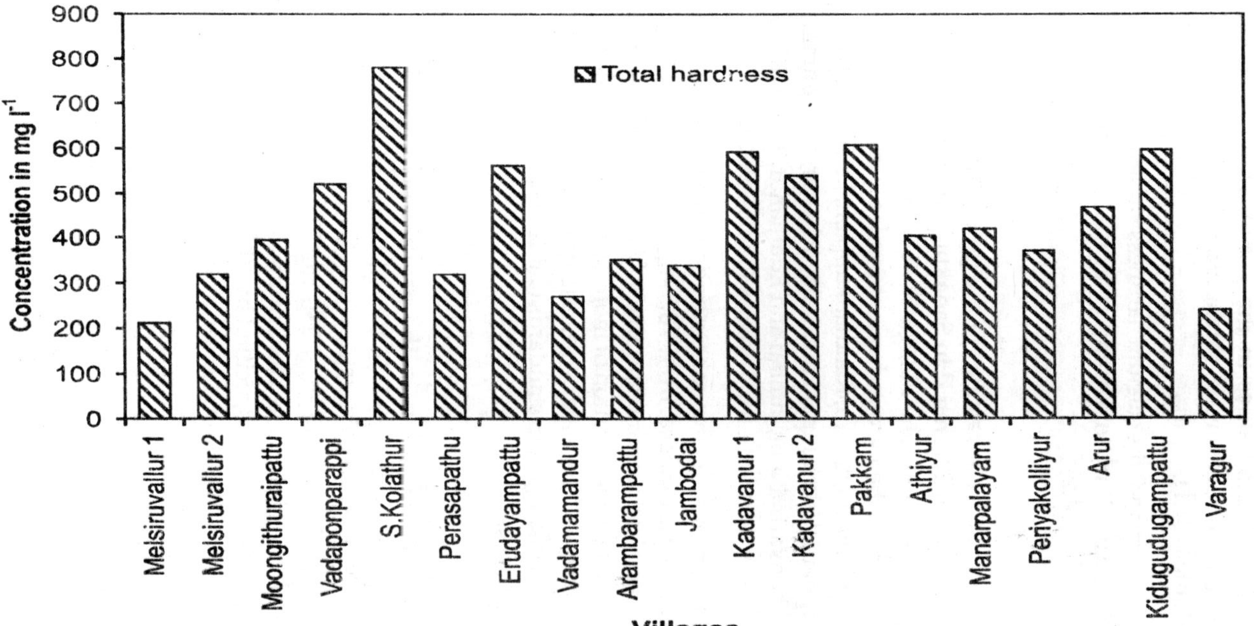

Fig. 14.8: Total hardness values in the ground water in SRBC command

Calcium Hardness

Calcium hardness values in the groundwater of SLBC comand ranged from 68 mg l^{-1} (Melandhal) to 392 mg l^{-1} in Palayanur (Table—14.1).

In SRBC command, the calcium hardness values varied from 196 mgl^{-1} (Melsiruvallur) to 420 mg l^{-1} in the groundwater of Kadavanur (Table—14.2).

Calcium

The concentration of calcium in the groundwater of SLBC command ranged from 22.4 mg l^{-1} in Sadakuppam to 157 mg l^{-1} in Palayanur (Figure—14.9). The villages that recorded more than 75 mg l^{-1} of calcium in their groundwater samples were Palayanur, Allikondapattu, Thachampattu, Kandiankuppam, Periyakallipadi, Devariyarkuppam, Velayampakkam, Valavachanur, Vanapuram, Melandhal, Thenkarimbalur, Jambai and Edathanur. The water of these villages was thus unfit for drinking (Table—9.23).

In SRBC command, the concentration of calcium in the groundwater samples ranged from 78.5 mg l^{-1} to 158.6 mg l^{-1} in Melsiruvallur and Kadavanur respectively (Figure—14.10). All the villages registered a concentration of more than 75 mgl^{-1} of calcium in their groundwater. The water with calcium content more than 75 mg l^{-1} as per drinking water standards (Table—9.23) is thus unfit for drinking purpose.

Chloride

Chloride concentration in the ground water of SLBC command varied from 97.5 mg l^{-1} in Valavachanur to 297.9 mg l^{-1} in Melandha! (Figure—14.11). The villages that recorded the chloride concentration more than 250 mg l^{-1} were Palayanur, Melandhal and Edathanur. The concentration of chloride higher than the permissible limit as specified by BIS for drinking water (Table—9.23) makes the groundwater in these regions unfit for drinking.

The chloride values in groundwater samples of SRBC command ranged from 57.7 mg l^{-1} to 435.8 mg l^{-1} Erudayampattu (Figure—14.12). Villages with chloride concentration more than 250 mg l^{-1} in their groundwater were Vadaponparappi, S. Kolathur,

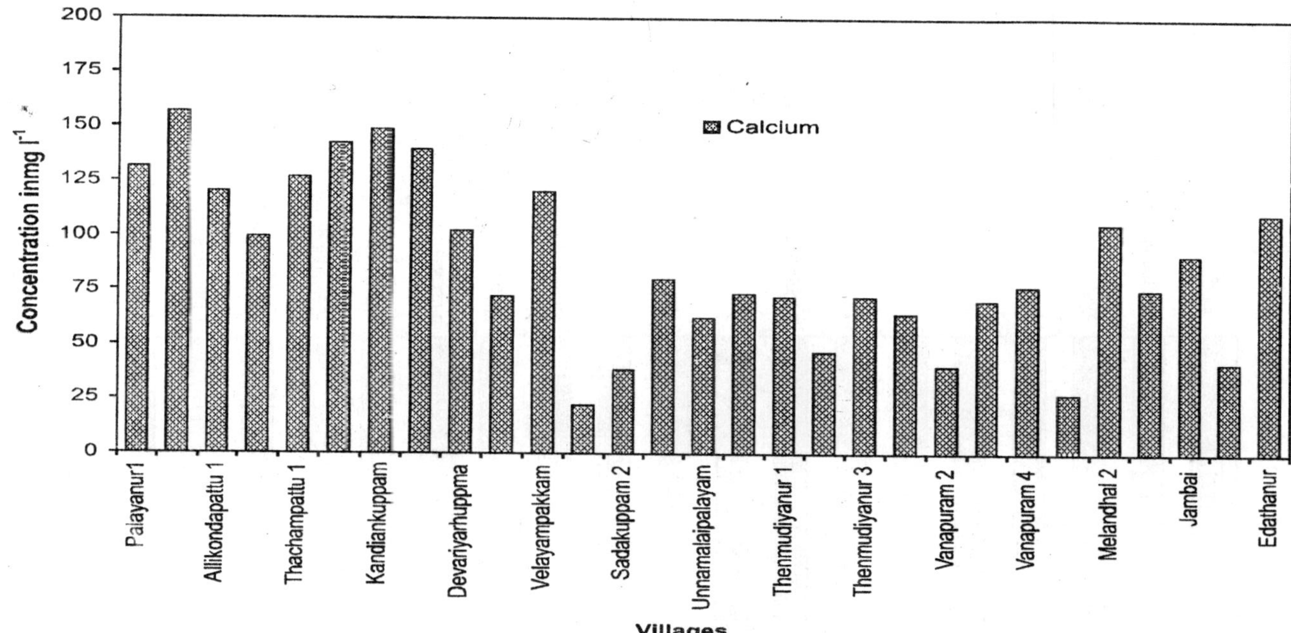

Fig. 14.9: Calcium values in the ground water in SLBC command

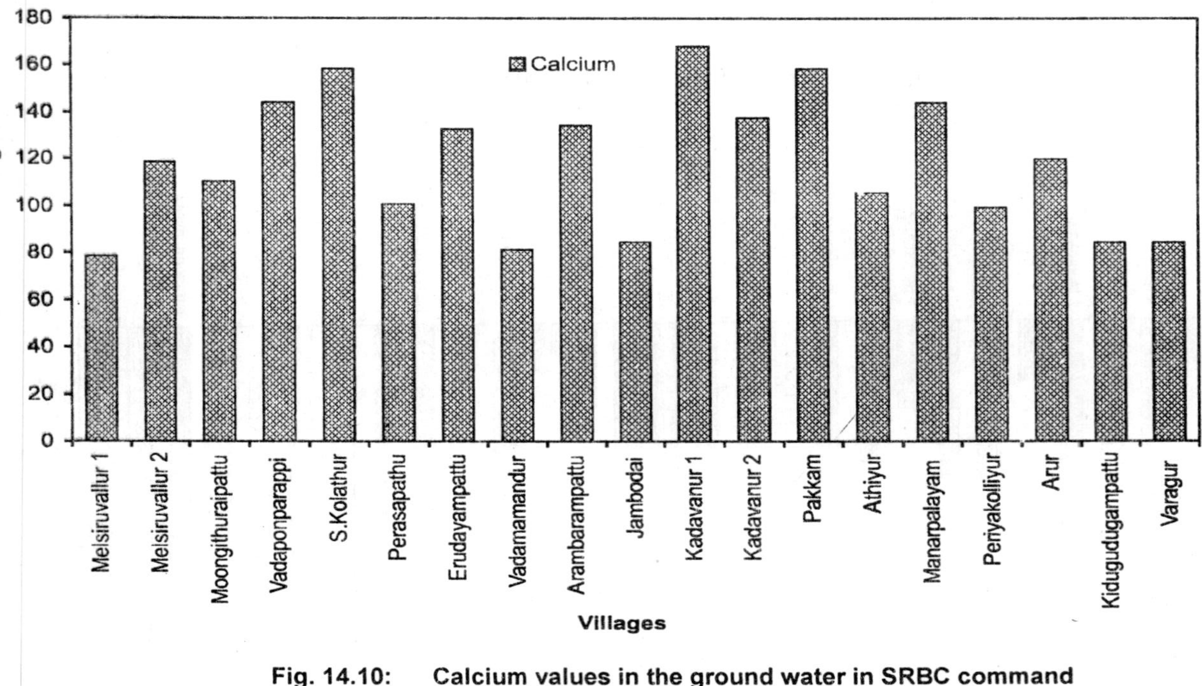

Fig. 14.10: Calcium values in the ground water in SRBC command

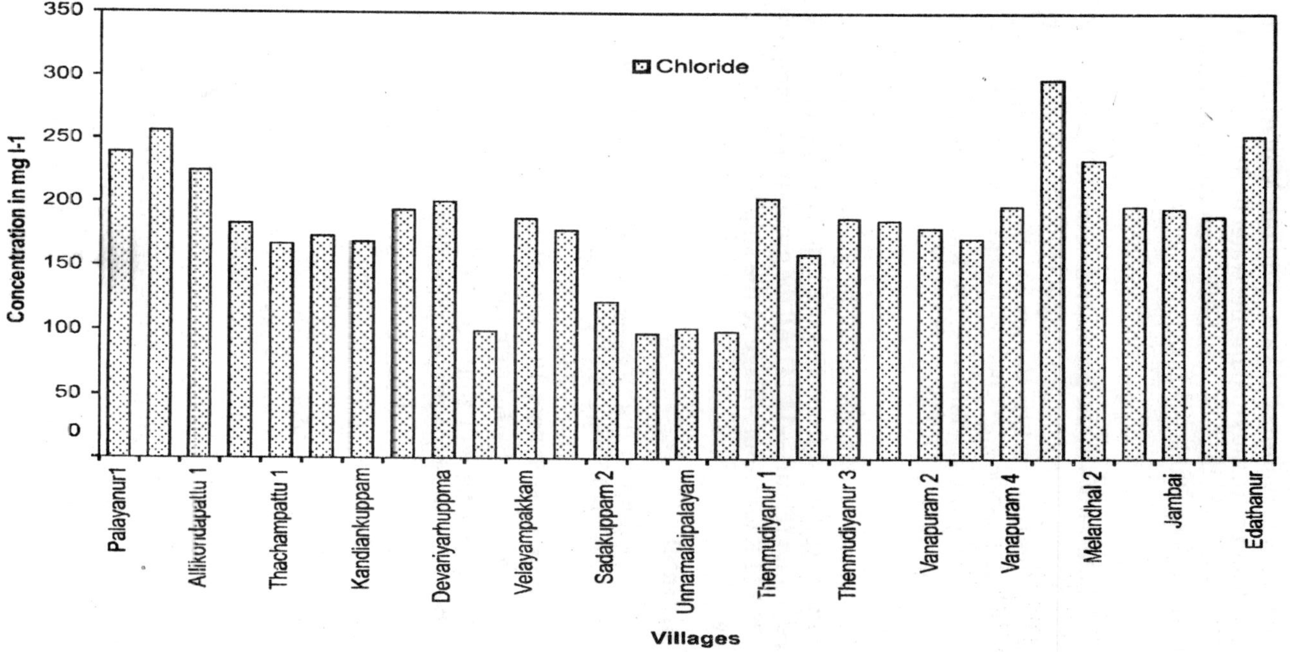

Fig. 14.11: Chloride values in the ground water in SLBC command

Erudayampattu, Kadavanur, Pakkam, Athiyur, Manarpalayam and Kidagudayampattu, thus making the water unfit for drinking purpose. The water in both SLBC and SRBC commands is suitable for irrigation purpose (Table—9.23).

Sulphate

The sulphate values in the groundwater of SLBC command ranged form 38 mg l^{-1} in Unnamalaipalayam to 184 mg l^{-1} in Sadakuppam (Figure—14.13). Values of all the groundwater samples were less than 200 mg l^{-1}-the permissible level as laid down by BIS for drinking water (Table—9.23). Kandiankuppam and Sadakuppam had more than 180 mg l^{-1} of sulphate values in their groundwater.

The concentration of sulphate in the groundwater samples of SRBC command ranged from 21.4 mg l^{-1} in S. Kolathur to 254 mg l^{-1} in Kadavanur (Figure—14.14). All the groundwater samples except for that of Kadavanur village re the sulphate content less than 200 mg l^{-1}.

Nitrogen (nitrate)

The nitrate values in the ground water samples of SLBC command ranged from 0.5 mg l^{-1} in Devariyarkuppam to 28 mgl^{-1} in Unnamalaipalayam (Figure—14.15). The water samples with nitrate concentration more than 10 mgl^{-1} belonged to Palayanur, Velayampakkam, Sadakuppam, Valavachanur, Unnamalaipalayam, Agarampallipattu and Vanapuram villages, thus rendering water unfit for consumption (Table—9.23).

In SRBC command, the nitrate concentration in the ground water varied from 1.3 mg l^{-1} in Arur to 19 mg l^{-1} in Pakkam (Figure 14.16). The villages that recorded more than 10 mg l^{-1} of nitrate in their groundwater were Moongilthuraipattu, S. Kolathur, Erudayampattu, Kadavanur and Pakkam. The water of these villages was thus unfit for human consumption (Table—9.23).

Phosphate

The phosphate concentration in the groundwater of SLBC command ranged from 0.6 mg l^{-1} (in Vanapuram and Thenkarimbalur) to 3.1 mg l^{-1} in Palayanur (Figure—14.17). In the villages of SRBC command the values varied from 0.5 mg l^{-1} in Melsiruvallur to 2.4 mg l^{-1} in Arur (Figure—14.18).

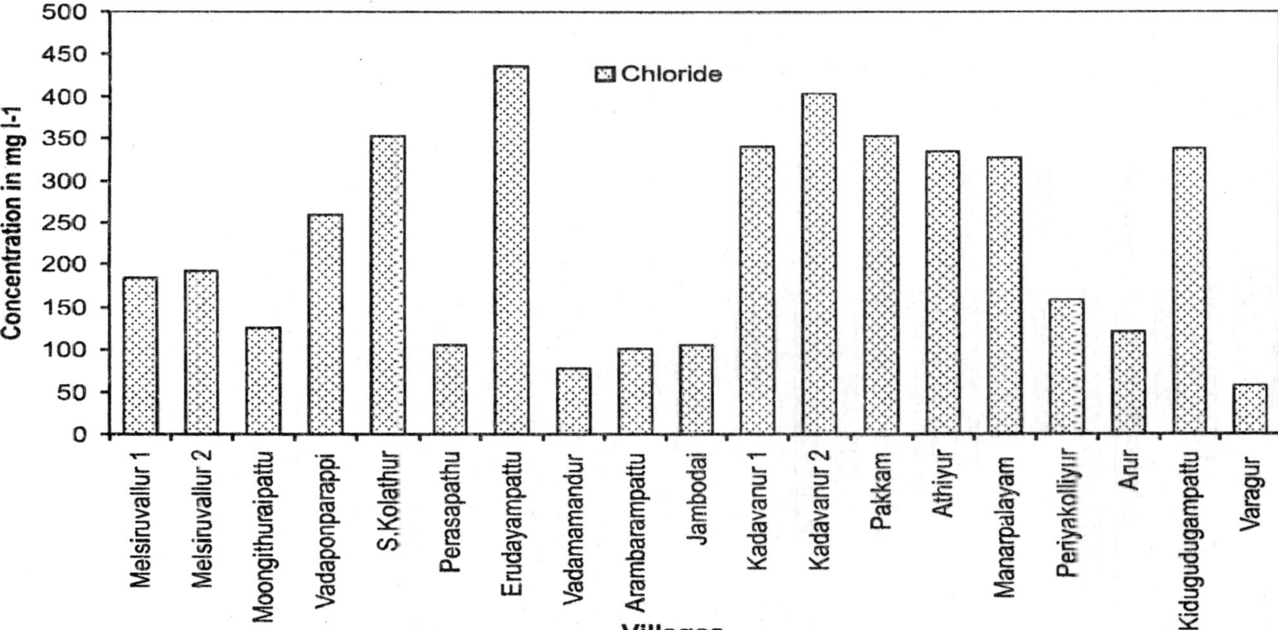

Fig. 14.12: Chloride values in the ground water in SRBC command

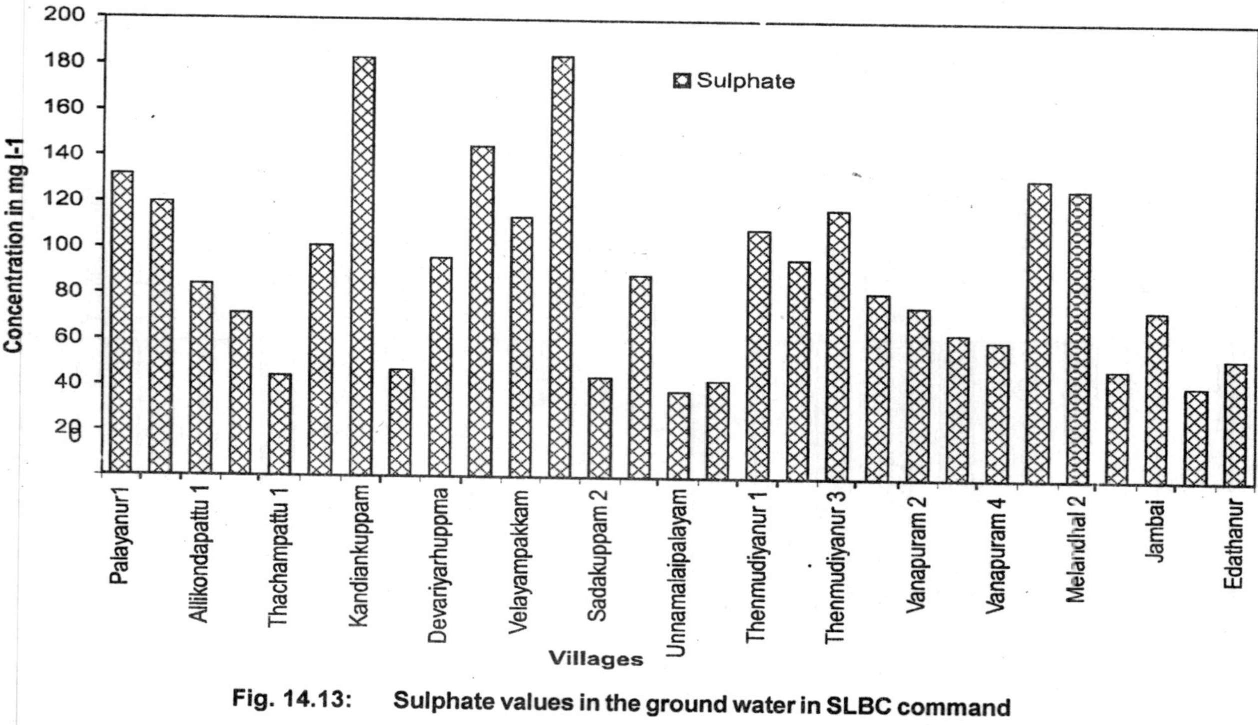

Fig. 14.13: Sulphate values in the ground water in SLBC command

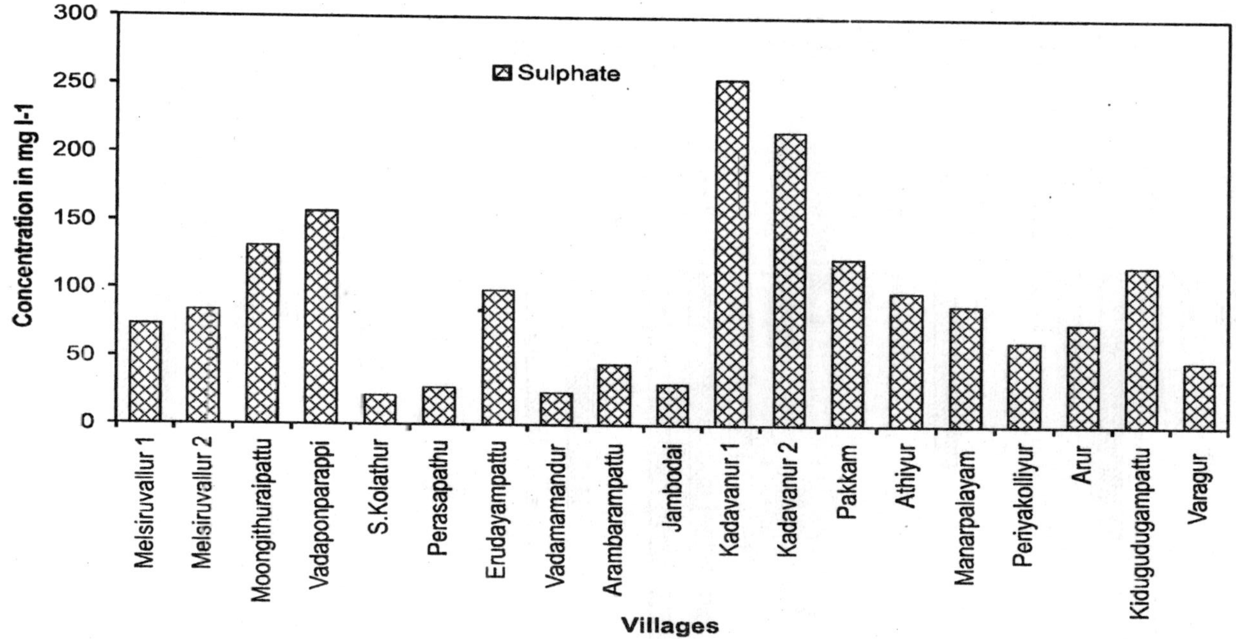

Fig. 14.14: Sulphate values in the ground water in SRBC command

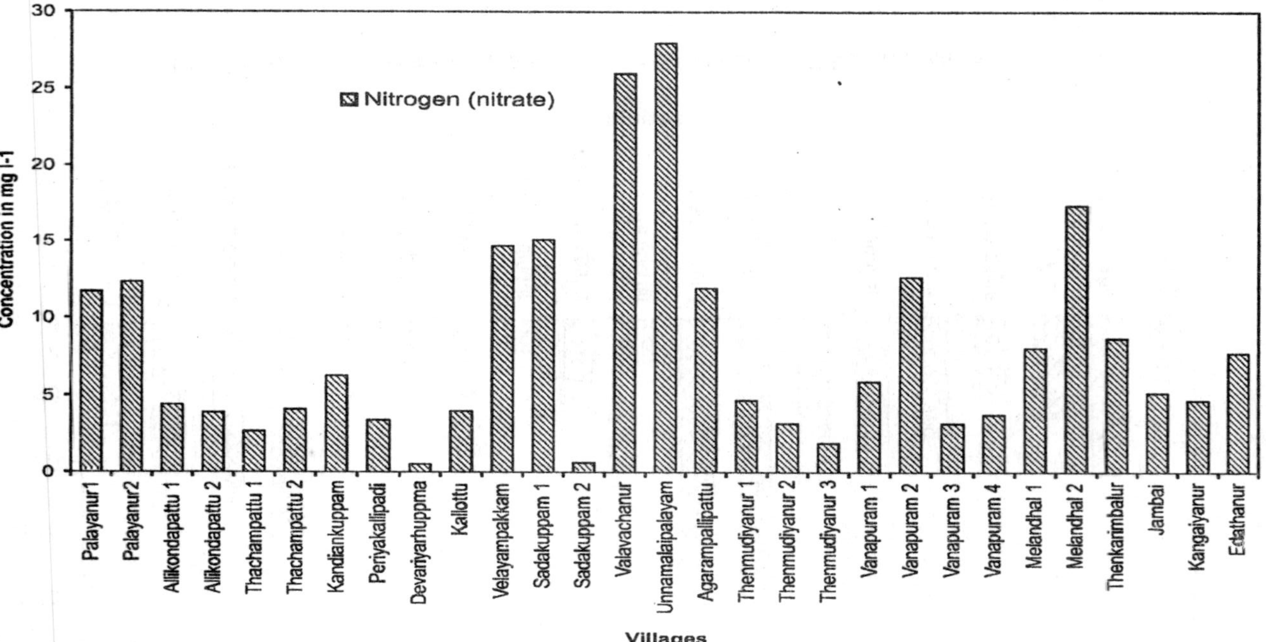

Fig. 14.15: Nitrogen (niterate) values in the ground water in SLBC command

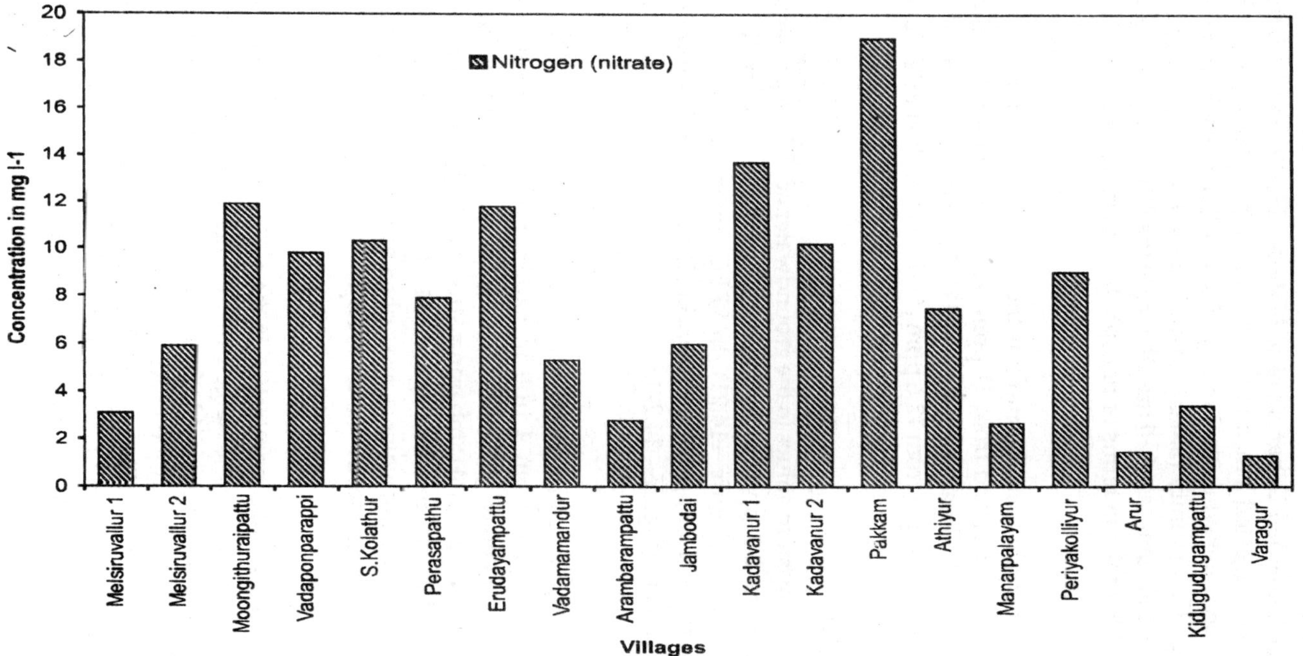

Fig. 14.16: Nitrogen (nitrate) values in the ground water in SRBC command

The Canadian Department of National Health and Welfare (1969) has suggested a maximum limit of 0.2 mg l^{-1} for phosphate in drinking water while the limit specified by that of the European Economic Community (Smeats and Amavis, 1981) is 0.54 mg l^{-1}. Groundwater samples of all the villages in the SCA exceeded these permissible limits in terms of phosphate concentration.

Sodium

The sodium concentration in the groundwater samples of SLBC command varied from 21 ppm (Agarampallipattu) to 47.5 ppm in Periyakallipadi (Figure—14.19). In SRBC command the concentration ranged from 12.6 ppm (Vadamamandur) to 38 ppm in Kidagudayampattu (Figure—14.20).

Potassium

The Potassium values in the groundwater of SLBC command ranged from 1.4 ppm in Agarampallipattu to 21.6 ppm in Periyakallipadi (Figure—14.21). In SRBC command the values ranged from 1.6 ppm in Periyakolliyur to 15.5 ppm in Manarpalayam (Figure—14.22).

Copper

The copper concentration in the ground water of SLBC command ranged from 4 ppb in Sadakuppam to 19 ppb in Devariyarkuppam (Figure—14.23). In SLBC command the values ranged from 9 ppb in Arambarampattu and Kidagudayampattu to 39 ppb in Erudayampattu (Figure—14.24). The values are within the range for drinking water as prescribed by the WHO and CPCB drinking water standards (Table—14.3).

Chromium

The chromium concentration in the groundwater samples of SLBC was recorded in the range 1 ppb in Agarampallipattu to 16 ppb in Edathanur (Figure—14.25). The values in the groundwater of SRBC ranged from 2 ppb in Arur and Kidagudayampattu and 100 ppb in Athiyur (Figure—14.26). Except for Athiyur, the concentrations of chromium in all other groundwater samples were within the prescribed limit for drinking water (Table—14.3).

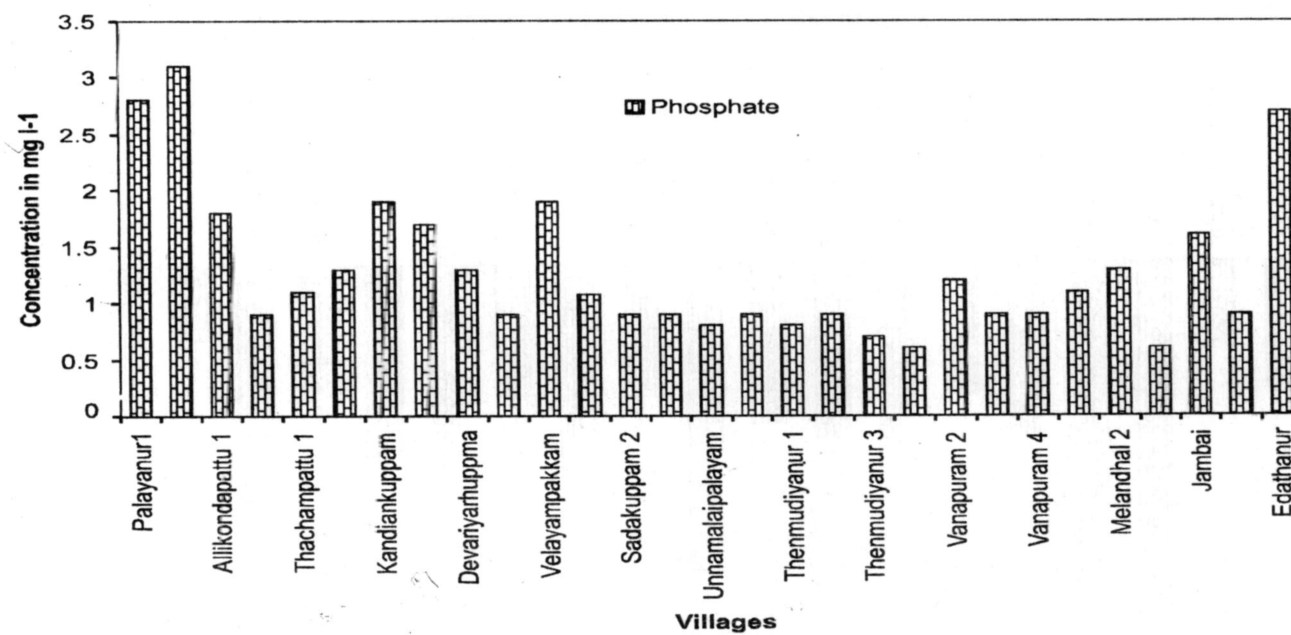

Fig. 14.17: Phosphate values in the ground water in SLBC command

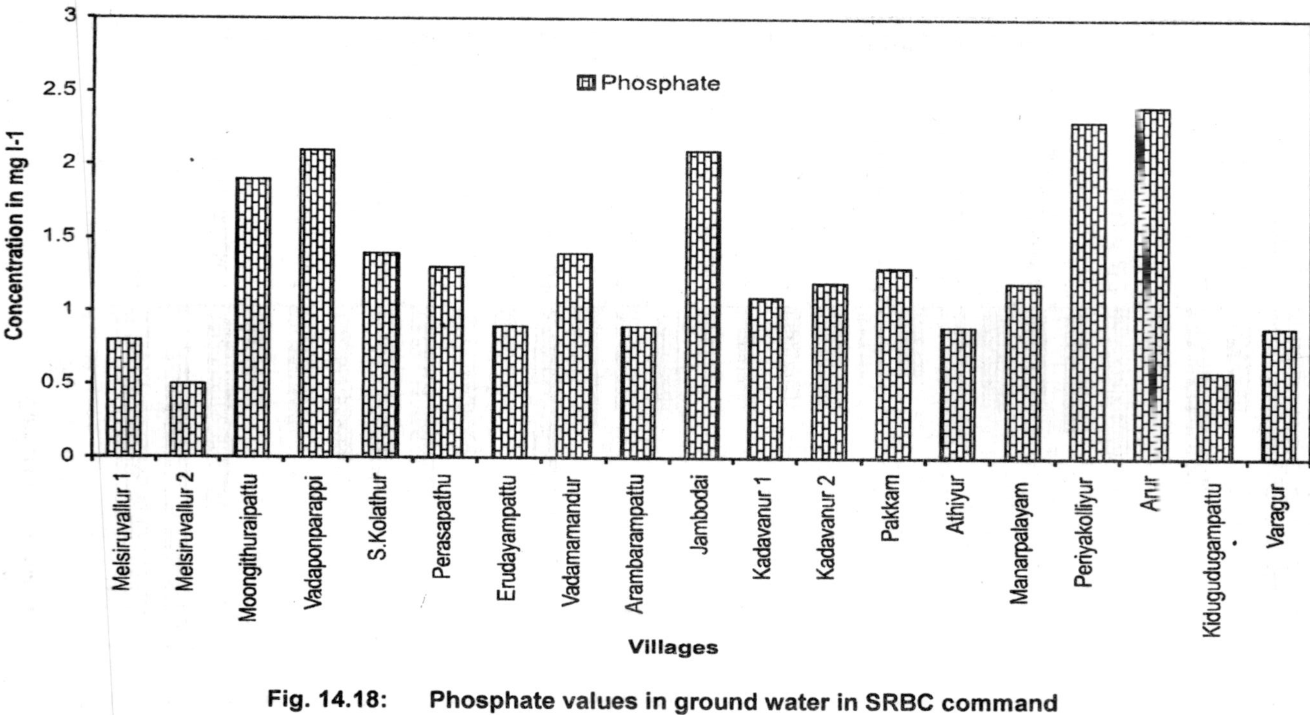

Fig. 14.18: Phosphate values in ground water in SRBC command

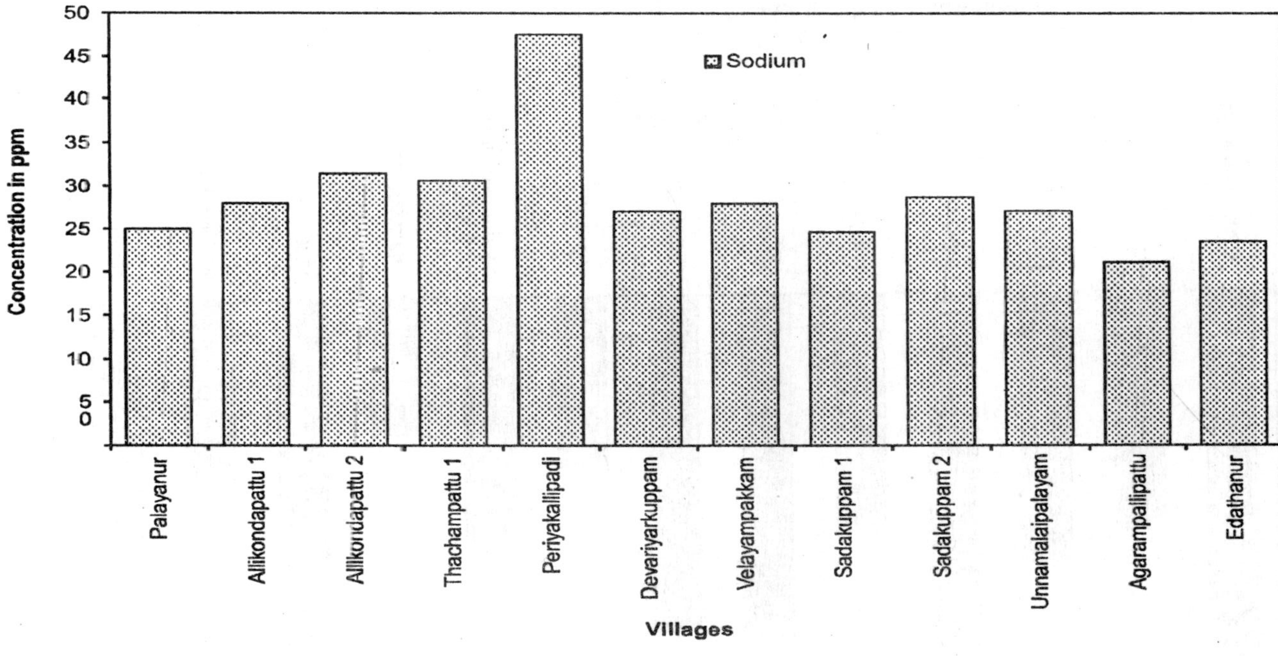

Fig. 14.19: Sodium values in ground water in SLBC command

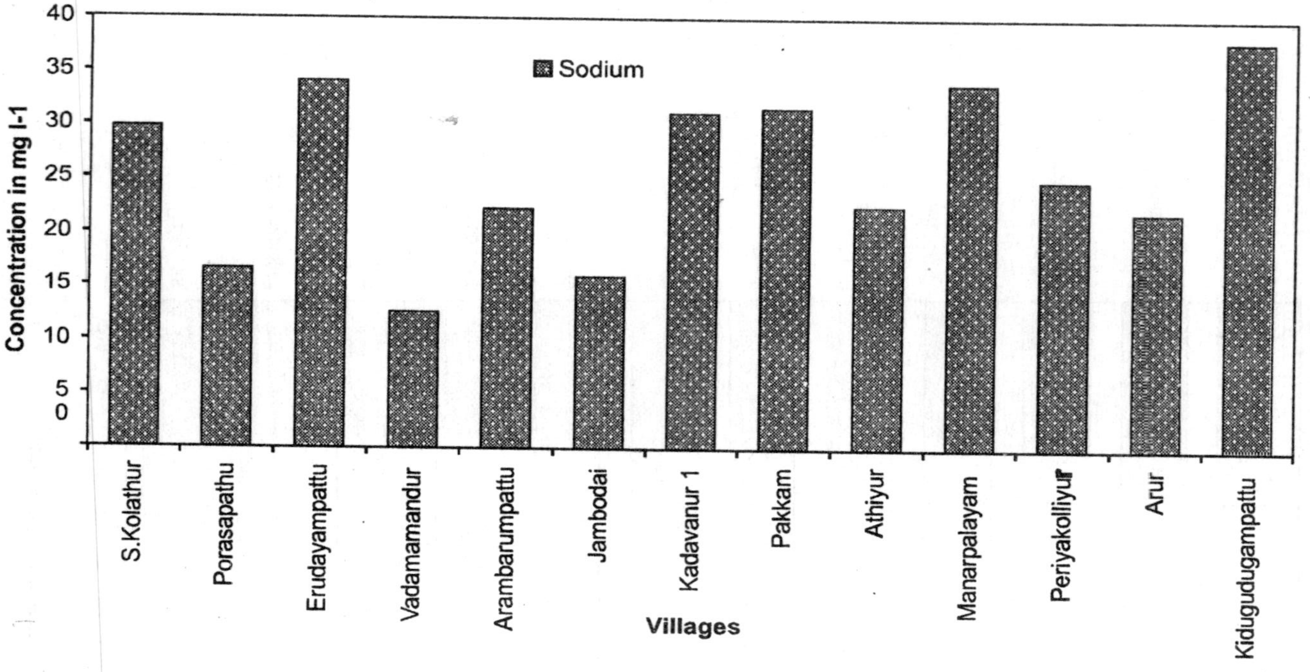

Fig. 14.20: Sodium concentration in groundwater in SRBC command

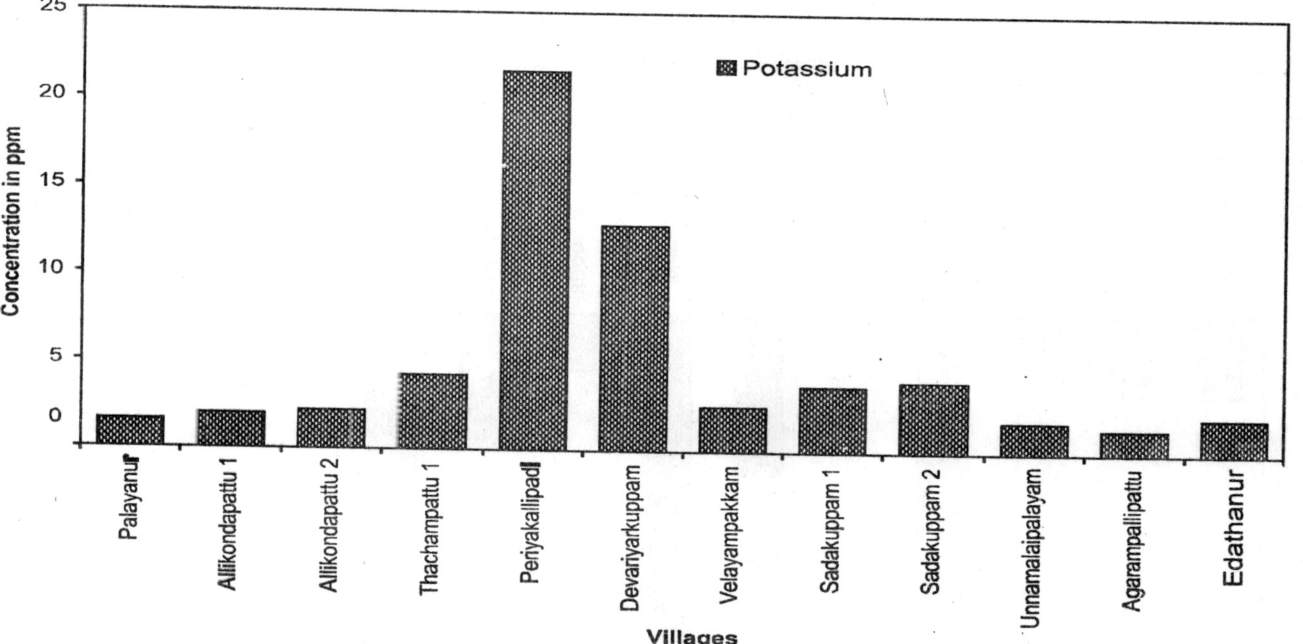

Fig. 14.21: Potassium concentration in ground water in SLBC command

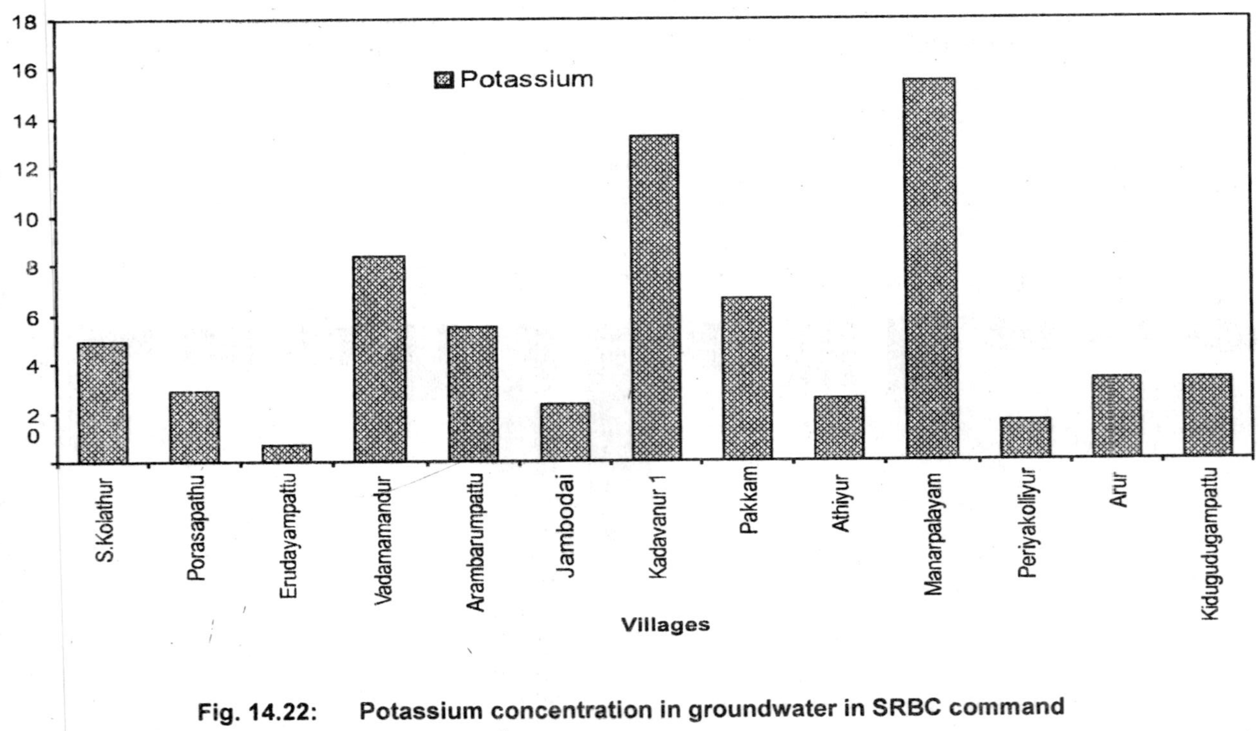

Fig. 14.22: Potassium concentration in groundwater in SRBC command

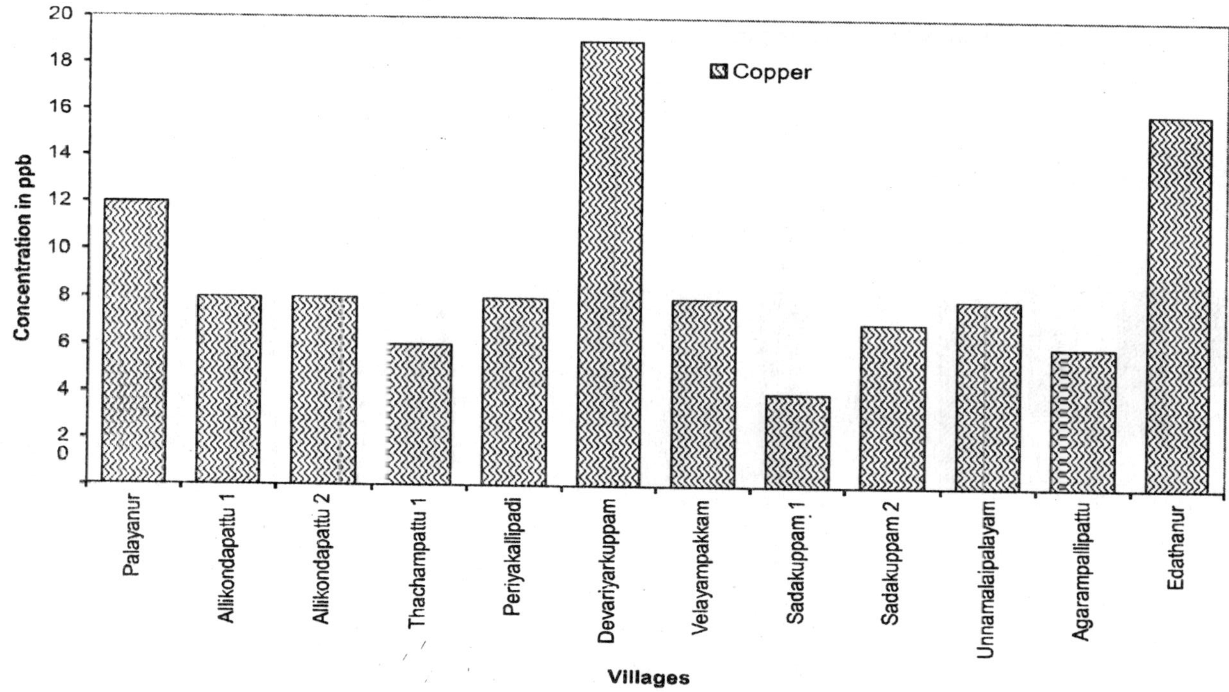

Fig. 14.23: Copper concentration in groundwater in SLBC command

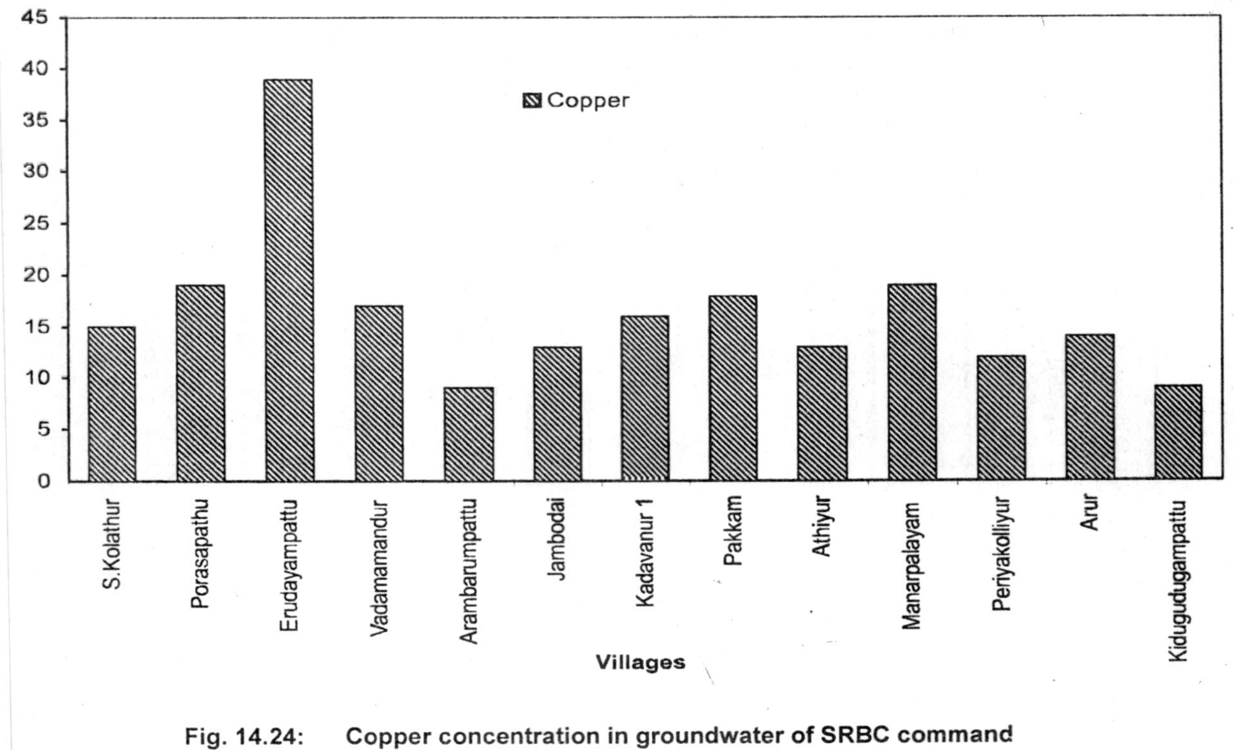

Fig. 14.24: Copper concentration in groundwater of SRBC command

Zinc

Zinc concentration in the groundwater samples of SLBC command ranged from 56 ppb in Unnamalaipalayam to 2550 ppb in Palayanur (Figure—14.27). The values in the samples of SRBC command varied from 48 ppb in Porasapattu to 2184 ppb in Kadavanur water samples (Figure—14.28). The concentration of zinc in Porasapattu, Kadavanur and Arur are higher than the permissible limit for drinking water (Table—14.3).

Cadmium

The Cadmium concentration in the groundwater samples of SLBC command ranged from trace amount to 16 ppb in Palayanur water sample (Table—14.1). In SRBC command, the values recorded in the Kadavanur and Arur groundwater samples were 30 ppb and 16 ppb respectively-way above the permissible limit for drinking water (Table—14.3).

The Tables—14.4 and 14.5 present the groundwater quality of few villages in SLBC and SRBC commands respectively for the year 1989-90. A comparative study indicates that there has been an increase in the values of all the chemical parameters in the groundwater over the years except pH. The alkaline pH is reduced to acid to neutral to slightly alkaline pH in most of the villages.

High concentration of nitrates, phosphates, fluorides and chlorides in the groundwater, can be attributed to the phenomenal growth in agriculture and fertilizers application in the SCA after the introduction of canal irrigation. This is supported by many such studies done by other authors (Gaumat et.al, 1992, Hamilton and Shedlock, 1992; Handa, 1983, 1990; Helgeson et. al, 1994; Jha and Jha, 1982; Klimas and Paukstys, 1993; Raju et. al, 1979). Super phosphates and NPK fertilizers act as an important source of nitrogen, phosphorus and potassium in the groundwater (Rao and Prasad, 1996). High concentration of chlorides and sulphates in groundwater can be due to the combined effect of non point sources such as bleaching agent for purifying water and sewage and also due to fertilizers (Pawar and Shaikh, 1995). Fertilizers and pesticides are also reported to be potentials source of heavy metal pollution such as cadmium and zinc (Arora et.al, 1975; Singh and Sekhon, 1977).

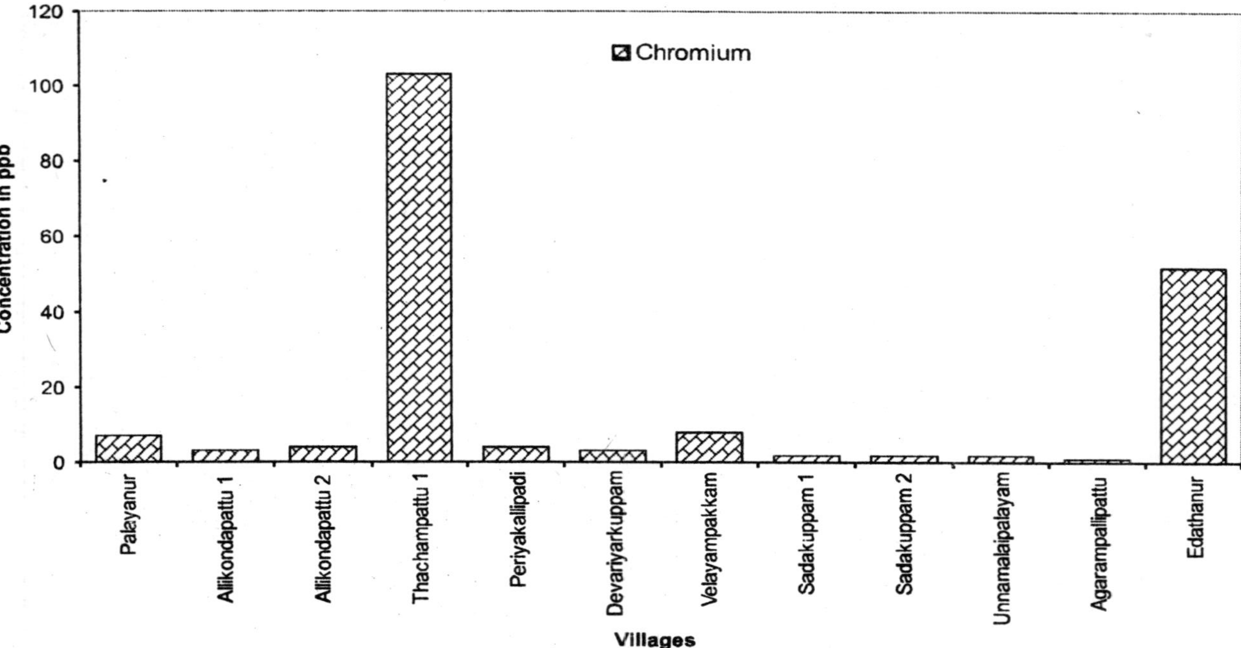

Fig. 14.25: Chromium concentration in ground water in SLBC command

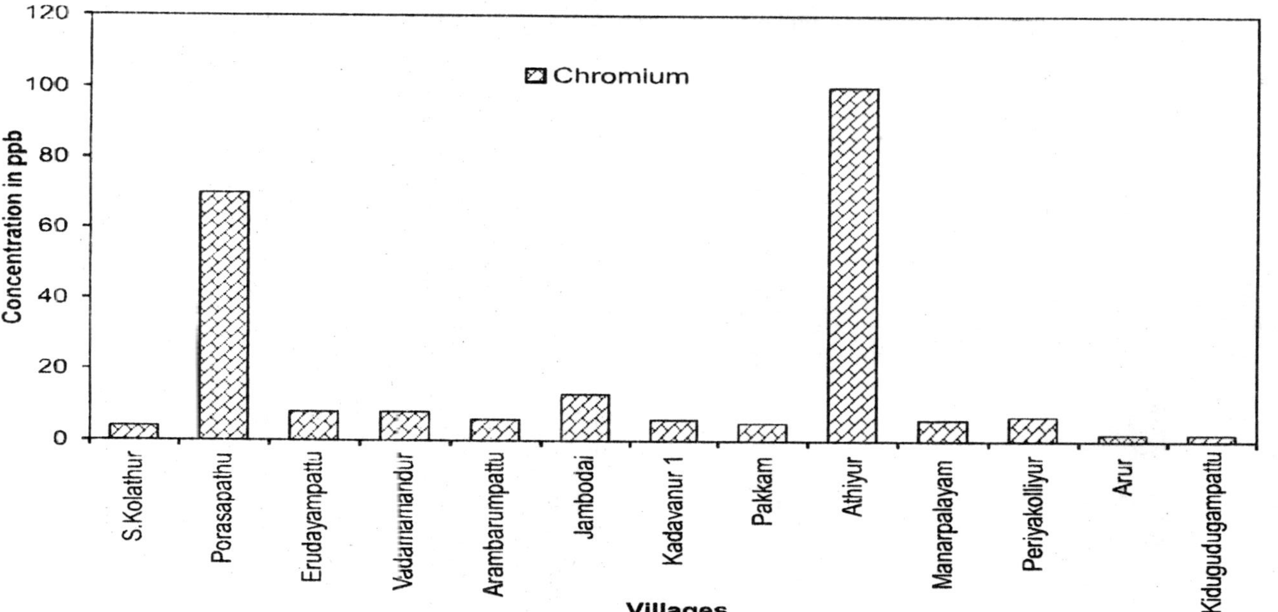

Fig. 14.26: Chromium concentration in ground water of SRBC command

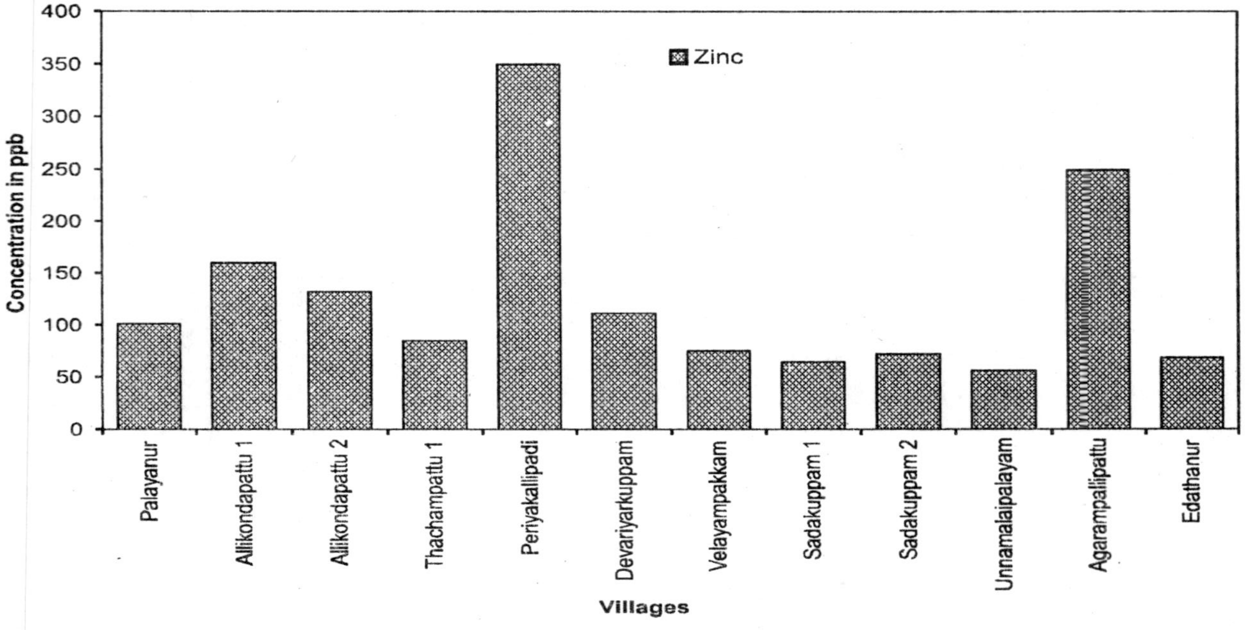

Fig. 14.27: Zinc concentration in ground water in SLBC command

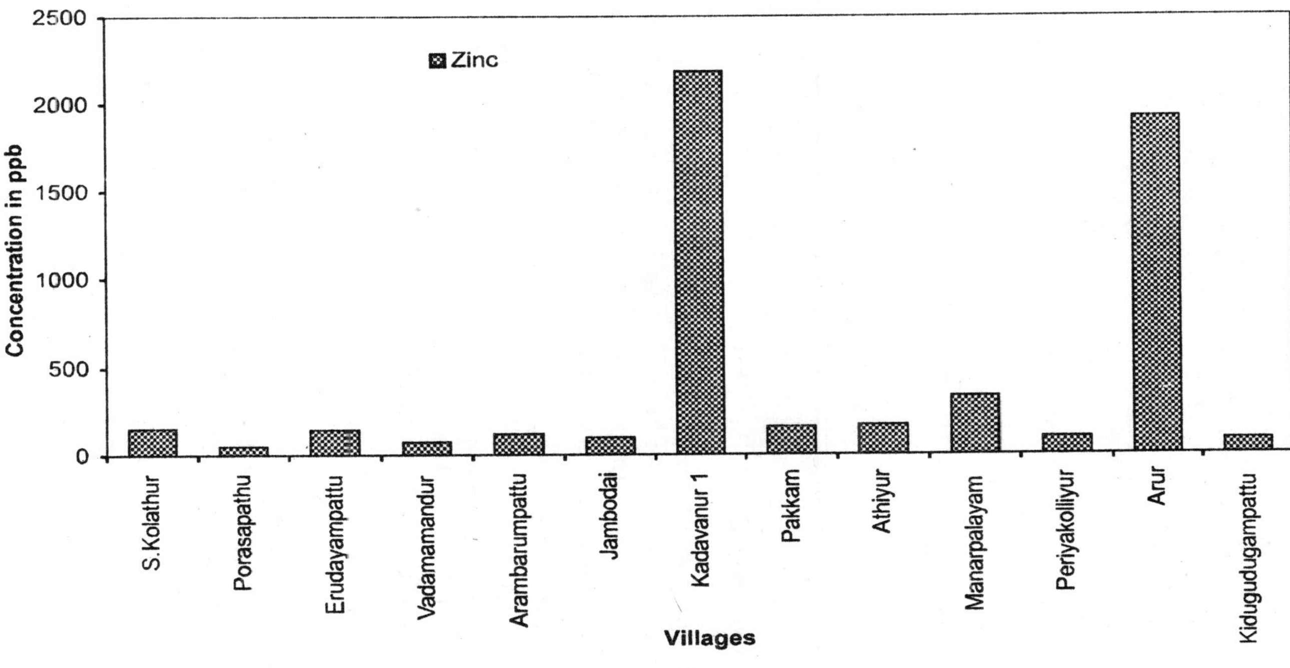

Fig. 14.28: Zinc concentration in ground water of SRBC command

15

Health

INTRODUCTION

Irrigation schemes—small, medium or large—have the potential of causing water-borne diseases in their command areas by providing conditions favourable for the breeding and transmission of disease vectors such as mosquitoes, snails and flies.

In this chapter we have made an attempt to evaluate the impact of the network of Sathanur canal system on the health aspects directly related to water – malaria and elephantiasis – in the command region. The impact of the measures employed by the health officials in countering these diseases has also been assessed.

STATUS

Data pertaining to the incidence of malaria in the population residing downstream of the Sathanur Project (SRP) is represented as Tables 15.1 and 15.2. The former covers the Sathanur Left Bank Canal (SLBC) command, and the later the Sathanur Right Bank Canal (SRBC) command. The data was collected form the offices of the Directorate of Public Health Services (DPHS), Thiruvannamalai and Kallakurichi. The terms *local, imported* and *'indicative populations'* have already been elaborated in section 7.1.1.

Table—15.1 Malaria incidence in SLBC command

Name of Village	Indicative Population	1994	1995	1996	1997	1998	1999
Vanapuram pudur	920	1 *	–	2	1	1	1
Vanapuram	3922	17 *	27 #	27	10	2	3
Serapapattu	961	–	–	–	1	–	–
Varagur	1719	–	2 *	1	–	–	1
Periyampattu	1927	4 *	–	4	2	–	1
Mazhuvampattu	910	4 *	3 *	–	3	–	–
Edakkal	1004	1 *	–	–	–	–	–
Valavachanur	1542	2* 9 7#	4#	3	–	–	–
Perunduraipattu	4428	1* 5 4#	5#	2	–	–	–
Kottaiyur	2526	1 *	–	–	1	–	–
Thenkarimbalur	2612	4 *	1 *	–	–	–	–
Sadakuppam	2551	13 #	2	1	1	-	-
Unamalaipalayam	225	2 *	–	–	–	–	1
Agarampallipattu	1762	1* 10 9#	5#	–	–	–	40
Thenmudiyanur	3993	10#	14# 16 2*	8	7	2	4
Edathanur	2439	88#	47# 51 4*	86	50	16	11
Allappanur	1025	11 #	1 #	6	–	2	5
Jambodai	443	–	–	3	–	–	–
Katampoondi	2185	11 *	10 *	6	7	–	1
Pappambadi	1524	–	–	–	–	–	1
Pavithuram	3525	4 *	5 *	–	–	–	
Navampattu	1245	–	–	–	–	–	1
Palayanoor	2904	2 *	5 *	2	5	–	–
Thachampattu	1640	1 *	–	–	–	–	–
Devanur	278	9# 2*	2#	75	12	3	–

(Table Contd...)

1	2	3	4	5	6	7	8
Devariyarkuppam	402	15 13#	1#	50	12	5	–
Pick-up-dam	12	2	1 #	–	–	–	–
Melandhal	2812	49	4	–	1	–	–
Kangaiyanur	3012	8	6	7	1	1	–
Jambai	2360	1	–	2	–	–	–
Murukkampadi	1951	–	–	4	1	–	–
Sithapatinam	1543	–	–	2	–	–	–
Manalurpet	6478	–	–	1	6	3	–

* Imported
# Local

Table—15.2 Malaria incidence in SRBC command

Name of Village	Indicative Population	1994	1995	1996	1997	1998	1999
Melsiruvalur	3061	-	-	-	3	-	-
Vadakiranur	1474	-	-	-	2	-	-
Mookanur	609	-	-	1	-	-	-
Olgalapadi	1344	-	-	-	-	1	-
Mangalam	1674	-	1	-	1	-	-
Arulambadi	2259	-	5	-	1	-	-
Vadaponparappi	2212	8	8	7	3	2	-
Vanapuralm	5119	-	-	-	1	2	-
A. Paudham	1456				-	3	-
Arur	1512	1	-	1	-	-	-
Arasampattu	627	-	-	2	-	-	-
Moongilthuraipattu	5659	135	46	14	61	25	-
Poravalur	454	2	1	-	2	3	-
Porasapattu	1704	16	1	-	2	-	-
Rayandapuram	2214	55 #	24 #	71	23	2	5
Thiruvadathanur	983	29 # 9*	6 # 2#	27	13	5	-
Illyankani	3645	11 2#	19 17* 2#	6	7	-	-
Puthurcheekady	671	3*	5 3*	-	6	-	-

* Imported
# *Local*

Table—15.3 lists out the total number of Primary Health Centres (PHCs) operating in the Sathanur Command Area (SCA), the villages and the corresponding populations covered by the individual health centres.

Information on the incidence of filaria for the year 1998-99 in the population residing in the SLBC command is presented as Table—15.4. Tables—15.5-15.11 present the profile of the various presumptive prophylactic measures taken by the health officials to check the spread of malaria in SCA Figures 15.18 represents the malaria and filaria endemic villages in SLBC command and Figure 15.18 illustrates the malaria endemic villages in SRBC command.

Annual Parasite Index

The definition, computation and the significance of the term *Annual Parasite Index* (API) have been dealt with in detail in section 2.7.1. API values as a function of time and annual precipitation in the region have been plotted individually for each village under SRBC and SLBC commands, and cumulatively for the individual command and the project. Impacts of precipitation and outflow along with other socio-cultural factors on the API have also been studied.

IMPACTS

At SLBC Command

There is no trend in the incidence of malaria encountered in the population residing in SLBC command as evident from Table 15.1. On one hand there are villages reported to have had high incidence of malaria among the natives while on the other hand in some villages viz Serapapattu, Edakkal, Kottaiyur, Jambodai, Pappambadi, Thachampattu, Unnamalaipalayam and Navampattu barely one or two isolated cases of malaria in the residents have been recorded over the past six years.

In still some other villages (not included in the Table—15.1) viz Varkillapattu, Kunglinatham, Sukhampalayam, Aradapattu, Athipadi, Kandiankuppam, Allikondapattu, Thalayamballam, Sakarathamadai, Nariyapattu, Pavampattu, Velayampakkam, Kalottu, Periyakallapadi, Chinnakalapadi and Manarpalayam, there has not been a single incidence of malaria in the native population during the study period.

Table—15.3 Villages along with the respective population under various Primary Health Centres (PHC) in SCA

Name of the PHC/Village	Population
1	2
Vanapuram PHC	
Vanapuram	3922
Vanapuram pudur	920
Serpapattu	961
Varkillapattu	772
Varagur	1719
Periyampattu	1927
Mazhuvampattu	910
Edakkal	1004
Kunglianatham	687
Valavachanur	1542
Perunduraipattu	4428
Kottaiyur	2526
Thenkarimbalur	2612
Sadakuppam	2551
Unnamalaipalayam	225
Agarampallipattu	1762
Thandarampet PHC	
Thenmudiyanur	3993
Edathanur	2439
Alappanur	1025
Rayandapuram	2214
Thiruvadathanur	983
Perungulathur PHC	
Ellyankani	3645
Puthurcheekady	671
Jambodai	443
Katampoondi PHC	
Katampoondi	2185
Sukhampalayam	745
Su. Pappambadi	1524
Pavithram	3525
Aradapattu	2388

(Table Contd...)

1	2
Palayanur PHC	
Navampattu	1245
Palayanur	2904
Athipadi	367
Kandiankuppam	1363
Thachampattu	1640
Allikondapattu	1472
Devanur	1262
Thalayamballam	2518
Sakarathamadai	775
Nariyapattu	705
Pavampattu	1478
Parayampattu	1556
Kallottu	863
Malamanjanur PHC	
Devanur	278
Devariyarkuppam	402
Pick up dam	12
Su Valavetti PHC	
Periyakallapadi	392
Periyakallapadi Pudur	335
Manarpalayam	198
Chinnakallipadi	1556
Vanapuram PHC	
Pakkam	1240
Kadavanoor	3386
Mangalam	1674
Arumbarampattu	1327
Vadamamandur	1463
Sirapathanalur	1753
Sirpanandhal	1756
Arkavadi	956
Athanoor	457
Edathanur	1178
Periyakolliyur	1267
Chinnakolliyur	1462
Sivapuram	702

(Table Contd...)

1	2
Tholuvanthangal	2614
Athiyur	3017
Ariyalur	2420
Nagalkudi	283
Vanapuram	5119
Erudayampattu	2153
Elayanarkuppam	3442
Kallipadi.S	249
Rishivandiyam PHC	
Jambadai	1695
Thiruvarangam	1439
Maniyandhal	829
Odiyandhal	1217
Sirupaniyoor	375
Kadambur	1328
Manalurpettai PHC	
Melandhal	2812
Kangaiyanur	3012
Jambai	2360
Pallichandal	1078
Murukkambadi	1951
Athiyandal	1196
Devaradiyarkuppam	1509
Sellankuppam	943
Sithapatinam	1543
Manalurpet	6478
Velayampakkam	1843
Vadaponparappi PHC	
Melsiruvalur	3061
Vadakeeranur	1474
Vadaponparappi	2212
Moongilthuraipattu	5659
Poravalur	454
Porosapattu	1704
Puthupettai PHC	
Mookannur	609
Olgalapadi	1344

(Table Contd...)

1	2
Arulampadi	2259
A. Pandalm	1456
S.Kolathur	2223
Arur	1512
Varagur	1041
Viriyur	4920
Arasampattu	627
Vadasiruvalur	2133
Ramrajapuram	755
Olgadayampat	841
Kidakudayampattu	700

Table—15.4 Annual parasite index of filaria in SLBC command

Village	API
Katampoondi	.5
Pavithram	1.5
Devariyarkuppam	3.4
Elayankani	.39
Kallottu	1.85
Thenmudiyanur	9.71

Table—15.5 Presumptive treatment

S. No.	Age group	Chloroquine dosage
1.	0-1	75mg
2.	1-4	150mg
3.	4-8	300mg
4.	8-14	450mg
5.	14 and above	600mg

Table—15.6 Radical treatment for malaria Single day treatment *Plasmodium falciparum*

Age	Chloroquine	Primaquine
0-1	75mg	Nil
1-4	150mg	7.5mg
4-8	300mg	15mg
8-14	450mg	30mg
14 and above	600mg	45mg

1. Chloroquine should not be given on empty stomach
2. Primaquine should not be given to infants and pregnant women

Table—15.7 Three days radical treatment for malaria *(Plasmodium vivax)*s

Age group	Choroquine	Primaquine		
		1st day	2nd day	3rd day
0-1	75mg	Nil	Nil	Nil
1-4	150mg	5mg	5mg	2.5mg
4-8	300mg	10mg	10mg	5mg
8-14	450mg	20mg	20mg	20mg
14 and above	600mg	30mg	30mg	15mg

Table—15.8 Radical treatment for chloroquine resistant *P.falciparum*

Age group	Sulfalene + pyremethamine	Primaquine
0-1	Nil	Nil
1-4	250mg + 12.5mg	7.5mg
4-8	500mg + 25mg	15mg
8-14	750mg + 37.5mg	30mg
14 and above	1000mg + 45mg	45mg

Table—15.9 Larvicide formulations and their dosages

	MLO	Temphos	Fenthion	*Bacillus spharicus*	*Bacillus thuringenisis*
Commercial formulation	100% petroleum products	50% Ec	100% Ec	Powder	Powder
Preparations of reeky to spray formulation	As it is 10 litres of water	2.5cc, in litres of water	5cc in 10 litres of water	500g in 10 litres of water	250g in 10 liters of water
Dosage m^2	20 cc	20 cc	20 cc	20 cc	20 cc
Dosage $50m^{-1}$	1 litre	1 litre	1 litre	1 litre	1 litre
Dosage ha^{-1}	200 litres	200 litres	200 litres	200 litres	200 litres
Frequency of application	Weekly	Weekly	Weekly	Weekly	Weekly

Table—15.10 Insecticide formulations and their dosage for indoor residual spray

	Malathion 2.5% wp	Delta methion 2.5% wp	Cyfluthrin 10% wp	Lambda chalothrin 10% wp	Feninthrothion 40% wp	Frimipho methyl 25% wp
Preparation of suspension in 10 litres of water	2 kg	400 g	125 g	1.25 kg	1.25 kg	2 kg
Dosage m^2	2 g	20 mg	25 mg	25 mg	1 g	2 g
Residual effect in weeks	6-8	10-12	10-12	10-12	6-8	6-8
No. of spray in weeks	3	2	2	2	3	3
Requirement/million population per round	300 MT	30 MT	9.38 MT	9.38 MT	93.75 MT	300 MT
Requirement/million population per annum	900 MT	60 MT	18.75 MT	18.75 MT	281.25 MT	900 Mt
Area to be sprayed by 10 litres of suspension for correct dosage	250 m^2	500 m^2	500 m^2	500 m^2	500 m^2	500 m^2

Table—15.11 Insecticide formulations and their dosage for space spray

	Pyrethrum extract	Malathion
Commercial formulations	2% extract	Technical malathion
Preparation of formulations	1:19 (1 part 2% pyrethrum extract in 19% of kerosene)	5 parts of technical malathion in 95% of oil
Equipment used	Flit pump or hand operated fogging machine	Vehicle mounted thermal fogging machine (km hr^{-1}).

The API values for the villages with high concentration of malarial cases among the natives are depicted in Figures 15.1 – 15.8. Impact of annual precipitation on the API values has also been studied.

Based on the API values of the villages of the SLBC command, the villages can be grouped under three categories:

1. Villages with API values ≥ 10

   Devanur, Devariyarkuppam, Edathanur, Alappanoor, and Melandhal.

2. Villages with API values ≥ 5 < 10

   Vanapuram, Valavachanur, Sadakuppam and Katampoondi.

3. Villages with API values < 5

Mazhuvampattu, Thenmudiyanur, Vanapurampuddur, Perunduraipatu, Periyampattu, Kangaiyanoor and Palayanoor.

Figure—15.1 presents the API values of Devanur and Devariyarkuppam (villages with highest recorded API values in the SLBC command). As is evident, the API values of both the villages indicate a strikingly similar pattern of rise and fall. The values marginally came down in 1995 before reaching the summit in 1996. The peak was higher for Devanur (API-269.7) compared to Devariyarkuppam (API -124.3). In 1997 the API values fell sharply by 83.99 per cent and 76.02 per cent of 1996 values respectively and this fall continued through 1998 till it ultimately reached the nadir in 1999. The API values for both the villages seem to follow the pattern of annual rainfall in the region till 1997 (being highest in 1996 when the rainfall too was maximum). But after this the patterns are dissimilar.

The Figure 15.2 illustrates the API values of Edathanur and Alappanur. The value for Edathanur declined marginally in 1995 before attaining a peak in 1996. There was a sharp decline in API in 1997; the trend continued through 1998 and 1999.The API-time curve for Alappanur reflects a similar pattern till 1997 after which there is a marginal increase in the API values. Both API curves peaked during the year 1996 which recorded the highest rainfall

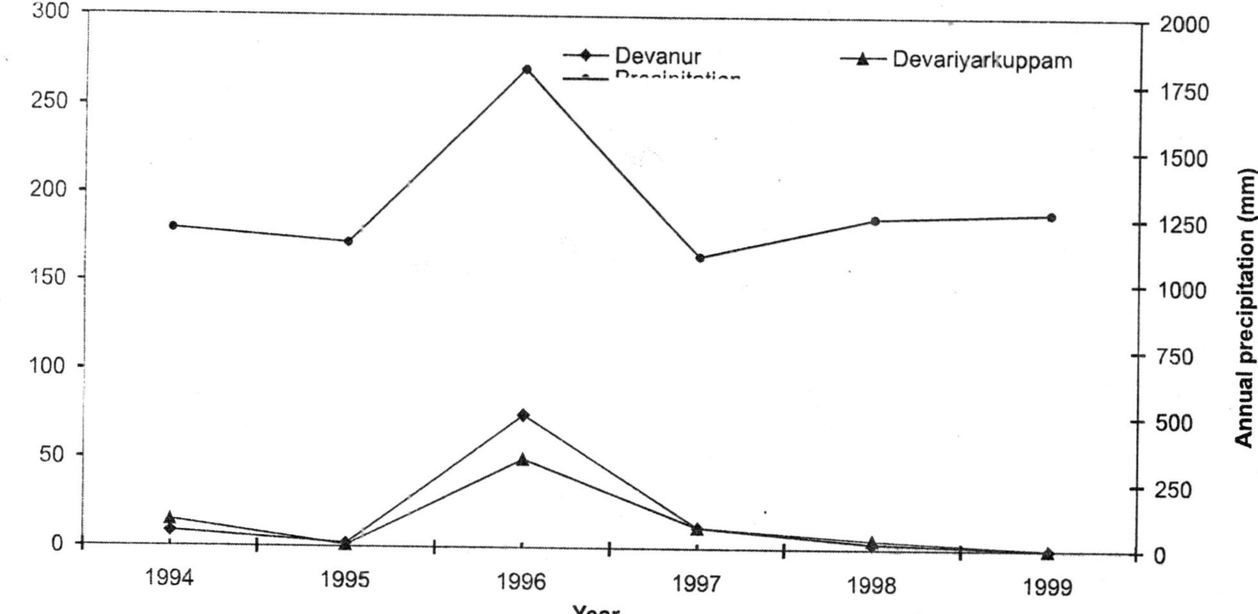

Fig. 15.1: Impact of precipitation on material annual parasite index in the villages of SLBC command

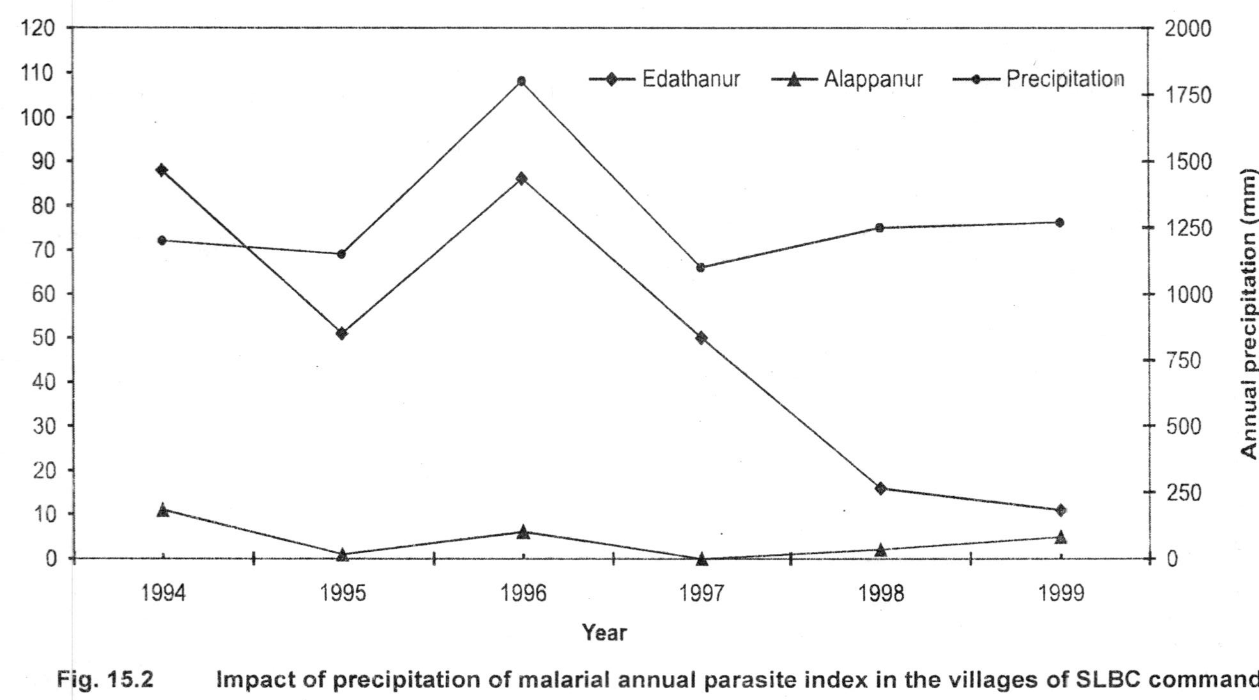

Fig. 15.2 Impact of precipitation of malarial annual parasite index in the villages of SLBC command

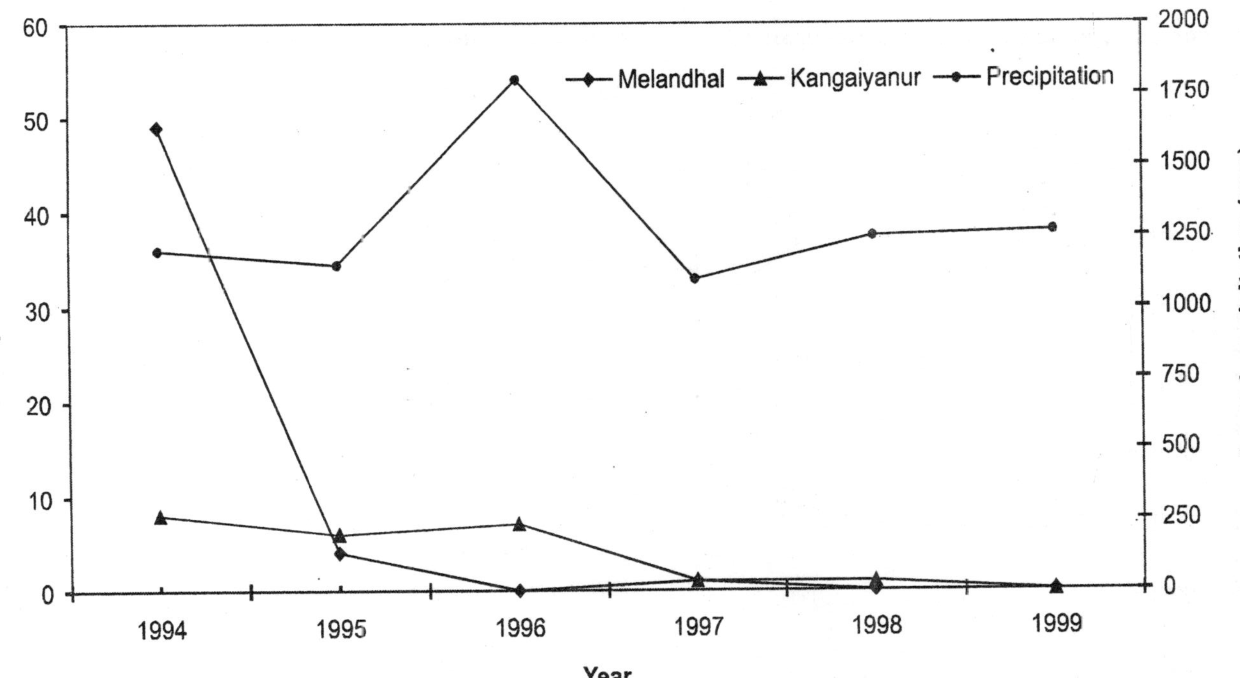

Fig. 15.3: Impact of precipitation on malarial annual parasite indes in the villages of SLBC command

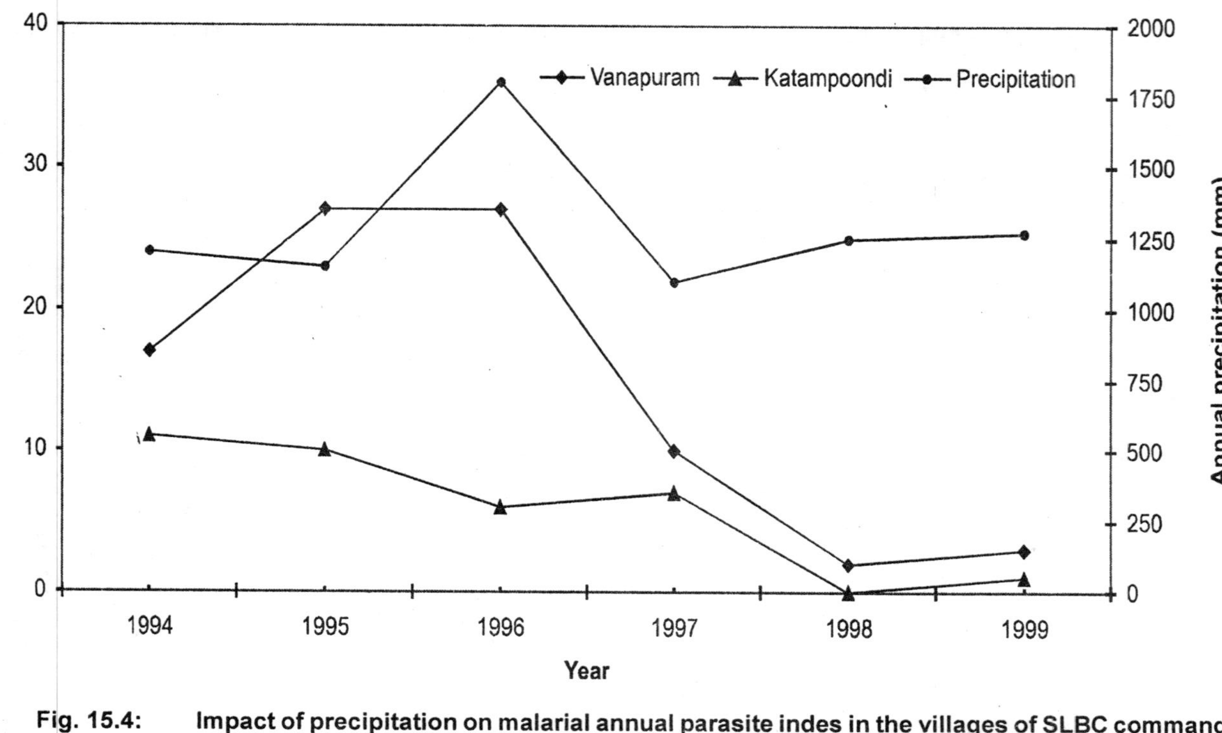

Fig. 15.4: Impact of precipitation on malarial annual parasite indes in the villages of SLBC command

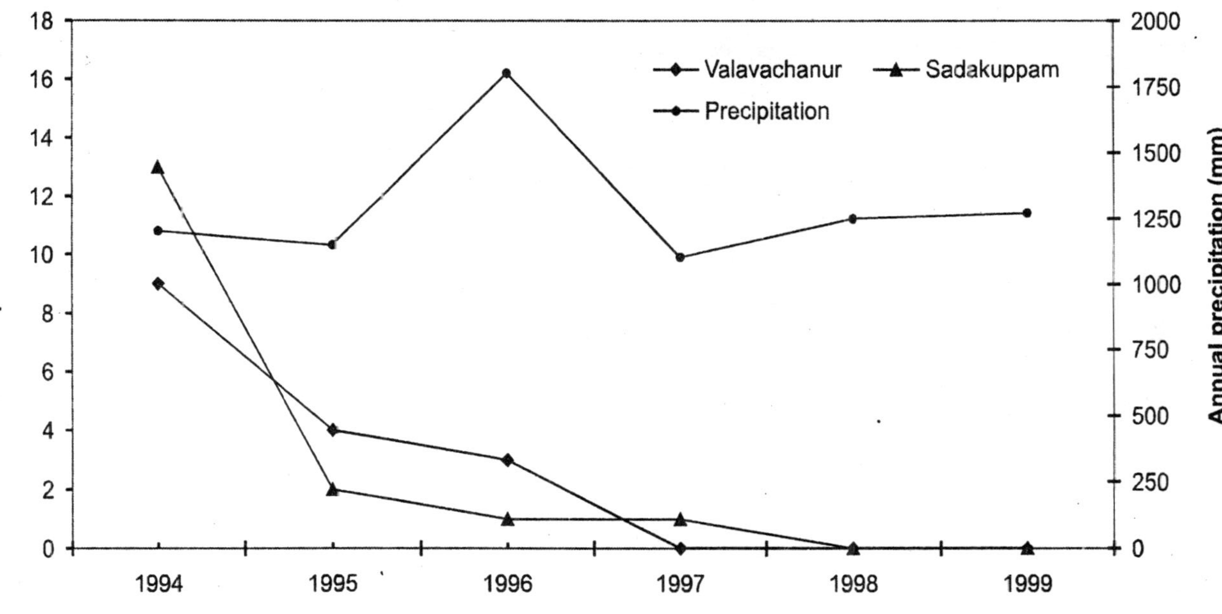

Fig. 15.5: Impact of precipitation on malarial annual parasite indes in the villages of SLBC command

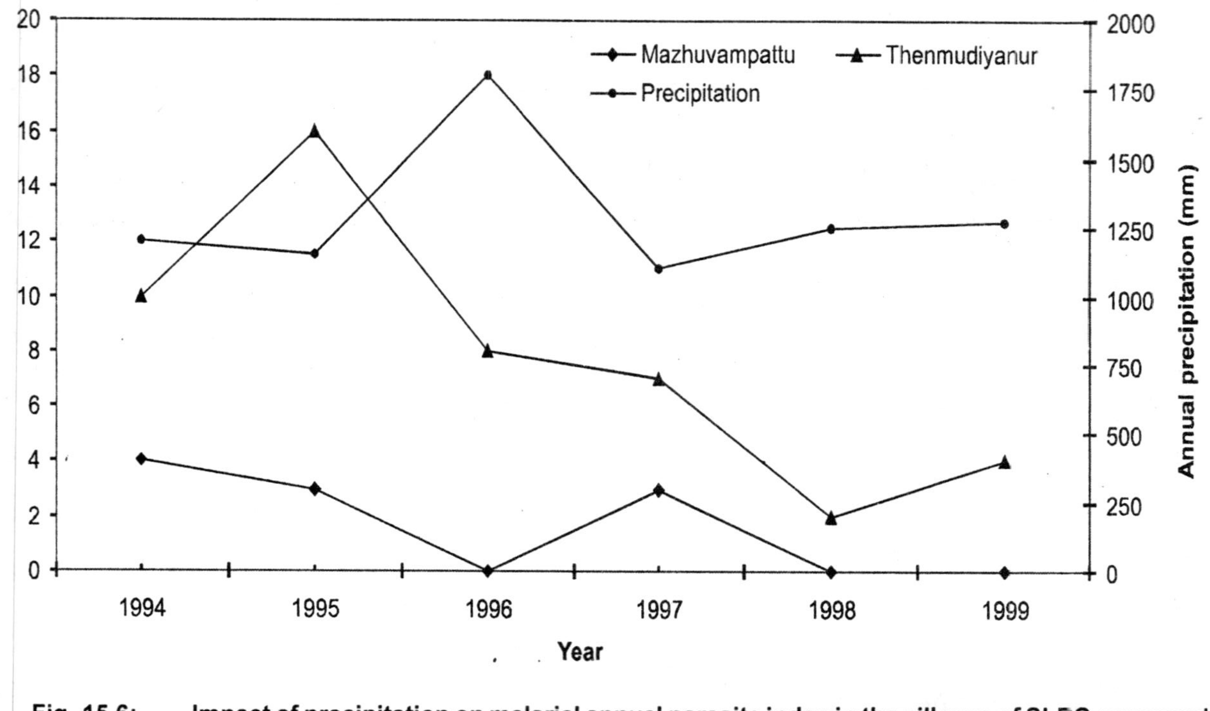

Fig. 15.6: **Impact of precipitation on malarial annual parasite index in the villages of SLBC command**

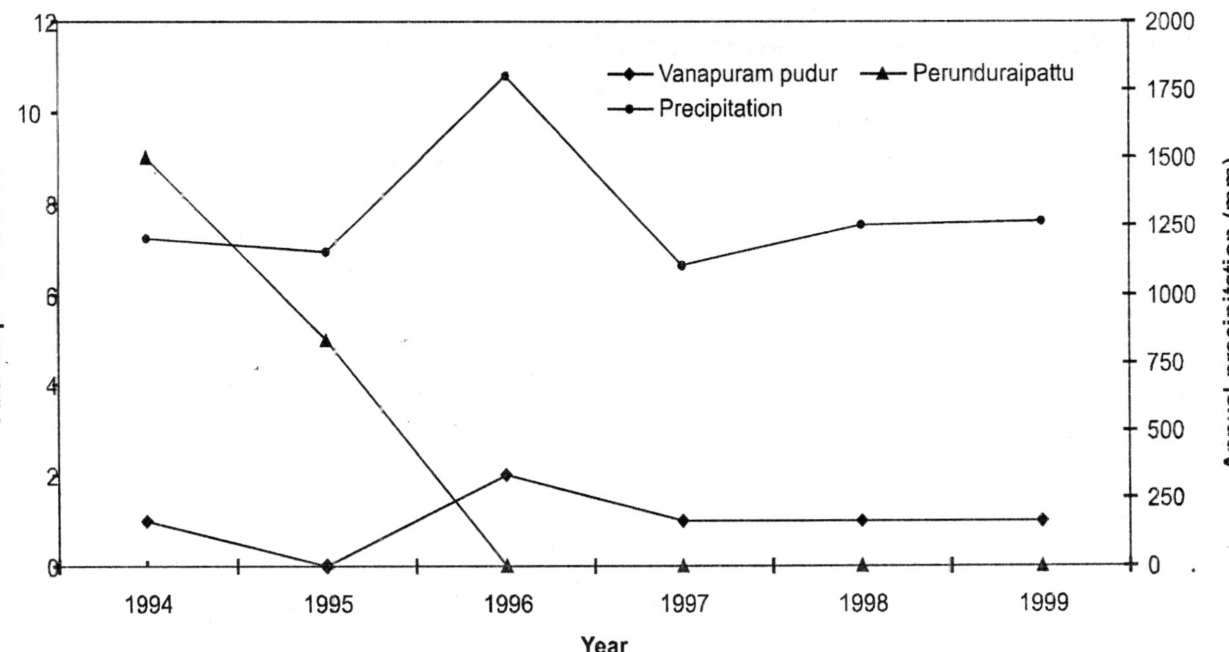

Fig. 15.7: Impact of precipitation on malarial annual parasite index in the villages of SLBC command

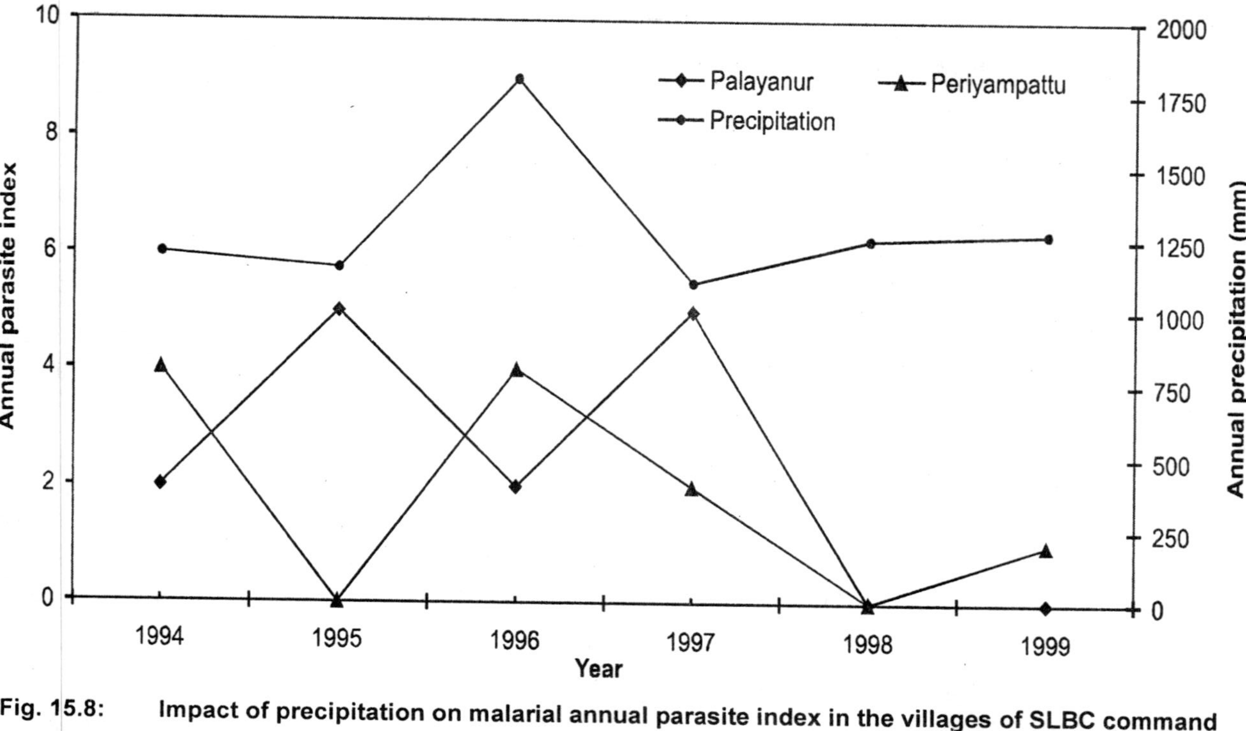

Fig. 15.8: Impact of precipitation on malarial annual parasite index in the villages of SLBC command

in the command. Alappanur curve varies in the same fashion as that of annual rainfall curve whereas the Edathanur curve manifests a change in the trend after 1997 (Figure—15.2).

The API values of Melandhal and Kangaiyanur are depicted in Figure—15.3. The noteworthy observation is that the API value of Melandhal that was 17.4 in 1994 declined by 91.83 per cent in 1995 and became nil in 1996. The pattern is represented by a steep fall of the curve in 1995. In 1997 there is again a marginal rise in the curve. There is absolutely no correlation between the API values and annual rainfall (Figure—15.3). The API value of Kangaiyanur peaked slightly in 1996 before coming down in the subsequent years. The API curve follows the similar pattern as that of rainfall (Figure—15.3).

Figure—15.4 illustrates the API values for Vanapuram and Katampoondi. There is a steep rise in the Vanapuram curve depicting API values in the year 1995. The curve remains constant for the year 1996 and falls sharply thereafter during the years 1997 and 1998. The curve rises marginally in 1999. The katampoondi curve declines in 1995, continues declining further in 1996 too. The curve rises in 1997 before falling steeply in 1998 when the API value was zero. The curve depicts a rising trend in 1999 similar to that of Vanapuram curve. The Vanapuram curve follows the pattern of annual rainfall from the year 1996, as evident from the Figure—15.4 while the Katampoondi curve imitates the rainfall curve from 1997 onwards. Vanapuram curve peaks in 1996 whereas the katampoondi curve shows a noticeable rise in 1997.

The API values for Sadakuppam and Mazhuvampattu are illustrated in Figure—15.5. There is a continuous fall in API values for Sadakupam. The fall was steep in 1995, which was marginal thereafter through the periods of 1996 and 1997. API values were reduced to nil in 1998 and 1999. In contrast, Mazhuvampattu curve, after an initial period of sharp declines in 1995 and 1996 (API-0) depicts a sudden rise in 1997 before falling steeply in 1998 touching the zero mark. No significant correlation is observed between the API curves and the annual precipitation curve. In fact, API of Mazhurvampattu was nil in 1996 when the command area received the highest recorded rainfall during the study period.

Figure—15.6 depicts the API values of Valavachanur and Thenmudiyanur. Valavachanur registered a sharp and steady decline in the API from 1994 to 1997 when it touched the zero mark. In 1999 there was a sudden jump in the API when it reached its highest ever value during the entire study period. Thenmudiyanur curve after illustrating a slight rise in 1995 shows a gradual and continual descent till 1998. After 1998 again a rising trend is noticed in the curve. Both, Valavachanur and Thenmudiyanur curves, don't confirm to the pattern of the annual rainfall curve.

Figure—15.7 indicates the trends of API values for Vanapuram puddur and Perunduraipattu. After an initial reduction in API in 1995, there was a, sharp and sudden increase in the value in 1996, as evident from the Vanapuram pudur curve. The value came down considerably in 1997 and attained a constancy thereafter during the periods 1998 and 1999. The API values for Perunduraipattu, fell noticeably in 1996 and thereafter in 1997 were reduced to zero, preceded by constant values in 1994 and 1995. The pattern of the rise and fall of the API values of Vanapuram puddur is consistent to that of the annual rainfall till 1997 and thereafter there is no consistency. There is no significant correlation between the API values of Perunduraipattu and that of annual rainfall in the command (Figure—15.7).

Figure—15.8 presents the API curves for Palayanur and Periyampattu plotted against the annual rainfall in the command area. As is observed, there is a wide range of fluctuations in API values over the years of both the villages. The API values for Palayanur rose considerably in 1995 and fell sharply in 1996 and repeated the similar pattern of sharp rise and sudden decline to zero in 1997 and 1998 respectively and remained nil thereafter in 1999. The API value of Periyampattu, from its highest value in 1994, was reduced to zero in 1995; increased sharply in 1996 and fell steeply in 1997. This fall continued in 1998 when it touched zero mark but 1999 experienced a slight rise in the index value. There is no similarity in the variations between the API values of Palaynur and annual rainfall. Periyampattu curve on the other hand follows the trend of the rainfall curve during most of the years with the exception of 1998, when the API was nil.

The noteworthy fact that emerges is that the index values of the villages and the annual precipitation in the command region apparently do not confirm to a similar pattern of variations over the study period. This observation points out towards the possibility of other parameters in governing the incidence of malaria in the SLBC command.

Figure—15.9 is a cumulative representation of the API values in the SLBC command and the impact of annual precipitation on the values. The API curve closely follows on the heels of the rainfall curve with the exception of the year 1998 when the rainfall curve depicts a rise while the API features a decline.

SRBC Command

The incidence of malaria among the residents of various villages in SRBC command for a period of five years (1994-98) is presented in Table—15.2. Of the forty-four villages in the SRBC command, only 18 villages had the incidence of malaria in the populace during this period. Of these 18 villages, for which the data is tabulated as Table—15.2, only 8 villages had high malaria incidence in the population. Rest of the villages had isolated cases of less than 5 in a particular year or two intervening years.

The API values of the villages with high concentration of disease in the populace have been plotted along with the annual precipitation in the region over a span of five years (Figures—15.9 to 15.12). As with the villages of the SLBC command, the villages in the SRBC command can be grouped on the basis of API values as follows:

1. Villages with API $\geq$ 10

   Rayandapuram, Thiruvadathanur and Moongilthuraipattu

2. Villages with API ^{3} 5 $\geq$ 10

   Elayankani, Puthurcheekady, Porasapattu and Poravalur

3. Villages with API < 5

   Vadaponparappi

The API values of Rayandapuram and Thiruvadathanur as functions of the annual precipitation in the command region are presented in Figure—15.10. There is a marked similarity in the trends of the API values for the two villages till 1998. The values came down sharply in 1995; peaked in 1996 before falling steeply in 1997 and falling further in 1998. The index value of Thiruvadathanur continued falling and touched the zero mark in 1999 while that of Rayandapuram rose marginally in 1999. The API curves of Rayandapuram and Thiruvadathanur closely follow the pattern of annual rainfall till the year 1997. All the three curves attain the crest in the year 1996.

Figure—15.11 depicts the API values of Vadamponparappi and Moongilthuraipattu. The Vadaponparappi curve illustrates a steady value of API over the years, with marginal reductions in 1997 and 1998. The API values of Moongithuraipattu registered a steep descent in 1995, which continued in 1996. There was a considerable elevation in the index value in 1997, which declined in 1998. There is no significant correlation among the API curves and the rainfall curve. In fact API value for Moongilthuraipattu was in ebb in 1996 when the rainfall recorded in the area, during the study period, was the highest.

The API values of Puthurcheekady and Elayankani have been plotted as Figure—15.12. The API curves of both the villages confirm to the similar pattern of crests and troughs. There was a marginal rise in the index values for both Puthurcheekady and Elayankani in 1995 followed by a sharp fall in 1996 when the index value of Puthurcheekady recorded zero. The value shot up in 1997 reaching its zenith in Puthurcheekady while in Elayankani there was a marginal increase. The values ebbed in 1998 to zero and continued to be so in 1999. There is absolutely no correlation between the index values and the annual precipitation. On the contrary, in 1996 there was a sharp fall in index values while the rainfall recorded an increase and in 1997 when comparatively low rainfall was recorded in the region the index values depicted a rise, quite extraordinarily high in case of Puthurcheckady.

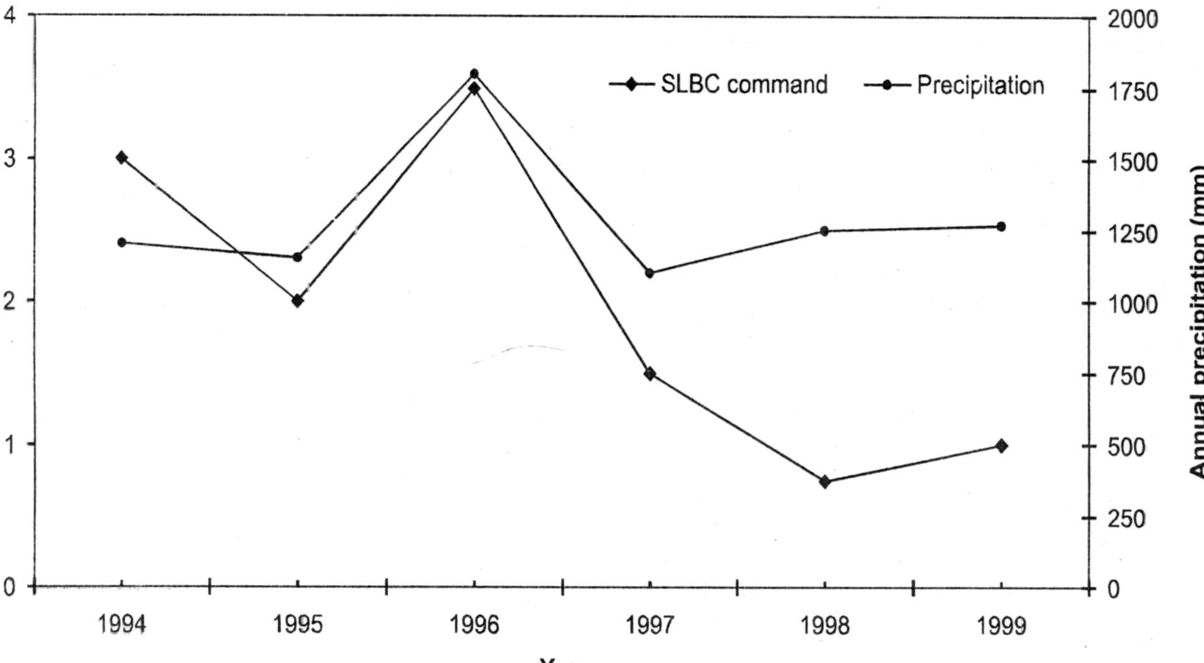

Fig. 15.9: Impact of precipitation on malarial annual parasite index in the villages of SLBC command

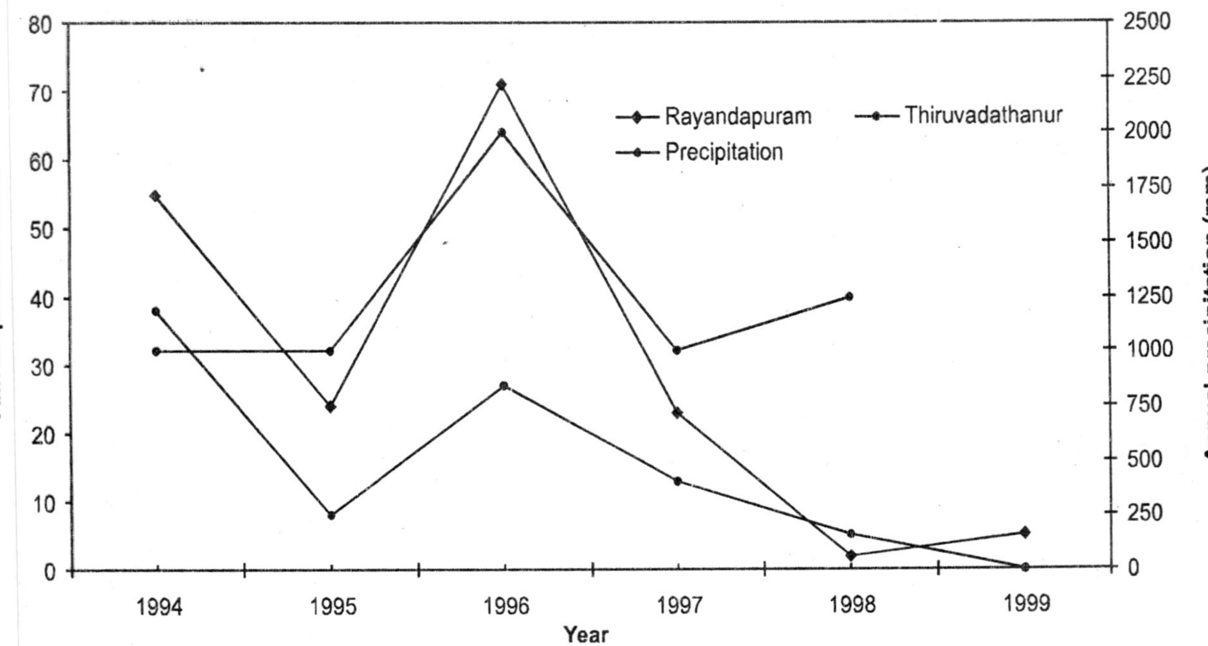

Fig. 15.10: Impact of precipitation on malarial annual parasite index in the villages of SRBC command

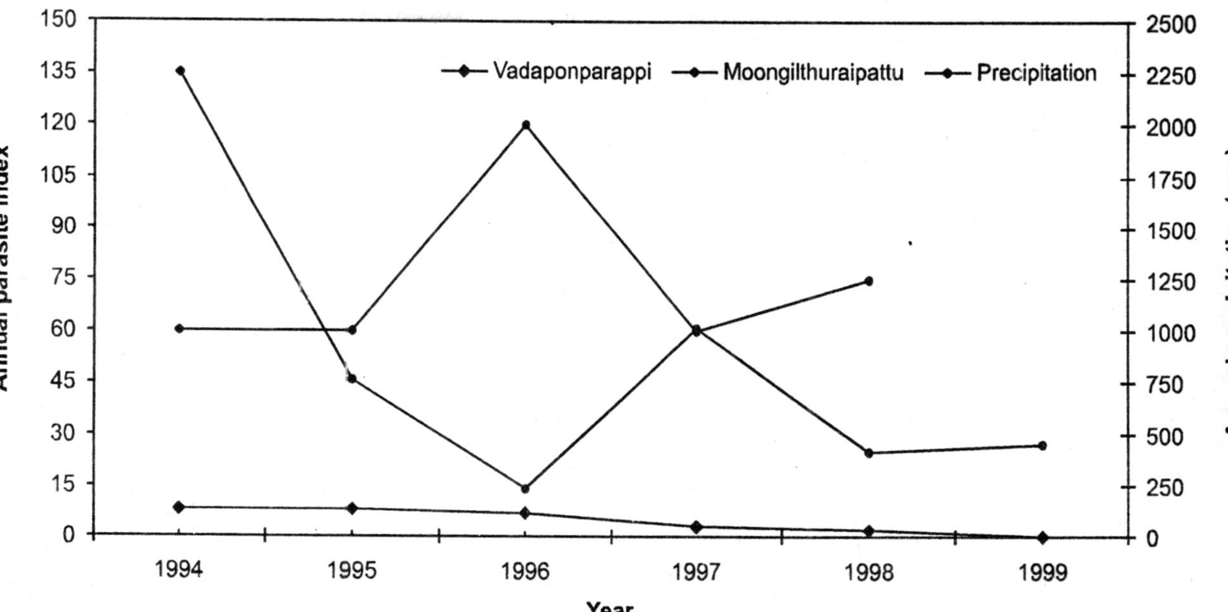

Fig. 15.11: Impact of precipitation on malarial annual parasite index in the villages of SRBC command

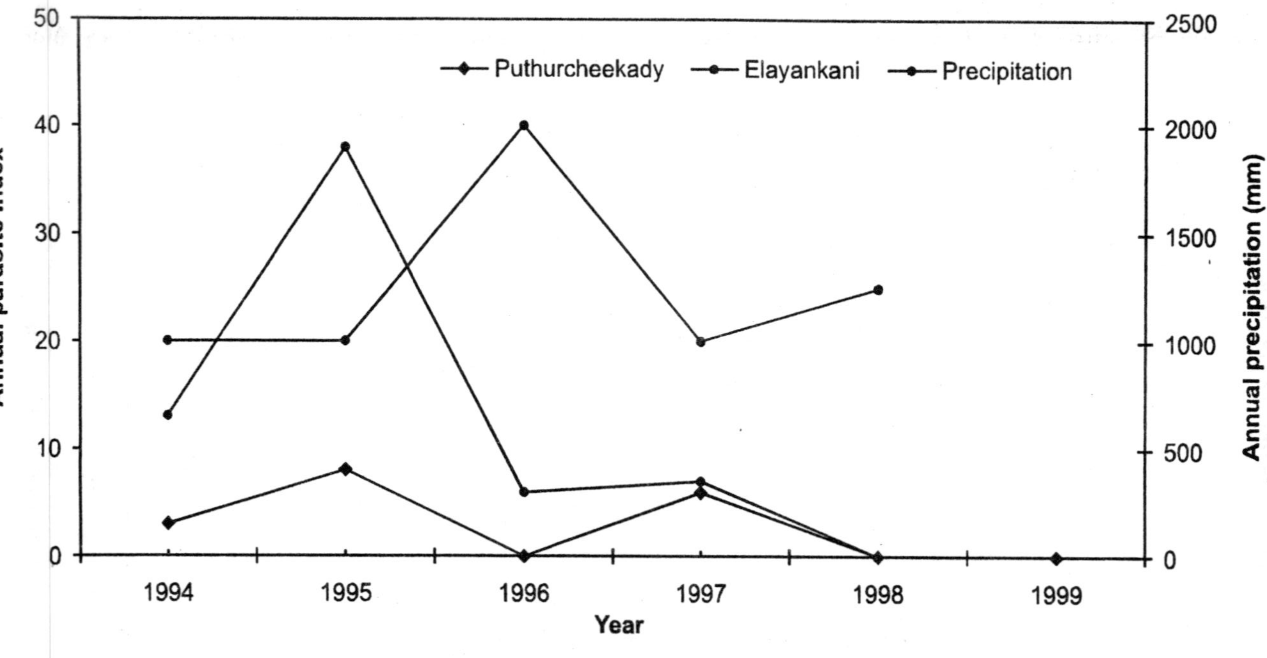

Fig. 15.12: Impact of precipitation on malarial annual parasite index in the villages of SRBC command

Figure—15.13 elucidates the fluctuations in the API values for Poravalur and Porasapattu over the years. A sharp plunge in the index value in Porasapattu was recorded in 1995 which ultimately dipped to zero in 1996. In 1997 there was a slight rise in the index value before it hit the zero mark again in 1998. The index values in Poravalur steadily kept declining till 1996 when it ebbed to zero. There was a sharp elevation in 1997, which continued further in 1998. Similar to the previous villages discussed under Figures—15.11 and 15.12 the API values of Poravalur and Porasapattu too do not correlate with the annual rainfall in the region. In 1996, the values were zero while the rainfall recorded was maximum, further, in 1997 when low rainfall was recorded in the command, the API values featured a rise.

A thorough comparative study of the Figures—15.10 to 15.13 highlights the fact that annual precipitation has little bearing on the API values in SRBC command. With the exceptions of Thiruvadathanur, Vadaponparappi and Rayandapuram, API values of no other village illustrate any similarity in variations to that of the annual rainfall in the region. Another point, which comes to light is the peak attained by the API values in the Poravalur, Porasapattu, Puthurcheekady, Elayankani and Moongilthuraipattu during the year 1997.

Figure—15.14 gives a cumulative account of API of SRBC command along with the annual precipitation in the region. There is no noticeable pattern between the API curve and the rainfall curve. The fluctuations in the rainfall over the years are not manifested in the API values which remained almost constant from 1995 to 1997 before dipping in 1998.

An attempt has been made to correlate the annual outflow from the reservoir with the API values in the SCA. Figure 15.15 illustrates the relationship between the annual precipitation received in the SCA (separately for the SLBC and SRBC commands) and the water released annually from the reservoir downstream. During 1994 and 1995 the outflow from the reservoir was very less which would have been due to the less rainfall received in the tract during those particular years. The trend changed in 1996 and 1997 when highest outflows from the reservoir were recorded for

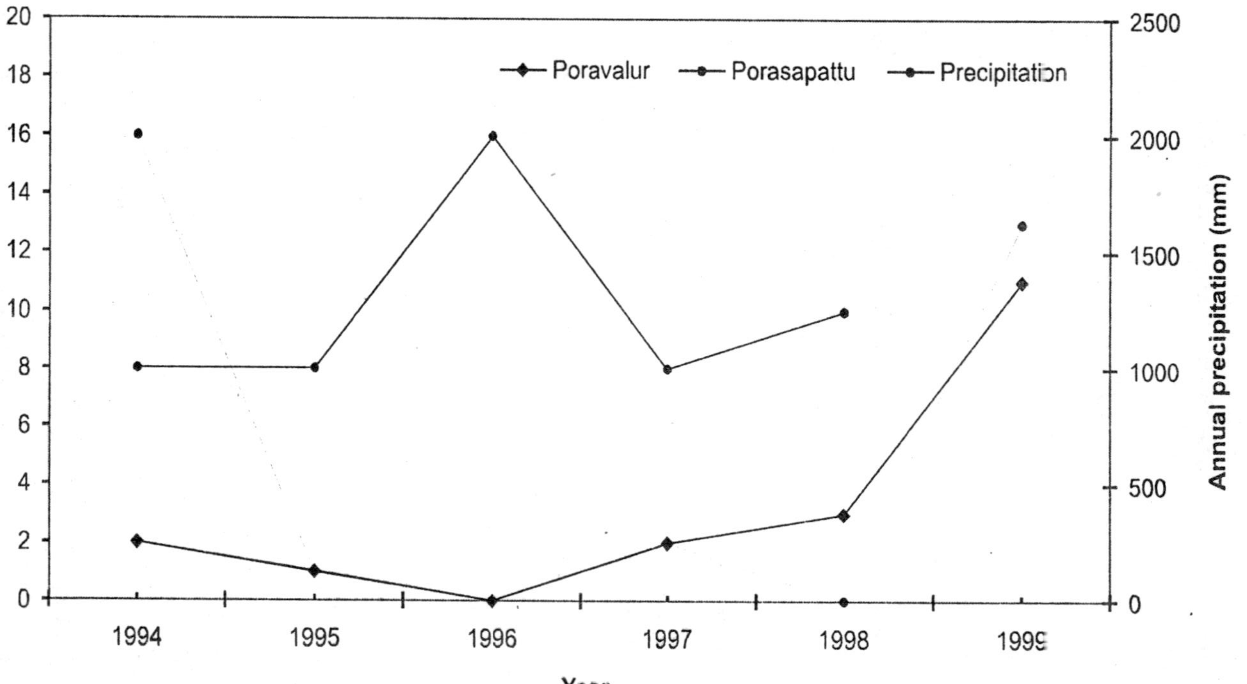

Fig. 15.13: Impact of precipitation on malarial annual parasite index in the villages of SRBC command

the study period. The outflow peaked in the year 1997 and this probably explains the peaks attained by the API values for various villages, as discussed in foregoing sections.

Figure—15.16 illustrates the impact of outflow from the reservoir on the cumulative index in SLBC command. The API curve closely follows the outflow curve till 1996. In 1997, there was an increase in the outflow from the reservoir but API values ebbed during that period.

Figure—15.17 presents the correlation between the API values of SRBC command and the annual outflow from the reservoir. The API curve confirms to the pattern followed by the outflow curve as comprehended from the Figure. Though the patterns are subtly similar in 1994 and 1995, there was a marginal decline in outflow in the year 1995 but API value registered a sharp decline. The trend reversed in 1996 when there was a sharp rise in the outflow but increase in API value was very slight.

Malaria doesn't owe its prevalence in the SCA to the rainfall or the outflow from the reservoir. It is the availability of water through SRP and its utilization in SCA that leads to the circumstances that cause the proliferation of vectors and transmission of the disease. Figures—15.18 present the malaria and filaria endemic regions in SLBC and SRBC commands respectively. An attempt is made to study their indirect impacts resulting in the spread of water-borne disease among the residents of SCA.

Irrigation and Agriculture

Farming is the main occupation of the people residing in SCA. It is practised with the conjunctive use of rainfall, surface water in the form of canals and tanks and groundwater in the form of tube wells and open wells. Following the year-round availability of water in the command reaches, a heavy shift in the cropping pattern in favour of paddy (*Oryza sativa*) and sugarcane (*Saccharum officinarum)* has been observed (Chapter 13, Agriculture). The net sown area is dominated by these two crops, which need extensive irrigation. The cultivators normally follow the traditional practice of flood irrigation.

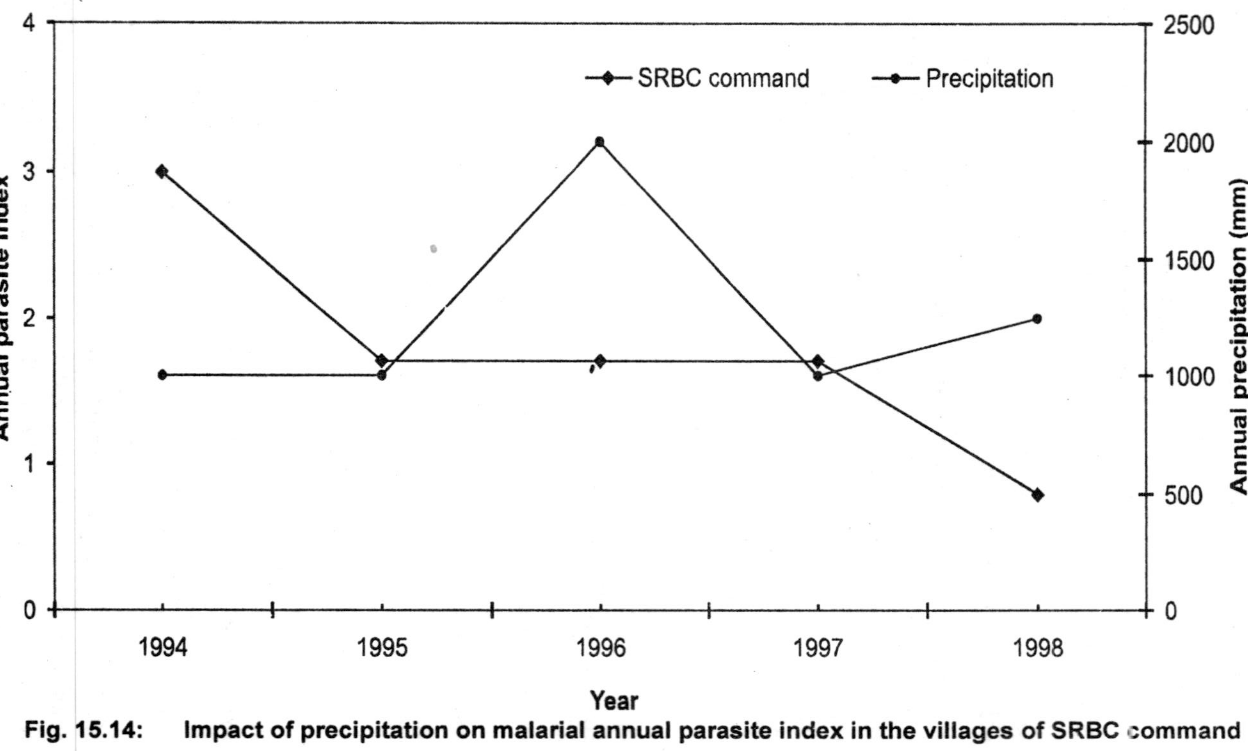

Fig. 15.14: Impact of precipitation on malarial annual parasite index in the villages of SRBC command

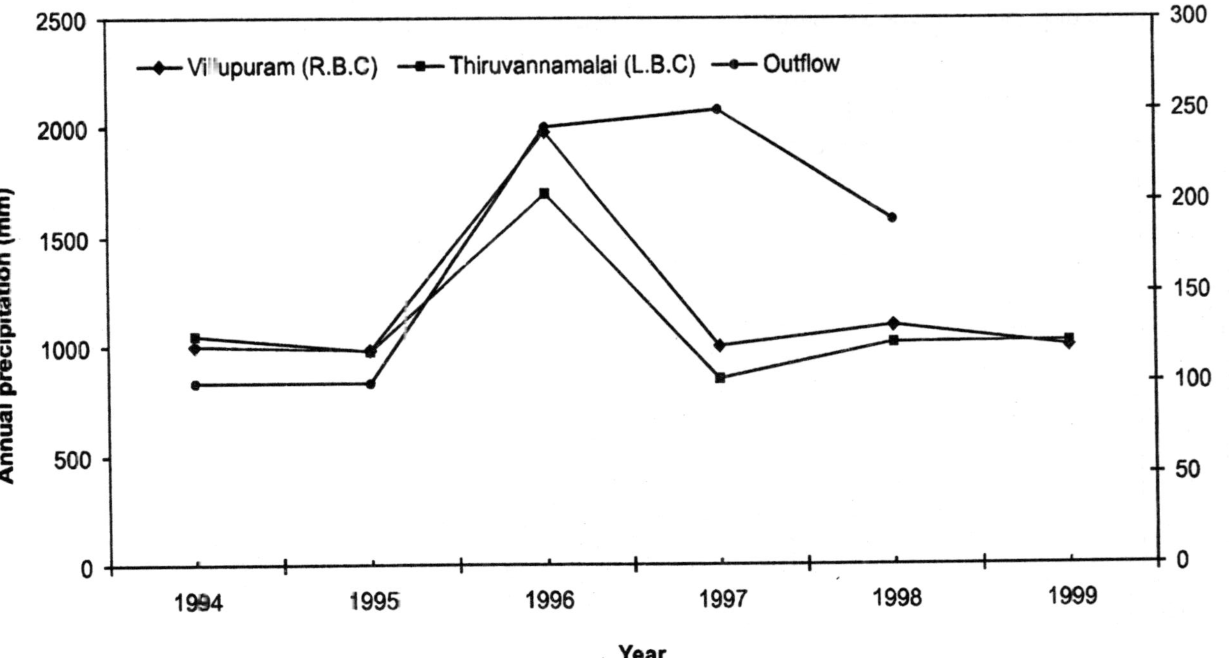

Fig. 15.15: Annual precipitation in SCA and outflow from the reservoir

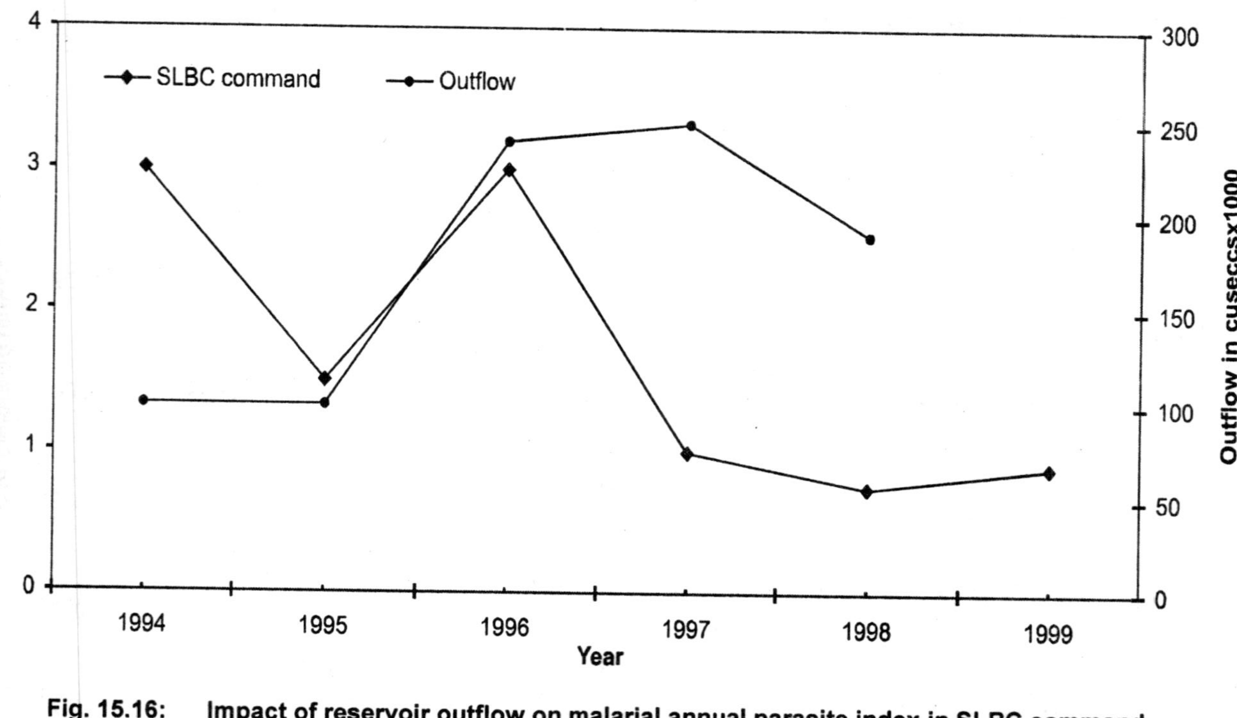

Fig. 15.16: Impact of reservoir outflow on malarial annual parasite index in SLBC command

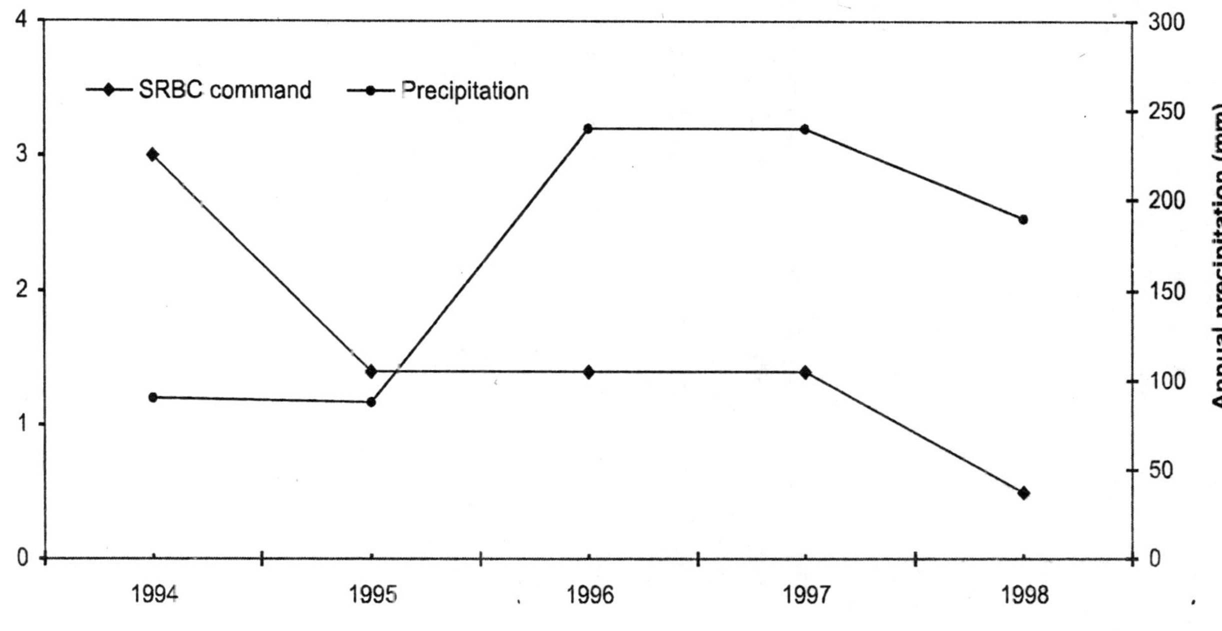

Fig. 15.17: Impact of reservoir outflow on malarial annual parasite index in SRBC command

A good monsoon generally ensures not only a good storage in the reservoir but also makes available plenty of water for rain-fed agriculture. As the main crops require extensive standing water, flood irrigation makes the situation worse. The presence of year-round water in the fields with almost nil dry period and no provision for the withdrawal of water for drying the fields, provides ideal conditions for the anopheline and culecine mosquitoes species to thrive and breed.

Status of Canals and Tanks

The status of canals and tanks as per the extensive field survey of the study area is described in chapters 10 and 16. The highlights of the findings are:

- The branch-canals and distributaries remain unlined due to the lack of adequate amount of funds;
- Most of the unlined branch-canals, drainage system and even the lined canals are in a dilapidated state in the absence of proper maintenance;
- The canal system and the tanks are silted and vegetation infested;
- The fields adjacent to the canals, tanks and the conveyance channels suffer from water logging problems as the water seeps out through them;
- The SRBC command gets less of water for irrigation through the canal system as per the regulations, which assign priority rights to the SLBC command.

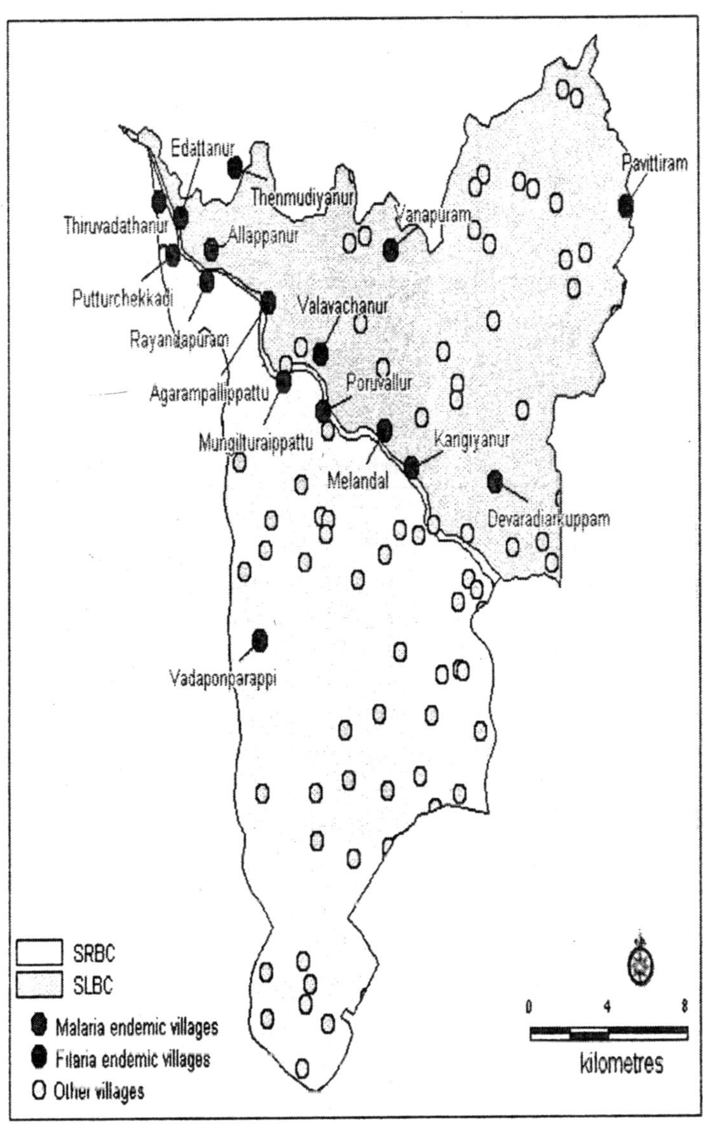

Fig. 15.18: Malaria and Falaria endemic villages in SLBC and SRBC command area

Malaria and filaria endemic villages in SLBC and SRBC

	SLBC	SRBC
Malaria endemic villages	Edattanur	Poruvallur
	Mungilturaipattu	Thiruvadathanur
	Melandal	Putturchekkadi
	Edattanur	Rayandapuram
	Allappanur	Agarampallippattu
	Vanapuram	Vadaponparappi
	Kangiyanur	
	Devaradiarkuppam	
Filaria endemic villages	Thenmudiyanur	
	Valavachanur	
	Pavittiram	

The defective, unlined and vegetation infested conveyance channels including canals, branch-canals, distributaries and the drainage systems play a crucial role in abating the mosquito menace. The vegetation and weeds provide niche to the mosquitoes and the water logged fields, soakage pits and seeping canal provide ideal environment for the pests to perpetuate. The fact that it is not irrigation *per se* but the faulty practices, which by misplacing water to the advantage of certain disease causing vectors leads to the proliferation of disease, has been supported by many studies across the world (Russel, 1938; FAO, 1997; Jain 1994).

Migration

About fifty percent of the workforce in the command region comprises on landless laborers who migrate to the nearby cities of Chennai, Bangalore, Thiruvannamalai and Mumbai in search of greener pasture (Chapter 16) where they earn their livelihood working sometimes in the unhygienic conditions.

The increase mobility of workers has accompanied the development of the command region with better transportation facilities. The migrants, to and from the villages, are the potential source of establishing malaria in the region as confirmed by the health officials. They work under unhygienic conditions in the malarial belts of the towns and come back infected with parasites.

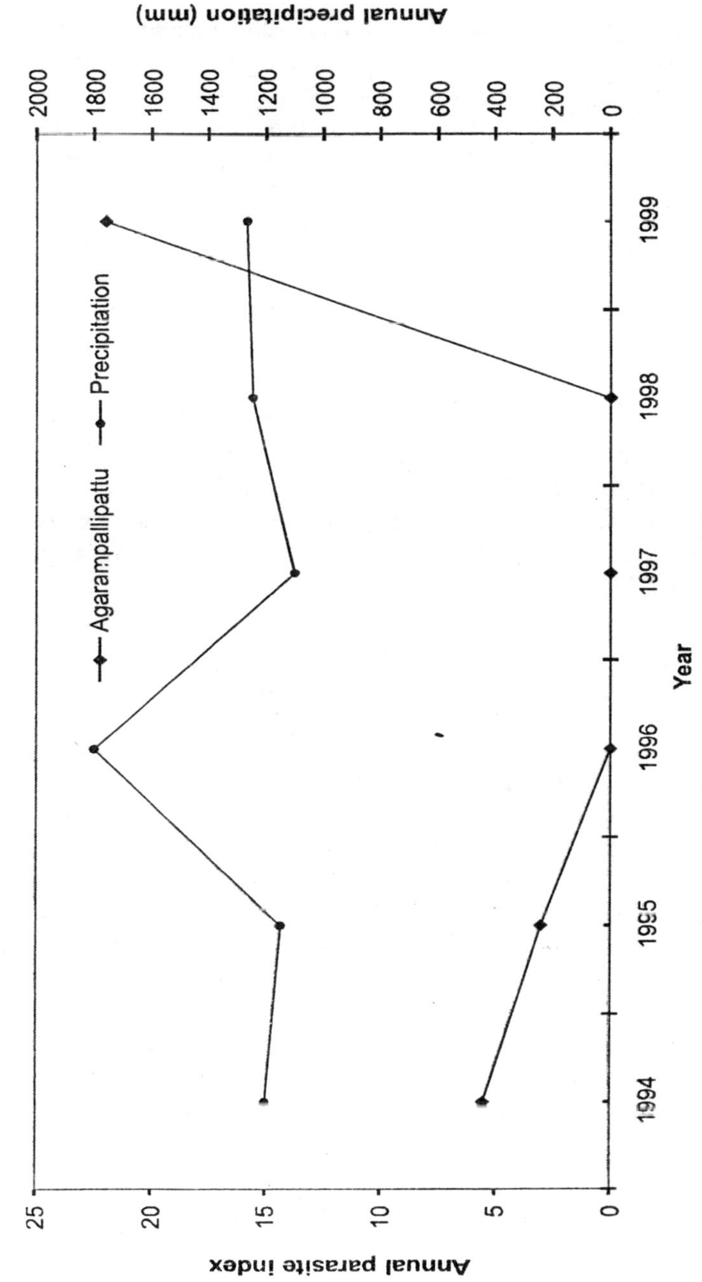

Fig. 15.19: Focal outbreak of malaria in Agarampallipattu village in SLBC command

An ideal combination of host, vector and parasites makes the region endemic to malaria, dengue fever and filaria. Figure—15.19 depicts one such case of focal outbreak of malaria in 1999 in Agarampalipattu village. After a lapse of three years malaria was encountered again in the village, which was brought in the village by the migrants.

The houses mostly built in the form of huts are situated near canals or fields and the male members of the houses sleep outside without any sort of protection against the mosquito bites. Studies conducted elsewhere (Hall et al, 1977) also support the fact that in majority of the irrigation schemes, the tendency has been to build the houses much closer to the canals and irrigated fields.

The SRBC command villages Melandhal, Porasapattu, Vadaponparappii, and Moongilthuraipattu are situated on the banks of the Ponnaiyar river apart from being near to the canals (Figure—15.18). These conditions facilitate breeding and propagation of mosquitoes. As indicated in Figures 15.3, 15.11 and 15.13 the above-mentioned villages recorded quite high API during the study period.

Industry

The Kallakurichi sugar mill, the only industry present in the SCA, is located in Moongilthuraipattu on the bank of Ponnaiyar river. The sugar mill releases its effluents in the river as per the complaints of the villagers downstream. These effluents substantially add to the health hazards along the entire stretch of the river. The poodles of water and other semi permanent pools formed in the river or along the riverbank are concentrated with the effluents and thus provide good breeding ground for the mosquitoes. The industry also brings in large number of hired labourers. Some of them come infected and trigger outbreak of diseases.

The perennial irrigation, agricultural practices of cultivating paddy and sugarcane, poor drainage, faulty and defective canals, waterlogged fields and soakage pits—all have profound bearing on the breeding and thriving of mosquitoes, the main disease vectors in the region.

Figure—15.20 presents the API values for SRBC, SLBC commands and SCA. The curves representing the API values of the three regions, follow the same pattern of fluctuations over the years. There is a decline in the curves in 1995 followed by a rise in 1996. The rise is particularly steep for SLBC command whereas the values of SRBC command have not changed from their previous years. The API values of SLBC and SCA came down sharply in 1997, but very slightly in SRBC command. A further decline was registered in the API values in 1998; this time the intensity of decline was almost similar for both SLBC and SRBC commands.

As discerned from the Figure, the API values for SLBC command were higher than the SRBC command throughout the study period. Wider fluctuations in the values in SLBC command are observed as compared to the values in SRBC command. The reason may be attributed to the fact that SLBC command receives more water for irrigation from the reservoir as compared to SRBC command due to the priority rights accorded in the regulations (Chapter 10). The intensity of irrigation is thus higher in SLBC command. More of water for irrigation through canals and tanks leads to more of water logging and seepage and therefore an increase in the habitats of disease vectors.

A socio-economic survey of the command region has revealed that some villages of SRBC command have not received water for irrigation for past ten plus years either due to the damaged canals or their location in the tail end or due to less water in the Sathanur reservoir (Chapter 16).

Figure—15.21 presents a comparative account of the malaria cases in the SCA and the districts and blocks where major villages of the command are located. The salient points, which emerge, are:

- Thiruvannamalai had the highest incidence of malaria. It is a malaria endemic region, followed by Villupuram district.
- The pattern of malarial cases in the residents of SLBC command closely follows that of the Thiruvannamalai which encompasses most of the villages of SLBC command.

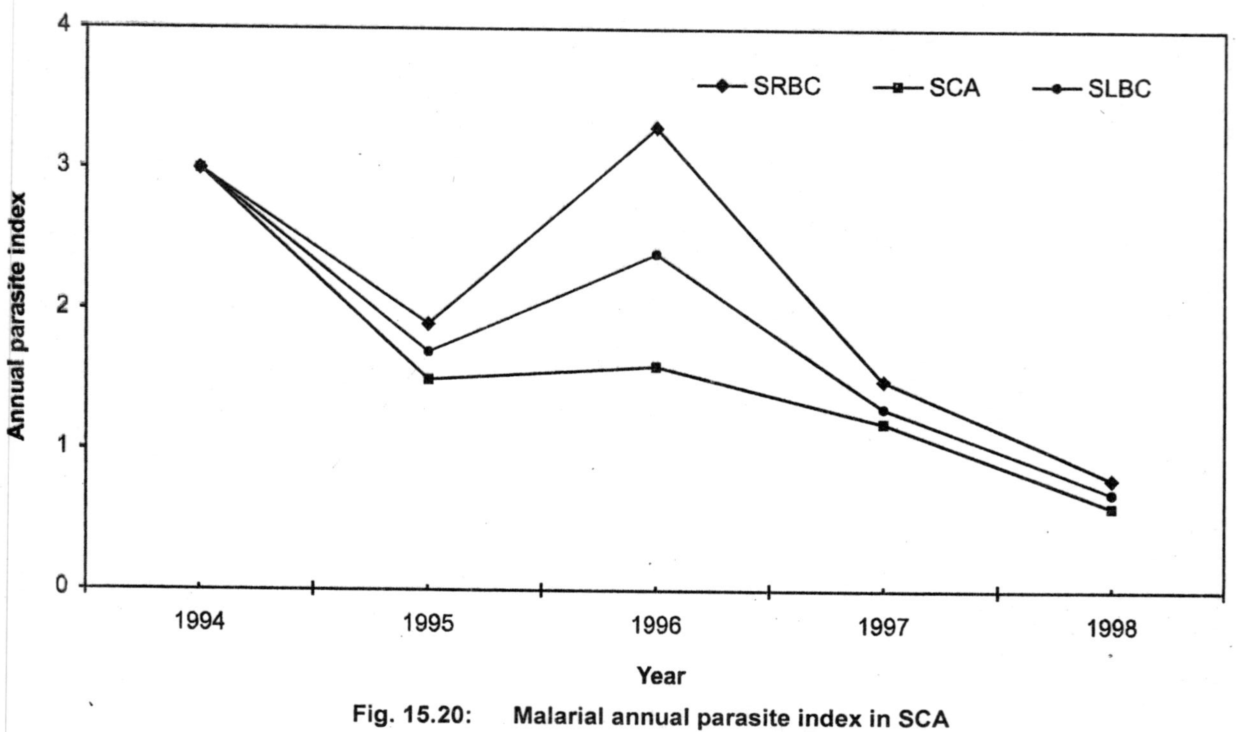

Fig. 15.20: Malarial annual parasite index in SCA

- All the curves in the Figure peaked in 1996, which was the year when maximum rainfall in the tract was recorded.
- The SRBC command falls in Villupuram district and some villages are in Kallakurichi block. The trend in malaria cases in the population of the SRBC command are therefore similar to that observed in Villupuram and Kallakurichi areas which present no wide fluctuations over the years.
- SLBC command had higher incidence of malaria than SRBC command.

Elephantiasis

The survey on the incidences of filaria was initiated only from the year 1998 in the SCA by the Thiruvannamalai DPHS. As per the 1998-99 survey, Palayanur PHC reported the maximum number of elephantiasis cases and Pavithram, Palayanur and Thenmudiyanur were reported to be the endemic villages (Figure 15.18). The API values of filaria for selected villages are presented in Table—15.4.

Current Status of Malaria

SLBC Command

A comparative study of Figures—15.1 to 15.8 reveals that after a period of wide fluctuations over the years, the API values in majority of the villages have come down heavily in SLBC command.

Figure—15.1 reveals a heavy downslide in API values for Devanur and Devariyarkuppam after 1996, the year when the API recorded very high values, 37.5 times and 50.1 times of 1995 values respectively for Devanur and Devariyarkuppam. In 1997 the API value of Devanur came down sharply by 84% and further reduced by 98.6 per cent of its 1996 value in 1999. In Devariyarkuppam, the API reduced to 76 per cent in 1997 and further to 96 per cent of its 1996 value in 1999.

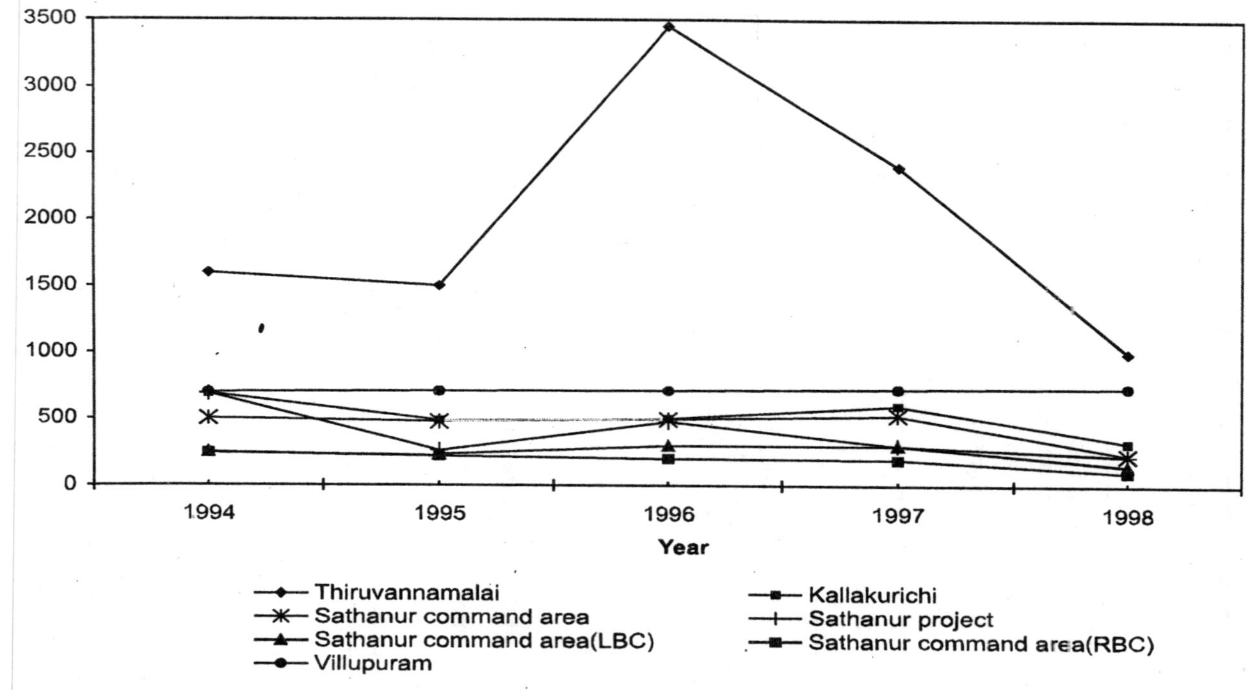

Fig. 15.21: Incidence of malaria

Similar trends of reduction in values were observed in Edathanur, Melandhal, Kangaiyanur, Sadakuppam, Mazhuvampattu, Perunduraipattu and Palayanur. In Edathanur (Figure—15.2) the percentage reduction in API achieved by the end of 1999 was 87.2 per cent of 1996 value while in Melandhal and rest of the villages no malaria incidences were reported in the year 1999.

Not all the villages recorded this declining trend in API. Some villages—Alappanur, Vanapuram, Katampoondi, Thenmudiyanur, Valavachanur and Periyampattu indicate an increase in the API in recent years. In Alappanocr (Figure—15.2) the API value had become zero in 1997 but has increased since then. The API in Vanapuram reduced to 92.8 per cent of its 1996 value in 1998 but in 1999 again rose slightly (Figure—15.4). The API in Katampoondi also registered a mild increase in 1999 after reaching zero in 1998 (Figure—15.4). The API in Valavachanur reached an all time high in 1999 after touching zero in 1997 and 1998.

The API values of SLBC command have fluctuated widely over the years (Figure—15.9). The API came down by 46.5 per cent in 1995 but rose sharply by 93.8 per cent of its 1995 value in 1996 before dipping to 58.6 per cent of its 1996 value in 1998. The drop continued to 80.9 per cent in 1998. An increase in API by 43.3 per cent of its 1998 value was observed in 1999.

SRBC Command

As indicated in Figures—15.10 to 15.14 there has been a general decline in the API values in the SRBC command over the years, with the exceptions in Rayandapuram and Poravalur villages.

The API value in Rayandapuram (Figure—15.10) registered a slight increase in 1999 after ebbing to 97.2 per cent of its 1996 value in 1998. In Poravalur, the API shot up in 1997 after reaching zero in 1996. The rising trend continued till 1998 (Figure—15.13). Thiruvadithanur and Elayankani registered nil API values in 1999 while in Vadaponparappi and Moongithuraipattu, the values kept reducing steadily till they declined by 75.1 per cent and 81.4 per cent of their 1994 values respectively in 1998 (Figures—15.10 to 15.12).

Cumulatively, API in SRBC command (Figure—15.14) is lesser in 1998 compared to its 1995, 1996 and 1997 values. The API of SCA has also been declining (Figure—15.19). The value dipped by 41.6 per cent of its 1996 value in 1997 and fell to 77.3 per cent of the 1996 API in 1998.

From the foregoing discussion it emerges that even though the menace of malaria in the SCA has not been wiped out completely, it has been checked to a great extent. The measures adopted by the health officials in controlling the disease vectors are summarized in the following section.

Preventive Measures Adopted by the Health Officials

The various measures implemented by the health officials in checking the propagation of malaria downstream SRP include:

- Routine blood smear tests on the villagers;
- Regular filling-up of the pits and depressions to prevent accumulation of stagnant water;
- Distribution of anti-malarial drugs, chloroquine and primaquine, free of cost amongst the villagers;
- Distribution of Diethyl Carbamazine Citrate (DEC) free of cost to the villagers suffering from filaria;
- Periodic spray of DDT, malathion and pyrethrum to kill mosquito larvae;
- *Public awareness:* frequent village to village tours of volunteers to make the villagers aware of the importance of personal hygiene in preventing water-related diseases.

The prophylactic measures adopted by the health officials are summarized in Tables—15.5 to 15.11.

The major constraints the health officials face are the poor socio-economic and educational status of the villagers, which provides them little opportunity of being groomed vis a vis personal hygiene. It takes a lot of effort on the part of the health officials to bridge their knowledge gap. The results are encouraging. The statistics indicate that in most of the places the officials have successfully prevented the diseases from becoming epidemics. After registering a peak in 1996 or 1997, the API has shown a declining trend. The focal outbreak of malaria in 1999 in Agarampallipattu was controlled within twenty-five days under the efficient supervision of health officials.

Even the villagers are more aware and vigilant now. Most of the migrants after coming back to their natives voluntarily go to PHCs to get themselves examined as per the information of health officials.

Findings of the Field Trip to SCA

Even as considerable success is indicated, as above, keeping malaria, filaria, and other water-borne diseases in check at SRP, in the perception of most of the villagers the governmental support in health care is woefully inadequate. They complain of lack of attention and lack of medicines. Some even allege that the medical records aren't maintained properly, obliquely suggesting that the claim of success in the control of water-borne diseases may be based on fabricated records. From other countries too, similar situations have been reported (Biswas and El-habr, 1991).

It is undeniable that the ratio of PHCs to the population in the command region is very low as depicted in Figure—15.22. Only thirteen PHCs exist, for over 100,000 people in SCA apart from the numerous villages outside the command area under each PHC. The distribution of the population of the villages in the command area under each PHC is presented in Table—15.5. It must be emphasized that inadequacy of health care in SRP is typical of thc rural India (Raghuram, 1996).

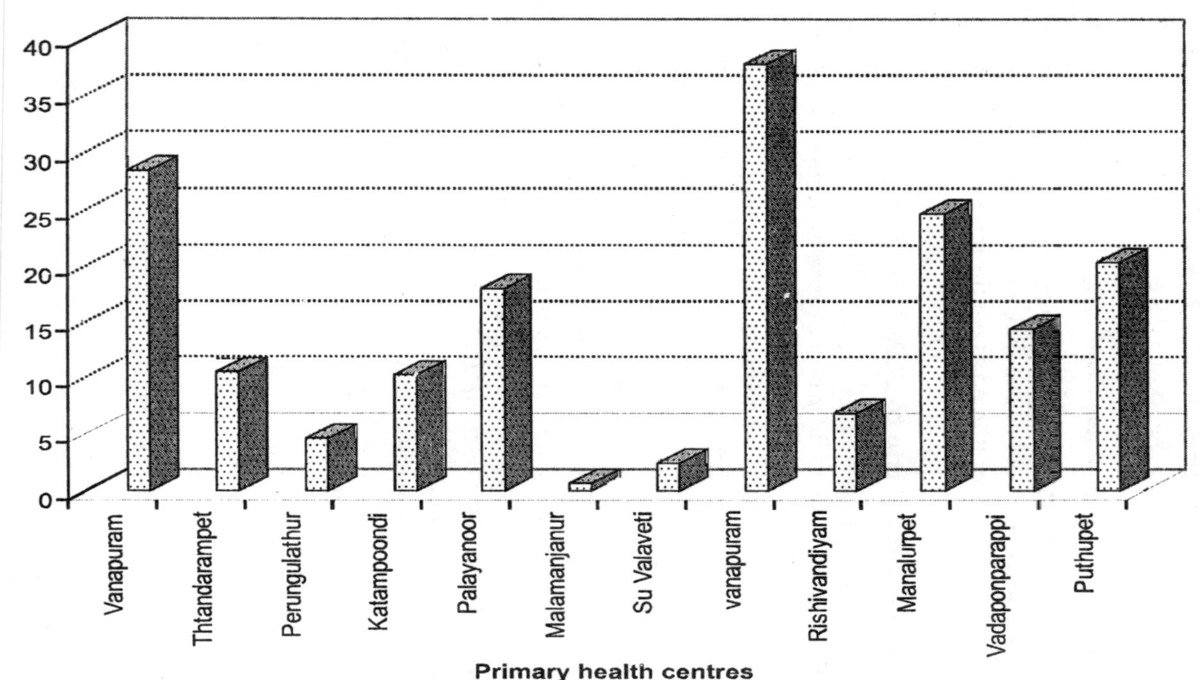

Fig. 15.22: Population distribution in various primary health centres in SCA

16

Socio-economics

INTRODUCTION

The *raison de être* of any water resources project is improvement of the socio-economic condition of its target population. This is the guiding theme, the ultimate objective. But assessment of the impact of a water resources project on the socio-economic environment is lot more difficult than assessment of impacts on various biotic/abiotic dimensions of the environment. To begin with, the definition of socio-economic progress is extremely difficult to establish as people belonging to different economic strata, social background, political beliefs, and levels of education can have very widely divergent perceptions of what constitutes socio-economic progress (Sen, 2000). There can be equally sharp differences in the perceptions of the project authorities and the people living in the project's command area. Further, unlike such parameters as water table, water quality, agricultural production or wildlife diversity, socio-economic impacts don't lend themselves to quantification with standardized equipment or by standardized procedures. And yet, assessment of socio-economic impacts is, perhaps, the most important component of any environmental impact assessment.

We have made an attempt to generate a body of information with which the perceptions of the end-users of Sathanur Reservoir

Project (SRP) can be assessed in a hopefully, quantitative manner. A scale was created which grades the people's responses from 'very good' to 'very bad'. The responses were then collected on the basis of an extensive survey assisted by an elaborate checklist. The matrix of type of response (positive or negative) and degree of response (very good to very bad) was then used to generate 'environmental impact units' (EIUs).

STATUS

An extensive socio-economic survey of the catchment and command area of the SRP was undertaken. The field survey included intensive interactions with the cultivators downstream. Fifty-four villages, twenty-seven in each of the Sathanur Left Bank Canal (SLBC) and Sathanur Right Bank Canal (SRBC) commands, were covered during the survey. The villages were selected so as to display representativeness in terms of area, distance from the Sathanur dam, area under canal irrigation, area indirectly under canal irrigation through tanks, location in the head, middle or tail reaches of the canals and the geo-climatic conditions. The respondents were also selected based on their socio-economic status, age and the size of their holdings – marginal farmers (possessing below 1.0 hectare of agriculture land), small farmers (possessing between 1.00 – 2.50 hectares of agriculture land), medium farmers (possessing between 2.5 – 5.0 hectares of agriculture land) large farmers (possessing more than 5.0 hectares of agriculture land) and the landless labourers.

The information was obtained in the form of a questionnaire (Annexure 16.1), which was designed so as to allow the respondents to express their views without bias or external pressure. The socio-economic indicators were placed under four broad categories namely agriculture, irrigation, maintenance of canals and tanks and socio-economic. Questions framed for each indicator were given seven qualitative choices ranging from excellent, to very good, good, no different, bad, very bad and worse. These choices were further given weightings on seven point scale ranging from – 3 to 0 to 3. A Delphi methodology was used in the final assessment of the qualitative data thus generated.

ANNEXURE

Name of village : Location :

Name of the person contacted :

Age :

Occupation :

Education :

Income :

Holding (in hectare) :

Major crops cultivated in various seasons :

AGRICULTURAL INDICATORS Status of	The situation is relatively						
	Worse	Very bad	Bad	No different	Good	Very good	Excellent
Area brought under cultivation							
Productivity per hectare							
Usage of chemical fertilizers							
Usage of bio fertilizers							
Problem of weeds in the fields							
Pests menace in the crops							
Usage of weedicides and pesticides							
Scientific approach towards agriculture							
Agricultural socictics and bank							
IRRIGATION INDICATORS							
Status of							
Sources of irrigation							
Mode of irrigation							
Water logging in the fields							
Salinity and alkalinity in the field							
Water-logging in the fields near to canals							
Drainage in the fields							
Canal Irrigation /tank irrigation							

Modern implements of agriculture and irrigation
INDICATORS OF MAINTENANCE OF CANAL AND TANK   **Status of**
Lining of the canals
Vegetation in the vicinity of canals
Vegetation in the vicinity of tanks
Siltation in canals
Siltation in tanks
Seepage from the canals
Optimal flow of water from reservoir through canals
Judicious distribution of water to the fields
Breaching and encroachment in canals and tanks
SOCIO-ECONOMIC INDICATORS   **Status of**
Income
Primary health centres
Water-borne disease
Schools
Forest
Roads
Drinking water
Power supply
Employment and migration

Though the thrust was on the canal and tank irrigation, but the approach was in such a way that the general questions related to each indicator were addressed first and then the implications of the canal irrigation were sought. This approach prevented the bias to creep into the methodology. Whenever the interviewees were in dilemma they were asked to compare and contrast the present scenario with the scenario fifteen to twenty years back.

The sample size of each village was kept constant – ten villagers. In some villages due to some unforeseen circumstances required number of the villagers were not interviewed. As, in a village due to death most of the men were unavailable as they were out attending the funeral ceremony. In such exceptional cases the data was converted to a ten-point scale. Similarly, the responses of twenty experts and the 540 villagers were also converted to a common unit of forty for comparison. Impact units were computed and compared so as to evaluate the impact of the Sathanur project on the various ecological parameters as perceived by the end-users and the project authorities. Regressions were performed on the ecological indicators so as to understand the significance of various interrelationships.

Tables—16.1 and 16.2 present village-wise cumulative impact units for various indicators in SRBC and SLBC commands respectively. Table—16.3 gives an account of the total impact units as assigned by the natives of SRBC and SLBC commands and cumulatively for the Sathanur Command Area (SCA). Tables—16.4 and 16.5 present the weightings assigned by the respondents to various ecological indicators in villages of the SLBC and SRBC commands respectively. Table—16.6 presents an overall account of the EIUs for various indicators in SLBC, SRBC commands and SCA. Table—16.7 presents a comparison between the EIUs as perceived by the dwellers in the SCA and the authorities connected with the various aspects of the SRP.

The results of the correlation analysis, which was performed to assess the interdependencies of indicators are presented in Table—16.8. The coefficients of the rank correlation between the various villages based on the weightings assigned as per the EIUs are presented in Tables—16.9 and 16.10 for SLBC and SRBC commands respectively.

Table—16.1 Weightings assigned by the respondents in SLBC command

AGRICULTURAL INDICATORS Status of	-3	-2	-1	0	1	2	3
Area brought under cultivation	–	–	–	23	131	113	3
Productivity per hectare	–	–	65	11	145	49	
Usage of chemical fertilizers	–	–		8	115	138	9
Usage of bio fertilizers	–	13	193	–	59	5	–
Problem of weeds in the fields	–	60	120	–	–	–	–
Pests menace in the crops	7	131	132	–	–	–	
Usage of weedicides and pesticides	–		14	20	199	37	–
Scientific approach towards agriculture	–	9	168	–	79	14	–
Agricultural societies and bank	–	7	110	–	103	50	–
IRRIGATION INDICATORS Status							
Sources of irrigation	–				112	107	51
Mode of irrigation	–	48	222	–	–	–	–
Water logging in the fields	–	4	57	–	176	33	–
Salinity and alkalinity in the field	–	24	44	–	135	67	–
Water-logging in the fields near to canals	21	57	149	–	34	9	–
Drainage in the fields	–		72	–	150	48	–
Canal Irrigation /tank irrigation	–	25	09		170	63	3
Modern implements of agriculture and irrigation	–	–	–	55	178	37	–

(Table Contd...)

INDICATORS OF MAINTENANCE OF CANAL AND TANK Status of							
Lining of the canals	9	55	51	-	37	97	21
vegetation in the vicinity of canals	-	41	147	-	72	10	-
Vegetation in the vicinity of tanks	30/180	39/180	76/180	-	23/180	12/180	-
Siltation in canals	-	2	136		122	10	-
Siltation in tanks	-	112/180	54/180	-	7/180	7/180	-
Seepage from the canals	-	63	171	-	32	4	-
Optimal flow of water from reservoir through canals	29	14	53	-	77	87	10
Judicious distribution of water to the fields	9	21	113	-	99	18	10
Breaching and encroachment in canals and tanks	11	89	28	-	119	23	-
SOCIO-ECONOMIC INDICATORS **Status of**							
Income	-	-	54	-	139	77	-
Primary health centres	-	-	60	-	155	27	28
Water-borne disease	8	17	83	-	132	30	-
Schools	-	-	22	-	141	91	16
Forest	13	8	170	-	67	12	-
Roads	-	-	22	-	71	167	10
Drinking water	-	8	45	-	125	92	-
Power supply	-	6	75	-	103	86	-
Employment and migration	-	90	134	-	41	5	-

Table—16.2 Weightings assigned by the respondents in SRBC command

AGRICULTURAL INDICATORS Status of	-3	-2	-1	0	1	2	3
Area brought under cultivation	–	–	–	7	72	191	–
Productivity per hectare	–	9	32	40	186	3	–
Usage of chemical fertilizers	–	–	–	12	87	171	–
Usage of bio fertilizers		40	94	–	136	–	–
Problem of weeds in the fields	–	113	157	–	–	–	–
Pests menace in the crops	–	57	213				
Usage of weedicides and pesticides	–			45	172	53	
Scientific approach towards agriculture	–	24	79	–	80	87	–
Agricultural societies and bank	–	99	47	–	25	99	–
IRRIGATION INDICATORS **Status of**							
Sources of irrigation				15	79	176	
Mode of irrigation		59	199	11	1		
Water logging in the fields		5	52	5	161	47	
Salinity and alkalinity in the field		20	48	25	110	66	1
Water-logging in the fields near to canals		35	73	22	84	56	
Drainage in the fields		13	77	–	123	57	
Canal Irrigation/tank irrigation	21	58	51	43	67	30	
Modern implements of agriculture and irrigation	–	–	–	81	169	20	

(Table Contd...)

INDICATORS OF MAINTENANCE OF CANAL AND TANK **Status of**							
Lining of the canals	78	72	40	–	3	62	15
Vegetation in the vicinity of canals	82	71	79	–	37	1	
Vegetation in the vicinity of tanks	46	136	58	–	30	–	–
Siltation in canals	21	83	134	–	24	8	
Siltation in tanks	50	106	114	–	–	–	–
Seepage from the canals	–	30	127	91	21	–	
Optimal flow of water from reservoir through canals	88	53	45	–	39	45	–
Judicious distribution of water to the fields	43	57	86	20	46	16	2
Breaching and encroachment in canals and tanks	95	12	29	-	100	34	-
SOCIO-ECONOMIC INDICATORS **Status of**							
Income	–	–	99	–	132	35	–
Primary health centres	–		94	5	129	32	10
Water-borne disease	–	12	84	9	123	42	–
Schools	–	–	–	–	160	100	10
Forest	–	–	231	15	18	6	–
Roads	–	4	7	5	99	192	43
Drinking water	–	1	29	–	107	133	–
Power supply	–	–	70	–	124	76	–
Employment and migration	–	–	99	171	–	–	–

Table—16.3 Cumulative weightings assigned by the respondents in SCA

AGRICULTURAL INDICATORS Status of	-3	-2	-1	0	1	2	3
Area brought under cultivation	–	–	–	30	203	304	3
Productivity per hectare	–	9	97	51	331	52	–
Usage of chemical fertilizers	–	–	–	20	202	309	9
Usage of bio fertilizers	–	53	287	–	195	5	–
Problem of weeds in the fields	–	173	367	–	–	–	–
Pests menace in the crops	7	188	345	–	–	–	–
Usage of weedicides and pesticides	–	–	14	65	371	90	–
Scientific approach towards agriculture	–	33	247	–	159	101	–
Agricultural societies and bank	–	106	157	–	128	149	–
IRRIGATION INDICATORS **Status of**							
Sources of irrigation	–	–	–	15	191	283	51
Mode of irrigation	–	107	421	11	1	–	–
Water logging in the fields	–	9	109	5	337	80	
Salinity and alkalinity in the field	–	44	92	25	245	133	1
Water-logging in the fields near to canals	21	92	222	22	118	65	–
Drainage in the fields	–	13	149	–	273	105	–
Canal Irrigation/tank irrigation	21	83	60	43	237	93	–
Modern implements of agriculture and irrigation	–	–	–	136	347	57	–

(Table Contd...)

INDICATORS OF MAINTENANCE OF CANAL AND TANK Status of							
Lining of the canals	87	127	91	-	40	159	36
Vegetation in the vicinity of canals	82	112	226	-	109	11	-
Vegetation in the vicinity of tanks	76/450	175	134	-	53	12	-
Siltation in canals	21	85	270	-	146	18	-
Siltation in tanks	50/450	218	168	-	7	7	-
Seepage from the canals		93	298	91	54	4	-
Optimal flow of water from reservoir through canals	117	67	98	-	116	132	10
Judicious distribution of water to the fields	52	78	199	20	145	34	12
Breaching and encroachment in canals and tanks	106	101	57	-	219	57	-
SOCIO-ECONOMIC INDICATORS **Status of**							
Income	-	-	153	4	217	112	-
Primary health centres	-	-	154	5	284	59	38
Water-borne disease	8	29	167	9	255	72	-
Schools	-	-	22	-	301	191	26
Forest	13	8	401	15	35	18	-
Roads	-	4	29	5	170	279	53
Drinking water	-	9	74	-	232	225	-
Power supply	-	6	145	-	227	162	-
Employment and migration	-	189	305	-	41	5	-

Table—16.4 Environmental impact units for various ecological indicators : SLBC command.

S. No.	Villages	Agricultural impact units	Irrigation impact units	Canal and tank maintenance impact units	Socio-economic impact units	Total impact unit
1.	Vanapuram	46	34	133	123	336
2.	Thenmudiyanur	36	32	98	117	283
3.	Perunduraipattu	28	40	72	62	202
4.	Jambai	22	36	45	73	176
5.	Periyakallapadi	7	64	44	58	173
6.	Sukhampalayam	7	45	56	49	157
7.	Kandiankuppam	22	60	-32	95	145
8.	Periyampattu	17	45	0	48	110
9.	Devariyarkuppam	10	45	-1	52	106
10.	Kangaiyanur	26	51	-34	57	100
11.	Sadakuppam	5	6	42	46	99
12.	Thachampattu	28	45	-73	95	95
13.	Palayanur	3	27	-37	105	98
14.	Pudur	11	52	-26	45	82
15.	Pappambadi	9	39	-30	42	60
16.	Agarampallipattu	-16	41	-45	69	49
17.	Pallirhandal	2	34	-3	-4	29

(Table Contd...)

1	2	3	4	5	6	7
18.	Valavachanur	15	41	-55	17	18
19.	Pavapattu	6	37	-76	48	15
20.	Parayampattu	12	23	-70	22	-13
21.	Melandhal	30	24	-85	14	-17
22.	Unnamalaipalayam	-18	13	-36	16	-25
23.	Arundhatiar Colony	-7	34	-87	29	-31
24.	Kunglinatham	14	24	-102	25	-39
25.	Sithapattinam	21	21	-112	21	-62
26.	Devanur	14	27	-121	14	-66
27.	Thenkarimbalur	-12	-39	-141	-40	-202

Table—16.5 Environmental impact units for various ecological indicators : SRBC command

S. No.	Villages	Agricultural impact units	Irrigation impact units	Canal and tank maintenance impact units	Socio-economic impact units	Total impact units
1.	Moongilthuraipattu	48	39	18	144	249
2.	Kadavanoor	50	18	27	101	196
3.	Vadamamandur	42	47	18	55	162
4.	Athiyur	51	60	-11	58	158
5.	Vadakiranur	30	18	-7	78	119
6.	S.Kolathur	34	69	-67	60	95
7.	Varagur	28	38	-10	21	77
8.	Arambarampattu	45	62	-100	67	74
9.	Arulampadi	25	-13	-26	58	44
10.	Edathanur	24	30	-50	32	36
11.	Tholuvanthaugal	24	11	-56	43	22
12.	Vadaponparappi	12	-5	-30	39	16.
13.	Sirpanandal	40	43	-133	36	-14
14.	Erudayampattu	36	36	-146	39	-35
15.	Jambodai	25	38	-150	19	-68
16.	Pakkam	14	4	-83	-6	-71

(Table Contd...)

1	2	3	4	5	6	7
17.	Melsiruvalur	1	25	-146	42	-78
18.	Olgalapadi	24	25	-144	17	-78
19.	Viriyur	7	54	-200	55	-84
20.	Poravalur	26	-10	-145	37	-92
21.	Arur	23	35	-193	39	-96
22.	Periyakolliyur	17	26	-220	59	-118
23.	Chinnakolliyur	4	24	-170	22	-120
24.	Porasapattu	7	-2	-140	6	-129
25.	Mangalam	25	20	-193	9	-139
26.	Kidagudayampattu	6	5	-180	-24	-147
27.	Thimendhal	14	14	-240	32	-180

Table—16.6 Cumulative environmental impact units for various ecological indicators: SCA

AGRICULTURAL INDICATORS Status of	*Environment impact units*		
	SLBC command	SRBC command	SCA
Area brought under cultivation	366	454	820
Productivity per hectare	178	142	320
Usage of chemical fertilizers	418	429	847
Usage of bio fertilizers	-150	-38	-188
Problem of weeds in the fields	-330	-383	-713
Pests menace in the crops	-415	-327	-742
Usage of weedicides and pesticides	259	278	537
Scientific approach towards agriculture	-79	127	48
Agricultural societies and bank	89	-22	67
IRRIGATION INDICATORS			
Status of			
Sources of irrigation	479	431	910
Mode of irrigation	-318	-306	-624
Water logging in the fields	-177	197	374
Salinity and alkalinity in the field	171	157	328
Water-logging in the fields near to canals	-280	53	-227
Drainage in the fields	174	134	308
Canal Irrigation /tank irrigation	246	-103	143
Modern implements of agriculture and irrigation	252	209	461

(Table Contd...)

INDICATORS OF MAINTENANCE OF CANAL AND TANK Status of			
Lining of the canals	105	-246	-141
Vegetation in the vicinity of canals	-137	-428	-565
Vegetation in the vicinity of tanks	-197	-438	-635
Siltation in canals	2	-323	-321
Siltation in tanks	-257	-476	-733
Seepage from the canals	-274	-165	-439
Optimal flow of water from reservoir through canals	113	-286	-173
Judicious distribution of water to the fields	-17	-245	-262
Breaching and encroachment in canals and tanks	-14	-170	-184
SOCIO-ECONOMIC INDICATORS			
Status of			
Income	239	103	342
Primary health centres	233	129	362
Water-borne disease	51	100	151
Schools	349	390	739
Forest	-130	-201	-331
Roads	413	438	851
Drinking water	248	342	590
Power supply	188	206	394
Employment and migration	-263	-369	-632

Table—16.7 Comparison of cumulative impact units of end users and experts: SCA

AGRICULTURAL INDICATORS Status of	*Environmental Impact Units*			
	SRBC command	**SLBC command**	**SCA**	**Experts**
Area brought under cultivation	67.25	54.22	60.74	104
Productivity per hectare	21.03	26.37	23.70	120
Usage of chemical fertilizers	63.55	61.9	62.74	120
Usage of bio fertilizers	-5.62	-22.2	-13.92	-40
Problem of weeds in the fields	-56.74	-48.88	52.81	-16
Pests menace in the crops	-48.44	-61.48	-54.96	-16
Usage of weedicides and pesticides	41.18	38.37	39.77	80
Scientific approach towards agriculture	18.81	-11.70	3.55	64
Agricultural societies and bank	-3.25	13.18	4.95	80
Cumulative impact units of Agriculture	97.77	49.7	73.77	496

(Table Contd...)

IRRIGATION INDICATORS				
Status of				
Sources of irrigation	63.85	70.96	67.40	152
Mode of irrigation	-45.33	-47.11	-46.22	-16
Water logging in the fields	29.18	26.22	27.70	40
Salinity and alkalinity in the field	23.25	25.33	24.29	32
Water-logging in the fields near to canals	7.85	-41.48	-16.81	-32
Drainage in the fields	19.85	25.77	22.81	96
Canal Irrigation/tank irrigation	-15.25	36.44	10.59	128
Modern implements of agriculture and irrigation	30.96	37.33	34.14	80
Cumulative impact units of irrigation	114.37	133.48	123.92	480
INDICATORS OF MAINTENANCE OF CANAL AND TANK				
Status of				
Lining of the canals	-36.44	15.55	-10.44	-32
Vegetation in the vicinity of canals	-63.40	-20.29	-41.85	-24
Vegetation in the vicinity of tanks	-64.88	-29.18	-47.03	-8
Siltation in canals	47.85	0.29	-23.77	16
Siltation in tanks	-70.51	-38.07	-54.29	-8
Seepage from the canals	-24.44	-40.59	-32.51	16
Optimal flow of water from reservoir through canals	-42.37	16.74	-12.81	40
Judicious distribution of water to the fields	-36.29	-2.51	-19.40	24
Breaching and encroachment in canals and tanks	-25.18	-2.07	-13.62	16
Cumulative impact units of canal and tank maintenance	-411.4	-100.14	-255.7	40

(Table Contd...)

SOCIO-ECONOMIC INDICATORS Status of				
Income	15.25	35.40	25.33	70
Primary health centres	19.11	34.51	26.81	40
Water-borne disease	14.81	7.55	11.18	32
Schools	57.77	51.70	54.74	56
Forest	-29.77	-19.25	-24.51	-16
Roads	64.88	61.18	63.03	54
Drinking water	50.66	36.74	43.70	44
Power supply	30.51	27.85	29.18	80
Employment and migration	-54.66	-38.96	-46.81	-20
Cumulative impact units of socio-economic	168.59	196.74	182.66	340
GRAND TOTAL IMPACTS	30.66	279.85	124.59	1356

Table—16.8 Correlation coefficients between various ecological indicators: SCA

	SRBC command	SLBC command
Productivity		
Area under cultivation	0.8	0.5
Usage of chemical fertilizers	0.5	0.6
Usage of bio fertilizers	0.5	-0.8
Problem of weeds in the fields	0.2	-0.3
Pests menace in the crops	-0.1	-0.4
Pesticides and weedicides application	0.1	0.5
Scientific approach towards agriculture	0.6	0.2
Sources of irrigation	0.5	0.5
Mode of irrigation	-0.4	-0.3
Salinity in the field	-0.4	-0.1
Water logging in the field	-0.5	-0.1
Modern implements of agriculture and irrigation	0.5	0.5
Canal irrigation	0.4	0.6
Optimal flow of water from reservoir into the canals	0.9	0.9
Judicious distribution of water to the fields	0.5	0.4
Income	0.6	0.6
Power supply	0.6	0.7
Roads	0.5	0.7
Lining of canal		
Water logging in the fields near to canal	-0.2	0.7
Water logging in the fields	-0.1	
Vegetation in the vicinity of canals	0.6	0.4
Siltation in the canal	0.5	0.4
Seepage from the canal	-0.3	0.3
Breaching and encroachment of the canals and tanks	0.4	0.4
Water-logging in the fields		
Salinity and alkalinity problems in the field	0.8	0.8
Drainage	0.5	0.2
Seepage from the canals	0.3	0.7
Canal irrigation/ tank irrigation	0.01	0.4
Optimal flow of water from reservoir through canals		
Water borne diseases	0.2	0.6

Table—16.9 Rank correlations between the villages based on the EIUs of ecological indicators in SLBC command

Indicators	Agriculture	Irrigation	Canal and tank maintenance	Socio-economic
Agriculture	1	0.23	0.26	0.47
Irrigation		1	0.31	0.30
Canal and tank maintenance			1	0.62
Socio economic				1

Table—16.10 Rank correlations between the villages based on the EIUs of ecological indicators in SRBC command

Indicators	Agriculture	Irrigation	Canal and tank maintenance	Socio-economic
Agriculture	1	0.39	0.60	0.55
Irrigation		1	0.01	0.21
Canal and tank maintenance			1	0.46
Socio-economic				1

IMPACTS

Agricultural Indicators

Area under Cultivation

There is a definite increase in the area under cultivation as envisaged from the EIUs of 366 and 454 of the total agricultural impacts units of 329 in SLBC and 660 in SRBC command respectively (Table—16.6).

In SLBC command 41.9 per cent of the respondents assigned the weighting 'very good' and 48.5 per cent 'good' to the status of land under cultivation. Of the respondents, 1 per cent found the situation 'excellent' while 8.5 per cent of the respondents opined that the situation was no different (Table 16.1).

In contrast to such homogenous pattern, 70.7 per cent of the respondents in SRBC command deemed the situation to be very good and 26.6 per cent to be good. Only 2.6 per cent of the respondents opined no change in the situation (Table—16.2).

The statistics substantiate the fact that after the introduction of canal irrigation in SCA, there has been an increase in the net sown area at the cost of the land meant for other purposes (Chapter 11). With the availability of plenty of water, there has been an increasing tendency to bring more and more of land under cultivation of food crops and the cash crops viz paddy (*Oryza sativa*), sugarcane (*Saccharum officinarum*), groundnut (*Arachis hypogea*) etc.

Natives of SRBC command, after going through the trauma of droughts years of water scarcity have reason to assign more weightings to this indicator thereby boosting the SRBC EIUs compared to the SLBC command.

Productivity Per Hectare

The EIUs of productivity of land are 178 and 142 for SLBC and SRBC commands respectively (Table—16.6). The opinions of the respondents were divided on this matter. While 53.7 per cent and 18.2 per cent of the interviewees in SLBC command considered the productivity to be good and very good respectively, about 24 per cent of them thought the productivity has gone down. Only 4% felt there was no difference (Table—16.1).

In SRBC command, 68.9 per cent of the villagers interviewed, expressed the yield to be good and 1 per cent as very good. About 14.8 per cent opined that there was no much of difference while 11.8 per cent and 3.3 per cent of the respondents deemed the situation to be bad and very bad respectively (Table—16.2).

The reasons given by the respondents for increased output were the usage of better variety of seeds, fertilizers, pesticides and, above all, the availability of water. Interestingly these were the very same explanations given by the interviewees who considered the situation to have become worse. According to them the usage of excessive chemicals has led to the loss of the natural fertility and moisture holding capacity of the soil. The near total absence of the use of traditional bio fertilizers such as compost might have contributed to the decline in the fertility of the soil. This situation is further worsened by the attack of pests and weeds. There is also a dearth of agricultural labourers to work in the fields as reported by the villagers. The landless labourers of the villagers migrate to nearby towns and cities to work.

The Agricultural Officers contacted by us were unanimous in that the productivity per hectare has increased. According to them the current status of the farmers is much better when compared to a decade back. They say that not only the people benefiting from the irrigation potential created by SRP but the entire state of Tamil Nadu has achieved self-sufficiency in food production. It is the lack of scientific and judicious approach towards agriculture that makes the productivity seem less.

Usage of Chemical Fertilizers

On the status of the usage of chemical fertilizers the responses veered towards a welcome trend. About 3 per cent of the respondents in the SLBC command considered the situation to be excellent, 51 per cent to be very good and 42 per cent to be good, (Table—16.1). In SRBC command 63 per cent of the interviewees expressed the situation as very good and 32 per cent to be good. Only 2 per cent and 4 per cent of the respondents in SLBC command respectively viewed the situation to be no different (Table—16.2).

Introduction of irrigation facilities, especially in the form of canal irrigation, invariably leads to the higher dependency on chemicals. Most commonly used fertilizers in SCA are DAP (Di ammonium phosphate), complex (17,17,17) urea, potash etc. Even the marginal and small farmers, who in normal circumstances can't afford to buy the fertilizers, somehow arrange for the loans and manage to use them with the blind perception that the usage of chemicals would invariably lead to better productivity.

The total EIUs garnered by this particular indicator are 418 (SLBC command) and 429 (SRBC command) (Table—16.6). Massive usage of chemical fertilizers has been deemed as a sign of progress and a direct impact of canal irrigation.

Usage of Bio Fertilizers

The negative EIUs of this particular indicator in SLBC and SRBC commands, –150 and –38 respectively, indicate its' sorry state (Table—16.6). The situation is better in SRBC command compared to SLBC command as evident by the impact units and also from the views of the respondents. In SLBC command 21 per cent of the respondents assigned the weighting of good to the usage of bio fertilizer to their fields whereas 71 per cent assigned bad and another 4 per cent very bad. A mere 1 per cent of the interviewers considered the status of application of bio fertilizers to be very good (Table—16.1). On the contrary 50 per cent of the respondents in SRBC command expressed the situation to be good, while 34 per cent as bad and 14 per cent as very bad (Table—16.2). Natives of SRBC command, being economically poorer, still stick to the traditional application of cow dung and composted manure along with the use of chemicals, which they normally buy in small quantities. On the other hand, the traditional fertilizers cow dung and composted manures have been completely replaced by the chemical fertilizers in the SLBC command.

The plausible reason for the reduced application of bio fertilizers is the decrease in the cattle population. Maintaining cattle have become very expensive as, there are no grazing lands left (Chapter 11). Also there are no good veterinary hospitals in the region. Villagers find it profitable to sell them for their flesh. The sight of cattle in the villages during the field trips was rare. As a result there is no dung manure available. The villagers consider buying fertilizers as better option than spending their money on buying dung manure from the other villages. Though some farmers did agree to actually paying and buying cow dung after their realization that soil was losing fertility fast with complete dependence on chemical fertilizers.

Problems of Weeds in the Fields

The high negative values of EIUs, –330 and –383 in SLBC and SRBC commands respectively, suggest the gravity of the situation (Table—16.6). There was a cent percent convergence of the opinions towards the negative scale; 77 per cent of the respondents in the SLBC command considered the situation to be bad and 22 per cent as very bad (Table—16.1), while 58 per cent of the natives interrogated in SRBC command assigned the weighting of bad and another 41 per cent as very bad to this indicator (Table—16.2).

The interviewers unanimously opined that there has been an increase in the weeds among the crops, a phenomenon rare before the introduction of the irrigation facilities. Though they could not connect the severity directly with the canal irrigation, there are chances of the weeds getting spread through the conveyance channels connecting the fields. Most of the canals and tanks are infested with weeds and other vegetation in the absence of proper lining. The most common weed encountered in the fields was *Ipomea spp*.

Pest Menace

The EIUs of pest menace have high negative values of –415 in SLBC command and –327 in SRBC command (Table—16.6). Of the respondents in SLBC command, 48 per cent opined the situation to be bad, another 48 per cent as very bad and 2 per cent as worst (Table—16.1). In SRBC command 78 per cent considered the status to be bad and 21 per cent to be very bad (Table—16.2). According to the interviewers sugarcane crops were getting more commonly infested with red-rot disease and early shoot borers. Paddy crops suffered from stem–borer, leaf-folder diseases, Infestation of crops with brown plant hopper and blast and tungro viruses were a common phenomenon. Villagers in Olgalapadi and Melsiruvallur reported the destruction of entire turmeric (*Curcuma domestic*) plantations owing to the pest infestation.

The reason attributed by the cultivators to the increase in the pests attacks as were extreme heat and poor performance of the hybrid verities of paddy and sugarcane crops supplied by the agricultural agencies. This situation is exacerbated with the improper applications of chemicals in the form of fertilizers, weedicides or pesticides as discussed in the following sections.

Application of Weedcides and Pesticides

The EIUs of the application of weedicides and pesticides in the SLBC and SRBC commands are 259 and 278 respectively (Table—16.6). About 73.7 per cent of the respondents in the SLBC command perceived the status to be good while 13 per cent as very good. 5 per cent of the interviewers considered the situation to be bad and 7 per cent as no different (Table—16.1). In SRBC command 63 per cent of the respondents deemed the situation to be good, 19 per cent very good and another 19 per cent considered the situation to be excellent (Table—16.2).

Application of weedicides and pesticides in the fields is a recent phenomenon in the SCA especially in the SRBC command, which was brought under irrigation very recently (Chapter 2). The autotroph weedicide is provided by the Kallakurichi Sugar Mill (Chapter 17) for the sugarcane plantations to the cultivators. In the wake of the recent pest attacks, which according to the cultivators have been on rise for the past five years or so, they find it convenient to use the pesticides. Most of them were quite unaware of the harm these chemicals cause through the crops especially food crops by getting incorporated in the food chains and food webs. Few cultivators did know this fact so they employ labourers to clear the weeds in the paddy crops instead of trying to kill the weeds using weedicides. Very recently cultivators have started opting for *Azadirachta imitea* extract cakes.

Scientific Approach Towards Agriculture

The EIUs for this indicator are negative, –79 for the SLBC command and –22 for the SRBC command (Table—16.6). The opinions of the respondents were diverse. In SLBC command 29 per cent of the respondents regarded the situation to be good while 5 per cent as very good, 62 per cent of them perceived the situation to be bad and 3 per cent as very bad (Table—16.1). In SRBC command, 29 per cent of the respondents viewed the status to be good and 32 per cent as very good while. 29 per cent of the interviewers considered it bad and another 8 per cent as very bad (Table—16.2). The situation seems to be better in SRBC command than the SLBC command as indicated by the impact units and the percentage of the respondents under the particular scale.

Canal irrigation is a recent phenomenon in the SRBC command as discussed earlier. Agriculture Officers concerned with this particular command area have taken steps to integrate science in the agriculture with the view to alleviate the general poverty prevailing in the region. Volunteers undertake trips to the villages explaining to them the modern techniques of agriculture and irrigation and scientific application of the chemicals.

Still, there are cultivators who resort to indiscriminate application of the chemicals in the absence of proper guidance or due to their perceptions that Agricultural Officers often mislead them. According to them, the quantities as specified by the Agricultural Officers are quite high and sometimes beyond their purchasing power.

Facilities of Co-operative Societies and Bank Loans

The net EIUs for this parameter in SLBC command and SRBC command are 89 and –22 respectively (Table—16.6). In SLBC command 38 per cent of the respondents viewed the situation to be good while 18 per cent to be very good. Another 40 per cent of the respondents deemed the situation to be bad and 2 per cent as very bad (Table—16.1). Extremes were observed in the responses of the farmers in SRBC command: 36 per cent considered the scenario to be very good and 9 per cent to be good. Again, 36 per cent regarded the situation to be very bad while 17 per cent fell it to be bad (Table—16.2). The extreme responses on either side of the scale can be attributed to the poor economic conditions prevailing in the SRBC command. The marginal and small cultivators get lesser support compared to medium and large farmers, as revealed in the justifications given by the respondents. The co-operative societies and banks seem to favour the richer farmers according to most of them.

In the SLBC command, the responses were not so negative and strong because of the general prosperity of the region. Still, the reasons given by the farmers for assigning the negative weightings were the same. The cooperatives were also alleged to sell the stocks at market prices to the rich farmers and so were often out of stock. Frustrated poor farmers thus stopped approaching them and instead preferred buying their requirements from open market.

The net agriculture impact units (AIUs) for all the villages surveyed in the SLBC command are positive with the exception of Agarampallipattu, Unnamalaipalayam, Arundhatiar colony and Thenkarimbalur (Table—16.4). Vanapuram, Thenmudiyanur and Perunduraipattu are the highest ranked in descending order, in terms of AIU. Vanapuram and Thenmudiyanur are situated along the main canal.

The cumulative AIUs for all the villages in the SRBC command are positive. Athiyur, Kadavanoor and

Moongilthuraipattu, in that order, have the highest impact units. Melsiruvalur has the lowest-1 (Table—16.5). Athiyur and Moongilthuraipattu are situated across the main branch canals.

Irrigation Indicators

Sources of Irrigation

The impact units under this indicator are the highest in the Irrigation category with 479 for SLBC command and 431 for the SRBC command (Table—16.6). There was a convergence of the views among the respondents towards the brighter side of the scales; 41 per cent of the respondents in the SLBC command considered the situation to be good, 39 per cent as very good and 18 per cent as excellent (Table—16.1). In SRBC command such extreme reactions were missing. About 29 per cent regarded the facilities to be good and 65 per cent as very good: 5 per cent considered the status to be no different.

As discussed in chapter 10, the irrigation, in SCA is accomplished with the conjunctive use of canal, tank and ground water. The sources of irrigation have significantly increased after the introduction of canal irrigation resulting in year round availability of plentiful of water. The residents of the SRBC command, because of the general agro climatic conditions prevailing in the region, second priority rights to the canal irrigation and damaged canals (Chapter 10), had no extreme reaction. But in general the current situation of sources of irrigation in SCA is relatively good.

Mode of Irrigation

The traditional mode of irrigation in which the paddy fields are flooded is prevalent in the SCA. The EIUs assigned by the interviewees are –318 and –306 in SLBC and SRBC commands respectively (Table—16.6). Surprisingly the villagers consider the situation to be bad as evident by the weightings assigned by them: 17 per cent in the SLBC command regarded the flooding mode of irrigation to be very bad and another 82 per cent as bad (Table—16.1). Similarly in SRBC command the majority (73%) of the respondents considered the situation to be bad and 21 per cent as very bad. A mere 3% regarded the situation to be good.

The cultivators are aware of the benefit of drip, furrow and lift irrigations but they have not started practising them in their fields. Paddy and sugarcane are the major crops grown in the area; the cultivators find it convenient to flood their fields with water. Of late there has been a realization of the harmful impacts of flood irrigation as evidenced by the rising problems of salinity, alkalinity and water logging in the agricultural fields. The farmers tend to utilize the extra water to flood their fields and grow the water intensive crops, even as they are aware of the pit falls associated with this method.

Salinity and Alkalinity in the Fields

The positive impact units of the indicator points towards the not so bad, status of salinity problem in the SCA. The SLBC command area has 177 units while the SRBC command region has 197 units (Table—16.6). There were mixed opinions among the respondents: 65 per cent in the SLBC command regarded the situation to be good in terms of non-existence of the problem in the field and 12 per cent as very good but according to 21 per cent of the respondents the situation is bad while 1 per cent deem it very bad (Table—16.1).

In SRBC command 59 per cent of the interviewees considered the situation has good, 17 per cent as very good, 19 per cent as bad and 18% as very bad (Table—16.2).

The fields with mostly red-soil didn't have such problems while the problems were encountered mostly in the fields with black clayey soil where water percolated slowly leading to water logging and salinity problems. The lakes and tank beds in most villages were found to have while encrustations indicating the alkaline nature of the soil. Most of the villagers were unaware of the development of alkalinity as they practiced agriculture year round, growing mostly paddy and never letting the fields dry.

Water Logging

Figure 16.1 illustrates the waterlogged areas in the SCA based on the ground truth studies during the field trip. The problem of water logging also garnered positive EIUs in both the command regions, 171 in SLBC command and 157 in SRBC command

(Table—16.6). 50 per cent of the farmers interviewed in the SLBC command considered the situation to be good and 12 per cent as very good. 16 per cent of them considered the status to be bad and 8% to be very bad (Table—16.1).

In SRBC command 40 per cent of the villagers interviewed regarded the status to be god and 24 per cent to be very good and .4 per cent to be excellent. About 17.7 per cent regarded the situation to be bad and 7.4 per cent as very bad (Table—16.2). Cultivators of the SRBC command didn't have complaints regarding the water logging and salinity because of the lack of sufficient water for irrigation as elaborated in Chapter 10. Water logged areas were of either low topography or had black clayey soil which was considered good by the cultivators since they needed standing water for their paddy and sugarcane plantations.

Similarly villages in SLBC command had waterlogging problems mostly in low-lying areas. The Tholuvanthangal, some areas are low lying, which remain water logged for 8-10 months as immediately after the rainy months, water is released in the canals. Similarly the tank in Periyakallipadi has water for all the 365 days because of immediate release of water via canal following rains. Sadakuppam, Valvachanur, Velayumpakkam, Parayampattu, Pavapattu, Palayanoor, Kallottu, Pudur, Pakkam, Arundhatiar colony and many other villages reported of water logging in the fields.

Water Logging in the Fields Near to Canals

There was a sharp divergence of views between the farmers of the SRBC and the SLBC commands, reflected in EIUs of opposing signs. The SLBC command had EIUs – 274 while SRBC command had EIUs 53 (Table—16.6). In SLBC command, (Table—16.1), 7.7 per cent of the respondents, considered the situation to be worse, 21.1 to be very bad and 55.1 to be bad. But 12.5 per cent regarded the situation to be good and 3.3 to be very good.

In SRBC command, 40.7 per centof the cultivators considered the situation to be good and 20.7 per cent as very good. But 27 per cent regarded the status to be bad and another 12.9 per cent to be very bad while 8 per cent of the respondents found the status to be no different.

Fig. 16.1: Water logged areas in Sathanur command area (as observed during the field survey)

The negative EIUs SLBC command may be attributed to either the non-lining of the canals or seepage from the canals. The water tends to flood the low-lying fields next to the canals thus leading to water logged conditions in those fields. Similar conditions prevail in SRBC command region too but water logging in the fields situated close to canals is not encountered. This is because of the fact that little water flows through the canal, as reiterated by the villagers.

Sithapatinam, the tank fed village, has 12 acres of the area opposite to the tank demarcated as waterlogged. No initiatives seem to have been taken by the authorities to desilt the tank and reclaim the fields. Vanapuram, Thenkarimbalur, Thenmudiyanur villages had the fields adjacent to canals water logged. Water logging was reported in the field near to Thachampattu tank (Palayaeri) as the tank is silted and there are no shutters to regulate the flow of water in the tank (Plate—16.1).

Drainage in the Field

The network of drains is, by and large in good condition as indicated by the positive EIUs of 246 and 134 for SLBC and SRBC commands respectively (Table—16.6). In the SLBC command 55.5 per cent of the respondents considered the drainage to be good, 17.7 per cent very good and 26.6 per cent bad (Table—16.1). In SRBC command 45.5 per cent of the interviewees considered the drainage to be good and 21.1 per cent to be very good. As against this, 28.5 per cent of the people considered the situation to be bad and 4.8 per cent to be very bad.

The positive responses of most of the respondents towards the drainage in the fields were a surprise since most of the drains encountered were earthen or made out of stones. Very few of the drainage channels were cemented. The cultivators themselves maintain their individual drainage, channels by occasional de-silting and de-weeding. They do wanted lined drainage networks but before that they wanted the canals, sub-canals and distributaries to be lined.

Canal Irrigation

On the aspect of canal irrigation mixed responses were elicited from the villagers. The EIUs are 246 for SLBC command

Plate 16.1: Water logged field next to Thachampattu tank

and –103 for SRBC command (Table—16.6). In SLBC command 62.9 per cent of the cultivators felt the status of the canal irrigation to be good, 23.3 per cent as very good and 1 per cent as excellent but 3.3 per cent of the regarded the situation to be bad and 9.3 per cent to be very bad (Table—16.1). In SRBC command 24.8 per cent of the cultivators regarded the situation to be good, 11.1 per cent to be very good, 18.8 per cent as bad, 21.5 per cent as very bad and 7.7 per cent as worse. The rest referred the situation to be no different (Table—16.2).

The dissatisfaction of the cultivators of the SRBC command with canal irrigation has been dealt in the previous chapter (Chapter 10). The SLBC command enjoys much greater benefits of the canal irrigation as elaborated in the same chapter. But, in general, most of the respondents agreed that canal irrigation is a boon to them as it boosts not only surface water availability but also the groundwater table.

Modern Implements of Irrigation and Agriculture

The EIUs for this indicator are positive in both commands: 252 and 209 for SLBC and right 252 and 209 for SLBC and SRBC commands respectively (Table—16.6). The positive impact units indicates acceptance of the modern implements. In SLBC command 65.9 per cent of the cultivators considered the situation to be good and 13.7 to be very good. Only 20.4 per cent of them regarded the situation to be no different. In SRBC command 62.6 per cent considered the situation to be good and 7.4 per cent to be very good, while 30 per cent of them felt the situation to be no different.

Tractors, pump-sets and bore wells are common in SCA. The implements have been purchased through loans. When the farmers can't afford to buy the machinery, they hire it. With the introduction of subsidized electricity, the majority of the farmers have installed pump-sets and motors. The farmers who regarded the situation to be no different belong the economically weaker strata. But with heavy subsidies and loan facilities, the use of modern equipment is increasingly common now. It is rare to see the traditional bulls ploughing the fields.

The irrigation impact units (IIUs) are positive in all the villages in SLBC command with the exception of Thenkarimbalur village, which garnered -39 units. Highest IIUs, were accomplished by Periyakallapadi followed by Kandiankuppam and Pudur villages (Table—16.4).

In contrast, SRBC command had several villages with negative IIUs. Arulampadi, Vadaponparappi, Poravalur and Porasapattu prompted negative IIUs. Highest impact units were that of S.Kolathur village followed by Arambarampattu and Athiyur (Table—16.5).

Maintenance of Canals and Tanks Indicator

Lining of the Canal

There is a marked variation in the EIUs of this indicator in both the command. While SLBC command had a score of 105 SRBC command scored –246 (Table—16.6). The responses of the villagers are widely divergent.

In SLBC command 13.7 per cent of the respondents assigned the weighting good to the lining of the canal, 35.9 per cent as very good and 7.7 per cent as excellent. On the other hand, 18.8 per cent regarded the status to be bad, 20.3 per cent as very bad and 3.3 per cent as worse (Table—16.1).

In SRBC command 1 per cent of the cultivators regarded the status to be good, 22.9 per cent to be very good and 5.5 per cent to be excellent. On the other side of the scale 14.8 per cent deemed the situation to be bad, 26.6 per cent to be very bad and another 28.8 per cent to be worse (Table—16.2).

The reason for such mixed and extreme responses is that the lining work of the canals, branch canals and distributaries is still to be completed (Chapter 10). Not all the villages have been covered under the various schemes of lining of the conveyance channels. Also, according to some villagers, the cement used in the lining of the canals is of inferior quality, which can't withstand the heavy flow of water, as a result, the lining comes off.

Most of the conveyance channels in SRBC command are in dilapidated state because of the absence of lining. In some places the canals have been totally broken down.

Vegetation Near Canals

The negative EIUs of this particular indicator in both the command regions indicate the severity of the situation (Table—16.6). The responses are precipitated towards the negative side of the scale: 15.2 per cent from the SLBC command regard the status to be very bad and 54.4 per cent to be bad while 26.6 per cent of them regarded the situation to be good and a mere 3.7 per cent very good (Table—16.1). In SRBC command 29.3 per cent of the respondents perceived the situation to be bad, 26.3 per cent to be very bad and 30.4 per cent to be worse; only 13.7 per cent and 0.4 per cent of the responses were for good and very good weightings (Table—16.2).

Most of the canals are weed -infested. Even the lined canals are being colonized by vegetation. The situation is worse in the SRBC compared to the SLBC command as most of the canals in the latter are unlined and completely damaged.

The unlined branch canals and distributaries of Valavachanur, Agarampallipattu, Pavapattu, Mangalam, Chinnakolliyur and scores of others were silted and infested with weeds and other vegetation. Even the lined canals of Velayampakkam, Palayanur, Pudur, Periyakallipadi, Periyakolliyur and many others are colonised by vegetation (Plates—16.2, 16.4 and 16.5).

Vegetation in the Tank

The negative impact units of – 917 and – 438 in SLBC and SRBC commands respectively point towards the unpopular status of tank maintenance (Table—16.6). The situation is practically bad in SRBC command where most of the tanks are full of vegetation. The tanks rarely get filled with water due to scarcity of water in the region even after the introduction of canal irrigation (Plates—16.3 and 16.6).

Siltation in the Canal

The EIUs for this indicator are negative, – 476, for SRBC command and positive, 2.2, for SLBC command. The responses almost are equally ranged on both sides of the scales in the SLBC command region with 45.2 per cent and 3.7 per cent of the

Plate 16.2: Peraiyampattu branch canal in a dilapidated state

Plate 16.3: Weed infested Peraiyampattu tank

Plate 16.4: Thenkarimbalur branch canal

Plate 16.5: Sorry state of Agarampallipattu branch canal

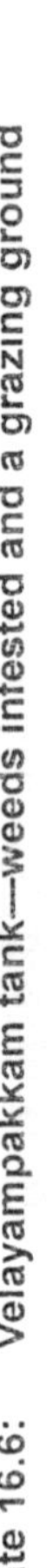

Plate 16.6: Velayampakkam tank—weeds infested and a grazing ground

interviewees considering the situation to be good and very good, respectively and 50.4 per cent and 4 per cent of them regarding the situation to be bad or worse respectively (Table—16.1).

Extremes in the responses are observed in SRBC command region yet again: 7.8 per cent of the respondents earmarked the situation to be worse, 30.7 to be very bad and 49.6 to be bad. A mere 8.8 per cent and 2.9 per cent of the respondents declared the situation to be good and very good respectively (Table—16.2).

The de-siltation has been accomplished only in those conveyance channels for which the lining work has been completed or is an ongoing process. Other channels are full of sediments. Even the lined channels at various villages have slowly started showing signs of acquiring silt (Plates—16.7).

Siltation in Tanks

The EIUs for SLBC command is –257 and –165 for SRBC command (Table—16.6). The depth of the tanks has reduced over the past several years due to siltation. In many villages the silt of the tank bed is scooped for use in construction.

Seepage from the Canals

The EIUs for the SLBC and SRBC commands are –274 and 53 respectively (Table—16.6).

About 63.3 per cent of the respondents considered the situation to be bad in SLBC command and another 23.3 per cent as very bad while 11.8 per cent perceived the situation to be good and 1.8 per cent as very good (Table—16.1). In contrast, 11.1 per cent of the interviewers in SRBC command considered the situation to be very bad, 47 per cent as bad, 33.7 per cent as no different and a mere 8.1 per cent as good (Table—16.2).

The reason for the varied responses in the SRBC command is that half of the channels remain unlined and in a bad shape. Water, if it does flow in such channels is quickly lost via seepage. Most of the time the channels remain dry and hence there is no water to seep!

On the other hand in the SLBC command all the main canals and the branch canals are lined and water flows regularly through

Plate 16.7: Status of the lined canal

them. This water seeps to the low-lying regions next to the canals and creates hazards of water logging as discussed earlier.

Optimal Flow of Water from Reservoir Through the Canals

The EIUs for SLBC command are 113, and that for SRBC command are –103 (Table—16.6). The opinions were diversified in both the command regions. In SLBC command 28.5 per cent of the respondents considered the status to be good, 32.2 per cent to be very good and 3.7 per cent to be excellent. On the other hand 19.6 per cent perceived the situation to be bad, 5.1 per cent as very bad and 10.7 per cent as worse (Table—16.1).

In SRBC command 14.4 per cent of the interviewees opined good status and 16.6 per cent. But 16.6 per cent considered the situation bad, 19.6 per cent very bad and 32.6 per cent worst (Table—16.2).

The problems associated with the flow of water in the SRBC command have been discussed in detail in chapter 10. Even the farmers of SLBC command have something to complain. According to them before the opening of the SRBC the water used to flow there in the canals for six months. But later this water was divided between the SLBC and SRBC commands. Though the SLBC farmers still enjoy priority rights over the SRBC farmers they consider the water to be insufficient for their needs.

The ire of the cultivators of both the commands was directed towards the officials concerned with reservoir and canal operation. It was alleged that during the harvesting period, when the crops need more water, the authorities do not release the water properly. As a result, many a times the crops get destroyed. Cases of bribery on the part of officials were alleged for getting the water released in the canals as reported by the villagers of Devariyarkuppam, Parayampattu, Varagur, Viriyaur, Erudayampattu, Melandhal, Kangaiyanur and many others.

Judicious Distribution of Water to the Fields

The EIUs for this particular indicator are negative in both the commands –17 for SLBC command and –103 for SRBC command (Table—16.6). The responses are varied as evident from Tables—16.1 and 16.2. The negative impact units suggest general dissatisfaction on the part of the end users. There seems to be a lack of cooperation among the farmers.

There are the villages like Thenmudiyanur, Vanupuram, Sitapathinum, etc. which are fortunate enough to be situated right across the main canals through which water flows uninterrupted for all the three months. Apart from using canal water for irrigation, those villagers use the same water for bathing, washing and other personal purposes (Plate 16.8). Such non-agricultural uses of irrigation water have been reported by Yoder (1981) and Small (1993) too. In contrast there are villages which either due to their topological location (Kalottu, Kunglinattham, Kangyanoor, Vadakirunur etc.) or due to faulty and mismanaged canals face acute shortage of water. There are villages in SRBC command like Mangalam, Chinnakolliyur, Sukhampaluyam etc., which have never reaped the fruits of canal irrigation.

Devarariyarkuppam, Pavapattu, Jambodai, Arur, Pakkam, Erudayampattu, Arundhatiar colony, Palichandal and scores other village's fall under the tail-end category. The water before reaching their respective canals gets exhausted as it is utilized by the villages in the head reaches. In most canal irrigation systems too much water is supplied to the head reaches and too little is supplied in an untimely and unpredictable manner to the tail end. Tail enders suffer multiple deprivation (Moore et al., 1983). Within the village itself (as reported by farmers of Pakkam, Kandiankuppam etc.) there is misunderstandings between them and generally the farmers having their fields at the tail end suffer from acute shortage of water as the farmers who have their fields situated at the head reach of the branch canal tend to use more water.

Encroachment and Breaching

The EIUs are –14 for SLBC command and 209 for SRBC command (Table—16.6).

In Paraympattu village, since some of the cultivators have encroached upon the land adjacent to the canal, the officials have reduced the width of the canal while lining as informed by the villagers.

In Kandiankuppam village, some farmers have done cultivation in the tank itself, as complained by the other villagers.

Plate 16.8: Thenmudiyanur main canal water being used for washing clothes

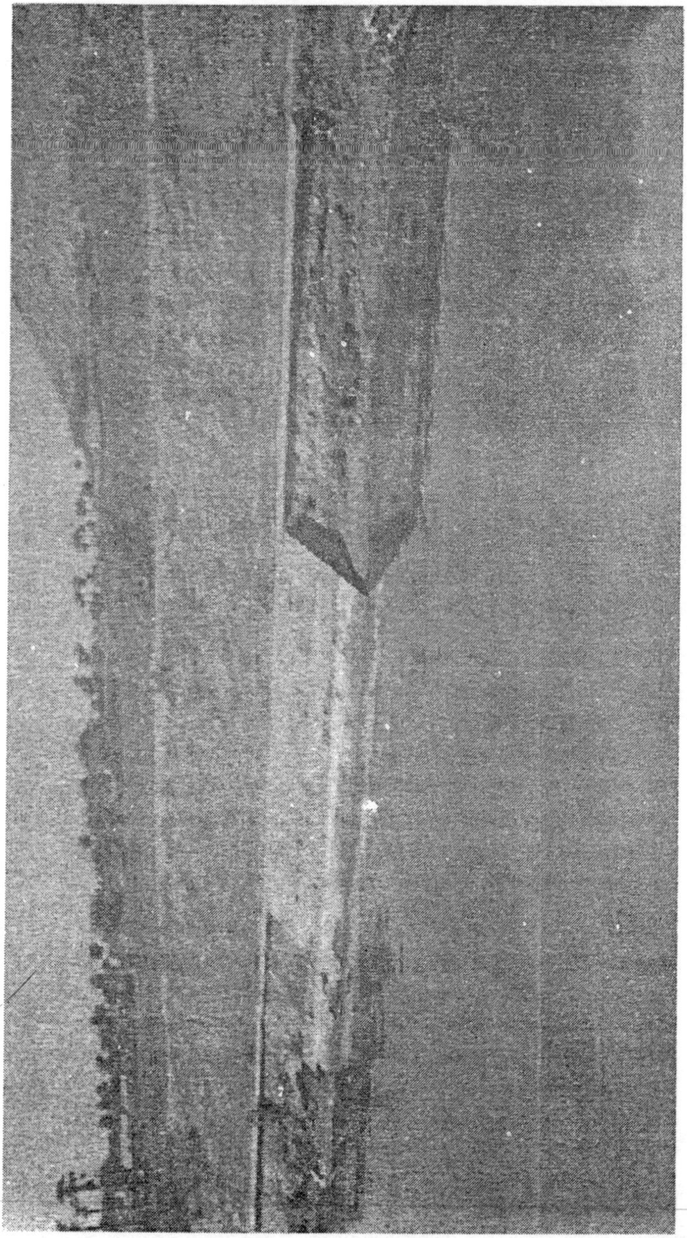

Plate 16.9: Cement slabs missing in Vanapuran canal

Thus, during the rainy months or during the time water is let out in the canal, those cultivators depriving others to get the water for irrigation remove the excess water of the tank. Similarly, farmers of some other villages for example Kunglinatham and Vadakeeranur, too, complained regarding the encroachment upon the canal and the tanks. The cement slabs of many of the lined canals were found to be broken. The area thus was encroached upon by the cultivators (Plate—16.9).

The impact units of the maintenance of canals and tanks in the SLBC command are mostly negative for all the villages with the exception of six villages viz Vanapuram, Thenmudiyanur, Perunduraipattu, Jambai, Periyakallapadi and Sukhampalayam (Table—16.4).

The situation is worse in SRBC command where, except Moongilthuraipattu, Kadavanoor and Vadamamandoor, all other villages garnered negative impact units (Table—16.5).

Socio-economic Indicators

Income

The EIUs of the indicator are 239 for SLBC command and 103 for SRBC command (Table—16.6). In SLBC command 51.4 per cent of the respondents deemed the status to be good while 28.5 per cent regarded the status to be very good, though 20 per cent regarded the status to be bad. In SRBC command region 48.8 per cent of the cultivators considered the situation to be good and 12.9 per cent to be very good. 36.6 per cent regarded the situation to be bad and 1.4 per cent to be no different.

Even though there has been an increase in the income of the individual farmer after the introduction of canal irrigation leading to intensification of the agriculture, the net savings are apparently nil in the wake of the introduction of new technologies of farming and procurement of expensive chemicals to supplement the agriculture. In general, marginal farmers were reported to earn Rs.4000 – 6000 per year, small farmers Rs.5000 – Rs.10,000 per year, medium formers Rs.10,000 – 40,000 and big farmers Rs.40,000 and above per year. The landless labourers earn their livelihood by working on the daily wages basis. Their wages vary from Rs. 50–60 per day.

Primary Health Centers

The EIUs for SLBC and SRBC commands are 233 and 129 respectively (Table—16.6). About 57.4 per cent of the villagers considered the status to be good, 10 per cent to be very good and 10.4 per cent to be excellent in the SLBC command. Of the respondents, 22.2 per cent of felt the situation to be bad. In SRBC command 47.7 per cent of the respondents considered the situation to be good, 11.8 per cent to be very good and 3.7 per cent to be excellent while 34.81 per cent of the respondents regarded the situation to be bad. There are thirteen PHCs functioning in the command area catering to the needs of one lakh people. The status of the PHC and other details has been presented in Chapters 7 and 15.

Water Borne Diseases

The EIUs are 51 and 100 for SLBC and SRBC commands respectively (Table—16.6). Since SRBC command gets less water for irrigation compared to SLBC command, there are fewer instances of water-borne diseases, as discussed in Chapters 7 and 15.

Schools

The EIUs are 349 and 390 for the SLBC and SRBC commands respectively (Table—16.6). Most of the responses are clustered towards the positive side of the scale as evident from Tables—16.1 and 16.2. All the villages have primary schools. Vanapuram, Kangaiyanur, Moongithuraipattu, Sadakuppam, Thachampattu, Palayanur, Pavapattu, Pavithram, Periyakallipadi, and scores others have high schools. For higher education the students go to Thiruvannamalai, Madras, Villupuram, Bangalore and other cities nearby.

Roads

The condition of the roads in the various villages in SCA is perceived to be good as judged from the positive impact units gained by this indicator: 413 in SLBC command and 438 in SRBC command respectively (Table—16.6).

In SLBC command 26.3 per cent of the villagers regarded the situation to be good, 61.8 per cent very good and 3.7 per cent excellent. Only 8.1 per cent of the respondents regarded the condition to be bad (Table—16.1).

In SRBC command, 36.6 per cent of the respondents considered the situation good, 41.5 per cent very good and 15.9 per cent excellent. Only 2.5 per cent of the people considered the conditions of the road to be bad and 1.4 per cent to be very bad while 1.8 per cent of the interviewers considered it to be no different. Most of the villages surveyed had pucca roads except for some that are situated in the interior. The Kallakurichi sugar mill has laid down roads in most of the villages for the easy passage of sugarcane plants to the mill. Yet, commutation is a problem since there are no frequent transport services, especially in the interior villages. Most of the villagers prefer walking to the nearby places or hire cycles rather than depending on the irregular bus services.

Forests

The status of the forests in the SCA is reflected from the negative EIUs of -130 and -201 in SLBC and SRBC commands respectively (Table—16.6). In SLBC command 62.9 per cent of the natives interviewed branded the condition to be bad, 2.9 per cent very bad and 4.8 per cent worse. Only 24.8 per cent of the residents interviewed considered the situation to be good and 4.4 per cent to be very good.

In SRBC command 85.5 per cent of the villagers regarded the status of forests to be very bad, 6.6 per cent to be good and 2.2 per cent to be good. Another 5.5% considered the situation to be no different.

The status of the forestry in the SCA is very poor as discussed in detail in Chapter 11. The villagers, especially women face a lot of difficulty in collecting firewood. Acacia (*Acacia nilotica)* is mostly used as fuel wood.

Drinking Water

The EIUs for drinking water were 248 and 342 for SLBC and SRBC commands respectively (Table—16.6). 46.3 per cent of the respondents in the SLBC command region considered the quality and facility of drinking water as good and 34 per cent as very good while 16.6 per cent regarded the condition to be bad and 2.9 per cent as very bad (Table 16.1).

In SRBC command 39.6 per cent of the residents regarded the drinking water facilities good and 49.3 per cent very good. But 10.74 per cent of the respondents considered it bad and 0.4 per cent very bad (Table—16.2).

Most of the villages have dug well and collect the water in a overhead tank for supply drinking water to entire village. The quality of drinking water has been assessed elaborately in Chapter 13.

Power Supply

The status of power supplies as indicated by the EIUs of 188 and 206 in SLBC and SRBC commands respectively can be termed as good (Table—16.6).

In the SLBC command 38.1 per cent of the respondents termedhe status to be good and 31.9 per cent to be very good, 27.7 per cent to be bad and 2.2 per cent to be very bad (Table—16.1).

In SRBC command, 45.9 per cent of the respondents regarded the situation to be good, 28.2 per cent to be very good and 25.9 per cent to be bad. All the villages in the command area are electrified and the houses are given free electricity supply for one bulb. But the supply is very irregular. The interruption in power, especially during the harvesting months, affects the agriculture adversely. Though the situation is getting better, uninterrupted power supply for running motors is yet to be achieved, as stated by the cultivators. Due to the late release of water into Commanur canal and irregular power supply paddy crop raised in 50 per cent of the area was reported to be withering in Nagarjunasagar project Karnataka, India (Rao, 2000).

Employment and Migration

The EIUs of this indicator -263 for SLBC command and-369 for SRBC command, reflect the disillusionment vis a vis employment opportunities. Nearly half (49.6%) of the farmers interviewed described the situation as bad and 33.3 per cent as very bad in the SLBC command (Table—16.1). Only 15.2 per cent of them deem the situation to be good and 1.8 per cent to be very good.

In SRBC command, 36.6 per cent of the interviewers regarded the status to be very bad and 63.3 per cent to be bad, in spite of the sugar mill being located in the command (Table—16.2).

The introduction of irrigation schemes and projects are believed to generate employment potential in the region by way of establishment of agro-based, and other small-scale industries. In SCA there is only one such industry, Kallakurichi Sugar Mill, located in Moongilthuraipattu. Lack of industries and scarcity of water for irrigation leads to large-scale migration of the villagers in the command region. Most of them migrate to the nearby towns and cities: Thirvannamalai, Villupuram, Kallakurichi, Chennai, Mumbai, Bangalore in search of greener pastures. They work in those places as skilled and semi skilled laborers and come back only during the periods when water flows through the canals. The situation is particularly bad in SRBC command which doesn't get sufficient water via canals thus leading to dearth of water for irrigation.

The bulk of marginal cultivators also work in the farms of bigger cultivators to supplement their incomes. Only the few prosperous villages, because of the main canal running through them, didn't have villagers complaining about the dearth of employment. As many as 80 per cent of the interviewers wanted their children to take up salaried jobs, preferably with the government, which would ensure regular income. None of the parents wanted their children to become cultivators. Downstream of Khasm el Girha dam in Sudan, the percentage of households involved in agriculture as a primary occupation fell from 92 per cent to 81 per cent while manual labour and trade increased (Acreman et al, 1999).

The net socio-economic impact units (SIU) for all the villages surveyed in the SLBC command are positive with the exception of Pallichandal and Thenkarimbalur villages. Vanapuram, Theinmudiyanur and Palayanur villages in that order had the highest SIUs in the region (Table—16.4).

With the exception of Pakkam and Kidagudayampattu villages, the net SIUs of all other villages are positive in SRBC command region as well (Table—16.5). Moongilthuraipattu and Kadavanoor have the highest SIUs in the region, 144 and 101, respectively.

The total impact units, for the villages are also illustrated in the Tables—16.4 and 16.5 for SLBC and SRBC commands respectively. The villages are arranged according to the total impact units, representing the descending order of prosperity in villages.

A comparative account of the cumulative impact units, as accomplished by the respondents in the SCA and the experts, is presented in Table—16.7. The wide gulf between the EIUs of the end users and the experts' points towards the failure in perceiving the magnitude of the impacts downstream by the project authorities. Both, the end users and the project authorities put the blame on each other for this disparity as discussed in various chapters and preceding sections. Illiteracy and poverty are the main hindrances in the prosperity of the region as maintained by the authorities. The literacy status, in SCA as per the 1991 census is depicted in Figures—16.2 and 16.3 and presented block-wise as Tables—16.11 to 16.17. The total literacy is dismally low, less than 35 per cent, in all the blocks in SCA as evident from the Figure—16.2. Gender wise, female literacy ranges from 35-40 per cent in various blocks and cumulatively in the SCA (Figure—16.3).

Literacy Status

The literacy status, as per the 1991 census, has been provided as Tables—16.11-16.13 for SLBC blocks, and Tables—16.14-16.16 for SRBC blocks. The cumulative literacy status for RS Sathanur catchment and command area has been provided as Table—16.17.

The pattern of literacy rate in the command and catchment area appears to be similar–with an average of 65 per cent made literacy (of the total literate population) and 35% female literacy (of the total literate population).

Inter-parametric Correlation

Regressions between various ecological indicators were performed to bring out the interdependencies, if any, among them. The correlation analyses were performed based on the ranks assigned by the respondents. The results are thus based upon the perceptions of the respondents in interrelating various indicators.

Productivity was correlated against area under cultivation, usage of chemical fertilizers, usage of biofertilizers, problems of needs in the fields, pests menace in the fields, pesticides and weedicides application, scientific approach towards agriculture, sources of irrigation, mode of irrigation, salinity in the fields, water logging in the fields, modern implements of agriculture and irrigation, canal irrigation, optimal flow of water from reservoir, judicious distribution of water, income, power supply, and roads.

Table—16.11 Literacy status:Chengam (SLBC command) block (1991 census)

	Literacy		
	Male	Female	Total
Jambadai	NA	NA	NA
Olgalapadi	NA	NA	NA
Thenmudiyanur	1117	637	1754
Edathanur	NA	NA	NA
Allappanoor	269	151	420
Vanapuram	1150	785	1908
Kunglinatham	167	104	271
Perunthuraipattu	507	372	879
Valavachanur	0	0	0
Kottaiyur	439	214	653
Agarampallipattu	334	146	480
Thenharimbalur	667	393	1060
Radhapuram	NA	NA	NA
Serapapttu	173	116	289
Varhillapattu	270	147	417
Sadahuppam	327	109	436
Unnamalaipalayam	162	46	208
Edakkal	269	105	374
Muzhuvampattu	192	75	267
Periyampathu	515	269	784

NA- *Not Available*

Table—16.12 Literacy status: Rishivandiyam (SRBC command) block (1991 census)

	Literacy		
	Male	Female	Total
Athiyur	483	278	761
Mangalam	302	174	476
Adathanur	913	565	1496
Erudayampattu	741	358	1099
Maniyandal	304	111	415
Vanapuram	991	502	1493
Odiyanthal	107	73	180
Edathanur	770	329	1099
Kadambur	419	263	682
Arambarampattu	219	98	317
Thiruvarangam	276	111	387
Sirpanandal	1395	770	2165
Jambadai	1394	846	2240
Periyakolliyur	74	34	108
Chinnakolliyur	668	323	991
Tholuvanthangal	431	276	707
Kadavanur	637	275	92
Pakkam	168	73	241
Vadamamandur	288	111	399
Nagalkudi	101	51	152

Table—16.13 Literacy status:Sankarapuram (SRBC command) block (1991 census)

	Literacy		
	Male	Female	Total
Rayasamudram	43	18	61
Arulampadi	125	50	175
Moongilthuraipathu	276	133	409
Porasapattu	325	195	520
Olgalapadi	424	235	659
Poravalur	467	232	699
Melsiruvalur	742	408	1150
Vadasiruvalur	707	218	925
Varagur	1025	473	1498
Arur	296	157	453
Mookanur	78	34	112
Thimmendhal	530	226	756
Chellankuppam	222	118	340
Viriyur	390	182	572
Arasampattu	335	173	508
S.Kolathur	208	141	349
Vadakiranur	NA	NA	NA
Vadaponparappi	475	230	705
Kidagudayampattu	569	295	864
Sivapuram	NA	NA	NA

NA- *Not Available*

Table—16.14 Literacy status: Chengam (SRBC command) block (1991 census)

	Literacy		
	Male	Female	Total
Rayandapuram	573	286	859
Thiruvadathanur	378	182	560
Thondamanur	225	72	297
Puthurcheekadi	247	160	407

Table—16.15 Literacy status: cumulative (1991 census)

Block	Literacy		
	Male	Female	Total
Thachampattu (SLBC command)	8851	5054	13905
Thirukoilur (SLBC command)	3715	2115	5830
Chengam (SLBC command)	6558	3642	10200
Rishivandiyam (SRBC command)	10699	5661	16320
Sankarapuram (SRBC command)	7237	3518	10755
Chengam (SRBC command)	1423	700	2123
SRBC command	19359	9839	29198
SLBC command	19124	10811	29935
SCA	38483	20650	59133

Table—16.16 Literacy status: Thachampattu (SLBC command) block (1991 census)

	Literacy		
	Male	Female	Total
Thachampattu	429	291	720
Allikondapattu	269	173	442
Katampoondi	824	535	1359
Periyakallipadi	495	303	798
Chinnakallipadi	480	264	744
Navampattu	398	216	614
Devanur	474	199	673
Velayampakkam	314	115	429
Kallottu	255	122	377
Athipadi	129	70	199
Palayanur	1030	515	1545
Kandiankuppam	335	159	494
Thalayampallam	646	391	1037
Chakkarathamadi	47	30	77
Nariyapattu	264	152	416
Perayampattu	417	239	656
Pavapattu	669	343	112
Pavithram	1019	789	1808
Aradapattu	357	148	505

Table—16.17 Literacy status: Thirukoilur (SLBC command) block (1991 census)

	Literary		
	Male	Female	Total
Melandhal	385	232	617
Kangaiyanoor	1091	518	1672
Pallichandal	111	32	143
Konganamur	625	440	1065
Murukkambadi	38	2	40
Athiyandal	265	99	464
Devariyarkuppam	258	134	392
Jambai	353	250	603
Chellankuppam	320	139	459
Sithapattinam	164	68	232
Manalurpet	105	38	143

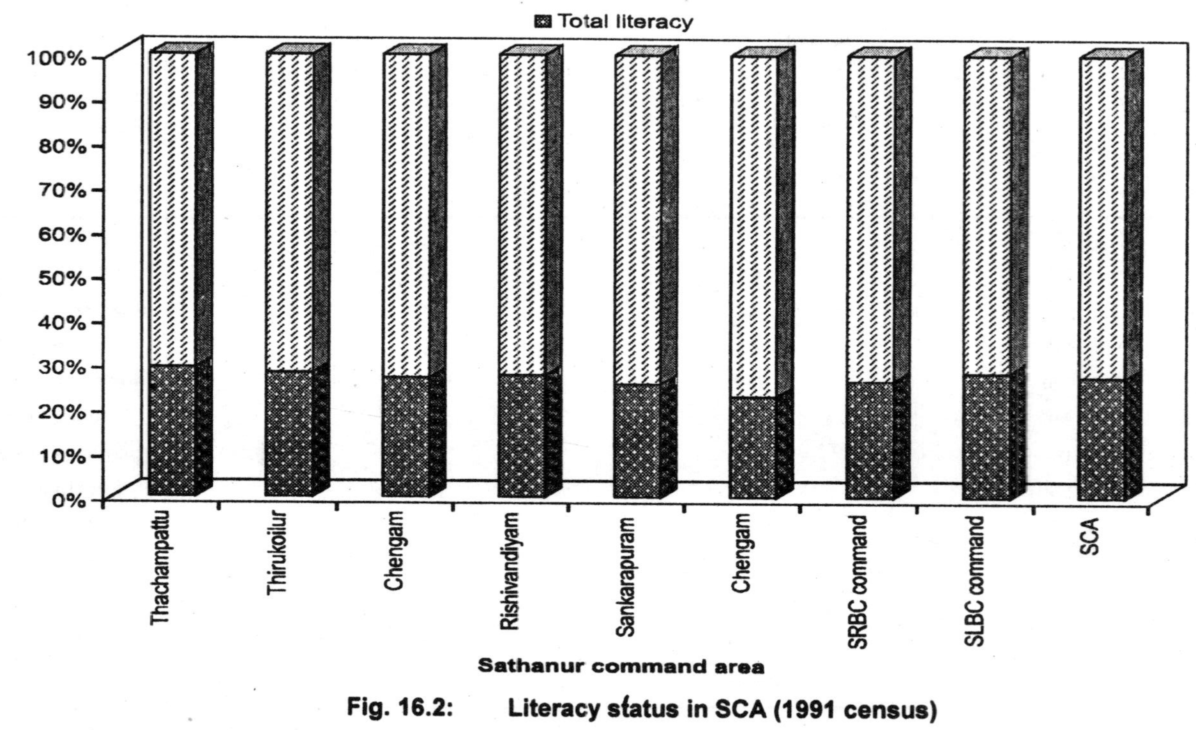

Fig. 16.2: Literacy status in SCA (1991 census)

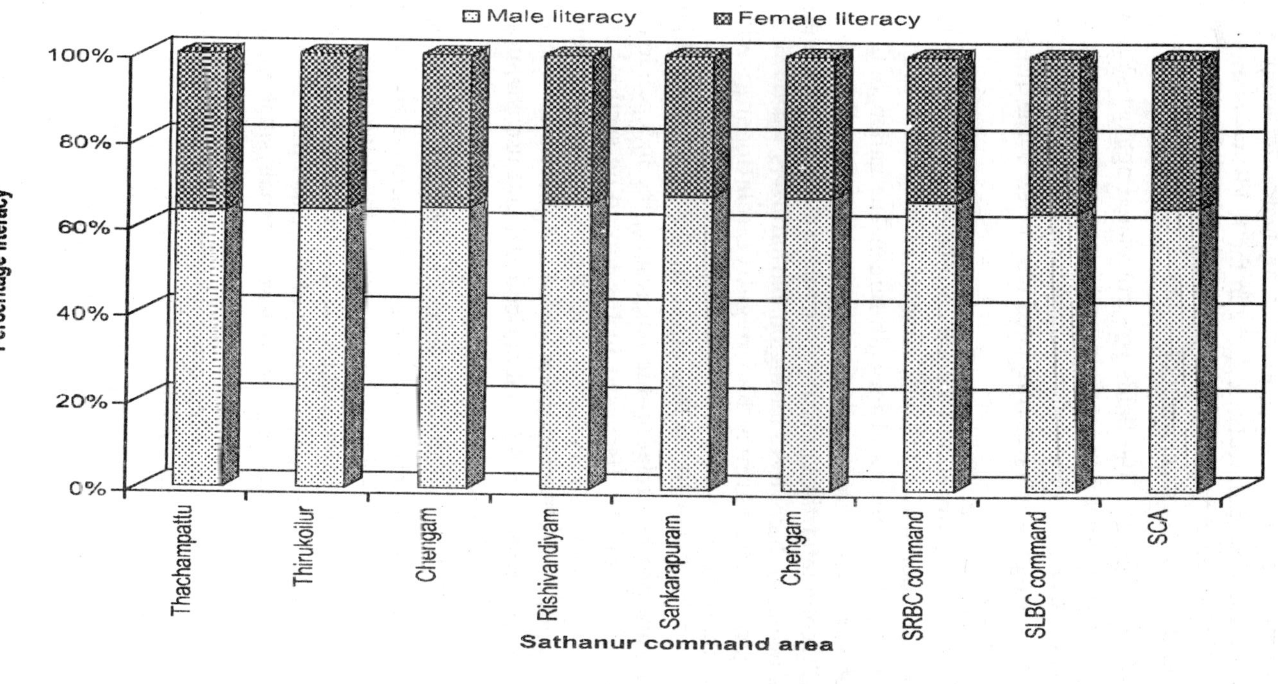

Fig. 16.3: Gender wise literacy status in SCA (1991 census)

Lining of the canal was correlated against water logging in the fields near to canal, water logging in the fields, vegetation in the vicinity of the canals, siltation in canals, seepage from canals and breaching an encroachment of canals.

Water logging in the fields was correlated against salinity and alkalinity problems in the fields, drainage, seepage from, the canals and canal irrigation and tank irrigation. Optimal flow from reservoir through canals was correlated with water borne diseases (Table—16.8).

The productivity shows a good correlation with area under cultivation, chemical fertilizers, sources of irrigation, modern implements of irrigation and agriculture, canal irrigation, optimal flow of water from reservoir, income, power supply and roads in the SLBC command, thus indicating the role of all these parameters in enhancing productivity of the region. Bio fertilizers, weeds mode of irrigation and salinity had negative relationship with productivity, as discerned from the coefficients of correlations (Table—16.8). Scientific application to the agriculture had insignificant positive relationship suggesting the independency of the parameter. As discussed in the foregoing sections the scientific approach is very less in the region. Cultivators don't generally pay much attention to the advice of Agricultural Officers. Instead they practice traditional agricultural and use chemicals out of their own choice.

Lining of the canal shows a positive correlation with water logging in the fields near the canal and negative with water logging in the fields. There is substantial seepage in the lined canals as well, thus leading to water logging as indicated in the positive correlation with the seepage. Lining of the canal also shows a positive relationship with vegetation near the canals, siltation, and breaching. The lined canals are infested with vegetation and are accumulating silt. The missing cement slabs and mutilating the lining of the canals have already been discussed.

Similarly water logging shows a positive correlation with salinity and seepage. Optimal flow of water in the canals correlates positively with water-borne diseases especially malaria.

Similar pattern of inter-parametric correlations is observed for SRBC command too. Lining of canals is negatively correlated with water logging in the fields near canals, and with seepage. This discrepancy is due to the fact that most of the branch canals in the SRBC command are unlined and villagers have a perception that once lined there will be reduction in water logging and seepage problems.

There is only a very weak relationship between waterborne diseases and optimal flow of water in the SRBC command since water very rarely flows in the SRBC system and villagers could not connect the two indicators. The correlation results are based on the responses of the villagers.

The coefficients of rank correlation between the villages based on the EIUs of the various indicators are presented in Tables—16.9 and 16.10 for SLBC and SRBC commands respectively. A significant correlation is observed between agriculture and socio-economic, and between canal maintenance and socio-economic indicators. In SLBC command (Table 16.9). In SRBC command there is a strong correlation between agriculture, canal and tank maintenance and between socio-economic indicators and again between canal and tank maintenance and socio-economic indicators, indicating the interdependencies of there indicators (Table—16.10).

Brief Notes on the Problems Encountered in the Selected Villages of SCA in Relation to the Canal Irrigation

- In Devanur village, 42 acres of the land was included in the canal irrigation area through tank. But the tank hence the water released by Sathanur reservoir in the tank is if no use to the farmers. The villagers refuse to be termed as the beneficiaries of the Sathanur canal irrigation project and have removed the signboard claiming thus.
- In Valavachanur village the canal starts from the low-lying land and subsequently goes to the higher fields. In the process the water does not reach the subsequent fields properly, spilling on the way. Thus, the villagers want the fields and lands, through which the canal passes, to be evened so that judicious distribution of water will be there.

- Another unique problem encountered in the Valavachanur was that near by the village there is a Government owned farm of 300 acres. The farm utilizes the water during the day and so the villagers get water for their fields only during nights. Thus there is a clash of interests between the village farmers and Government officials.
- In Velayampakkam, there are four tanks. Villagers complained that all the four tanks are always dry as water does not reach these tanks at all.
- In the Palayanur village, some farmers have put stones and boulders on the shutter of the canal, thus blocking the canal towards the shutter end, hence denying water to some parts of the fields (Plate—16.10).
- Thatchampattu is having two tanks but the channel connecting the two tanks is not lined. Also the tanks are silted so water scarcely comes to the 2nd tank.
- In Pudur village, farmers have broken the shutters and sides of the canal in order to get more of water towards their fields.
- Jambodai, which has provision to get water from Arumbarampattu branch canal via Sivanur, doesn't get much water through this canal. The reason being that Arumbarampattu has a comparatively high lying land and as the water comes through the canal from higher to lower land, silt and soil deposit in the canal and ultimately water doesn't reach the Jambodai village tank. Thus instead of double cropping the farmers have been restricted to only one crop per year.
- Periyakolliyur and Chinnakolliyur are connected by the same branch canal. Periyakolliyur branch canal (which ultimately goes to Chinnakolliyur) is lined but on the way the canal has been completely silted by the villagers themselves for facilitating easy commutation and hence the water doesn't reach the Chinnakolliyur Lake.

Plate 16.10: Canal mutilated by the villagers of Palayanur to divert more water towards on side

Plate 16.11: Hard to believe that the branch canal of Mangalam has been filled up for easy passage

- Villagers of the Sirpanandal too complained of no water reaching their village tank as the canal is positioned in a place from where water doesn't reach this upland village. So the villagers want the canal to be reconstructed through some other place or repositioned.
- In the Mangalam village too, since branch canal is not lined and is damaged, people have put mud and soil in order to fill it up and are commuting over it. The canal has been blocked in such a way that is hard to believe that canal existed there once (Plate—16.11).
- The tank in the Tholuvanthangal village is low lying whereas the fields and farms are located higher. So, the water from the tanks doesn't reach the fields.
- Arur village being situated at the tail end doesn't get enough water in the tank. The tank has been planted with Acacia trees now in the name social forestry.
- Poruvalur village hasn't got water for years. According to some villagers water came for just 2-3 years after the canal was opened. Many meetings with farmers were conducted by the officials to look in to the matter with no results whatsoever. Before reaching the poruvalur village the water gets utilized and on the way low lying lake comes where the watergets collected and doesn't come further till the village.
- Kidakudayampattu, a village with poor economic condition, is having 3 tanks to be fed under Sathanur reservoir project canal via Arur. But since Arur village itself doesn't get water there is no question of the tanks of this village getting filled.
- Thimmenendal village comes under tank irrigation scheme of the SRP but no water comes to the tank through the canal since the canal has not been constructed even after 17 years of SRBC coming in to existence, though the canal exists in the map.
- Viriyur village also never receives water in the canal unless the farmers bribe the officials.

17

Forestry, Industry and Water Supply

FORESTRY

Introduction

Apart from the aspect of dislocation of people, the loss of forests is arguably the most harmful of the impacts of large water resource projects. Unlike the few very opportune situations—such as the one existing with Hoover dam (Lake Mead) USA where the storage site had no forests at all-most water resources projects entail loss of natural forests. For a long while this loss was considered inconsequential. From 1970 onwards increasing attention was paid to deforestation caused by large water resource projects; yet the impact was, more often than not, undervalued by the authorities interested in pushing through new projects. In India, matters reached a head during late 1970s when the authorities, in the state of Kerala, which would have submerged large tract of a tropical rainforest, approved a major project. The region, called 'Silent Valley' has since become a part of the environmental folklore, more so because protracted public agitation eventually succeeded in stalling the project.

In this chapter we have presented an account of the status of the forests in the Sathanur Command Area (SCA). The impacts of the forest on the environment and the eco-restoration attempts on the part of the officials have also been evaluated.

STATUS

SCA has a meagre forest cover as evident from the rather bleak statistics of area under forest in the various villages (Chapter 11). Vanapuram, Kadavanur, Athiyur and Periyakallipadi have some of their land under the designation 'reserve forest' but in reality all that Kadavanur and Athiyur have by way of 'forest' is eucalyptus *(Eucalyptus globulus)* (Plate—17.1). The flora is a little more diverse in the forests associated with Vanapuram and Periyakallipadi, which consists of teak *(Tectona grandis)*, *Tamarindus indica, Acacia nilotica, Accacia planiform, Albizia amara* and *Azadirachta imitea* etc. The flora underrlines the southern umbrella thorn forest, typical of southern peninsular India. The 'reserve forest' is the responsibility of Forest Department functioning in the Tamil Nadu state. The villages as such have very scarce forest cover. The only trees encountered in the villages were that of *Acacia nilotica,* especially in the dry regions of SRBC command.

IMPACT

The scarce resources of trees in the SCA put tremendous pressure on the village folks in terms of fuel wood. The villagers use wood of *Acacia* sp. for this purpose. According to the villagers, the *Acacia nilotica* trees are becoming a nuisance now as they crop up along with the crops in the fields and also in the vicinity of canals and on the tank beds. Acacia trees are planted in some tanks of Arur and many other villages in the name of social forestry by the Forest Department. The villagers are not allowed to take wood products form the reserve forests. The villagers were also conscious of the water table conditions of their regions. Most of them had complaint regarding the plantation of Eucalyptus trees in the vicinity of their villages (Vanapuram, Varagur, Arulampadi) (Plate—17.1). Extreme reactions were encountered form the villagers of Arulampadi who insisted on the removal of those trees, which according to them were depleting the ground water in their village.

The follow up visits to the officials of Forest Departments (Thiruvannamalai, Villupuram and Kallakurichi) were made. The officials laid the blame for the scarce forest cover in the region squarely on the villagers. According to them the villagers in their

Plate 17.1: Eucalyptus plantation right across the varagur tank

greed to bring more and more area under cropping have started encroaching upon even the graveyards let alone forests. The programmes of social forestry and agro forestry have failed in the SCA partly because of the extreme heat and dry conditions prevailing in the regions but more strongly due to the non-cooperation of the villagers. The sturdy *Acacia nilotica* trees have been planted by the Forest Department to cater to the fuel wood needs of the villagers because no other species survives in the prevailing agro climatic conditions. Also there is a lack of space for planting other trees as the villagers refuse to allot their land for forestry purposes. The inclination of the villagers at large is to bring more and more of available land under the cultivation of cash crops viz paddy *(Oryza sativa)* and groundnut. The officials denied the role of Eucalyptus plantations in the depletion of ground water and maintained that no plantations have been undertaken for the past seven to ten years.

INDUSTRY

Introduction

Large irrigation projects often witness growth of agro-based industries in the command region, for utilizing the agricultural produce. These industries influence the socio-economics of the region by creating new employment and entrepreneurial opportunities.

Sathanur irrigation project led to the establishment of Kallakurichi Co-operative Sugar Mill in Moongilthuraipattu village (Plate—17.2). This is the only one industry operating in the Sathanur Command Region (SCA). Present study assesses the impact of the coming up of the sugar-mill on the environs of SCA.

STATUS

The sugar mill was built in the year 1962, four years after the commissioning of Sathanur Reservoir Project (SRP). The mill started with the sugarcane *(Saccharum officinarum)* processing capacity of 1000 tones and increased to 2500 tonnes in the year 1995. The mill caters to the demand of various villages of Sathanur Left Bank Canal (SLBC) and Sathanur Right Bank Canal (SRBC) commands.

Plate 17.2: Kallakurichi Cooperative Sugar Mill

Plate 17.3: Sugar mill effluent released into the Ponnaiyar—a drinking water source for the natives of melandal

Impacts

During the field trip it was revealed that majority of the cultivators had grievance against the sugar mill. The cultivators don't get their payments in time as the mill owners claim the mill to be running in loss for the past several years, as reported by the villagers downstream. The mill delays the cutting orders of the sugarcane plantations by six months to a year or more. The plantations lose their worth with time, accruing great loss to the cultivators. The farms remain blocked for two or more years with the plantations, as a result the cultivators are unable to grow any other crop. In fact these were the reasons attributed to loss in productivity and income by many cultivators during the survey.

The villagers also implicated the mill in spreading pollution in the SCA. It was alleged that the mill releases its untreated effluents into the Ponnaiyar River, which meanders through Jambai, Pallichandal, Melandhal and Kangaiyanur villages. The villagers have dug wells in the river and use the water for drinking and other domestic purpose (Plate—17.3) According to the villagers, the water turns black and emanates a pungent odour, whenever the effluent is released in the river. There were reports of fish kills also due to the toxic effect of the effluent. Interestingly an official in Valavachanur village related the effluents rich water as beneficial for the crops cultivated on for the Government owned 300 acres of farm in the village. The water was supposed to be rich in certain nutrients which enhanced the crop production.

A follow up visit was made to the mill office to ascertain the facts. The mill is situated right across the bank of Ponnaiyar River. An Effluent Treatment Plant (ETP) is installed inside the factory. Our attempts to photograph the plant were thwarted by the officials. The officials claimed that no untreated effluents were being released into the river. According to them the effluents were treated with the aid of ETP and then let into a small farm inside the factory premises. The form consisted of coconut *(Cocus nucifera)*, mango *(Mangifera indica)* and palm trees. The reasons regarding the late issuing of sugarcane cutting orders were attributed to the abundant cane production, a phenomenon being witnessed all over the country, since 1995 and not in SCA alone. The officials complained that the cultivators in their greed to earn more cash profusely grew sugarcane, which ultimately didn't find buyers. As to compensating the cultivators in terms of cash, the officials maintained that they had cleared all their pending dues till that date. In summary, as with other aspects of SRP, there was great difference in the perceptions of thc authorities and the end users vis a vis the sugar mill as well. The sugar mill has laid roads in many of the villages for easy transportation of the cane and also gets the soil quality of the cultivators tested in the Soil Testing Laboratory, Cuddalore. Based on the test results the mill authorities prescribe the amount of fertilizers and pesticides to be used for the plantations, as stated by the mill authorities.

WATER SUPPLY

Introduction

Sathanur Reservoir Project (SRP) meets the domestic demands of Thiruvannamalai district by supplying about 0.4 million litres of water per day from the reservoir. The present study gives the status of the water supply from the Sathanur reservoir.

Status

The details regarding the water demand and the various water supply schemes are presented below:

Water supply schemes executed before SRP:

1. 1939 Samudram pumping station;
2. 1969 Olgalapadi head works;
3. 1989 Olgalapadi head works World bank scheme.

Details of the Scheme

Year of original scheme	–	1934, 1969 and 1989		
Administrative sanction number and date	–	G.O. No. 839, MA & WS Dt. 6.9.90		
Estimate cost	–	1969 - Rs. 0.5 millions 1989-Rs. 14.01 millions		
		75% grant		
		25% loan		
Design year and design population	–	Year 2021 - Population 2,10,000		
Source details	–	1939 - Samudram iri, Infiltration Gallery		
		1969 and 1989 Thenpenni river Surface water		
Number and Capacity of SCS		Old scheme New scheme		
		1. 6.82 !akh litres		18.00 lakh litres
		2. 6.82 lakh litres		10.00 lakh litres
		3. 4.50 lakh litres		8.00 lakh litres
		4. 10.00 lakh litres		3.00 lakh litres
System provided		Existing	– 39.67 km	
		New scheme	– 45.70 km	
			85.37 km	
Number of HSCs given	–	9217		
Number of PFs provided	–	595		
Designed supply	–	90LPCD		
Month and year of completion and commissioning	–	May 1995		

(Contd...)

Present supply in MLD		–	12MLD
1991 and present population		–	109196 and 122500
Present service level		–	100 LPCD
Water supply tariff			
	Domestic	–	Rs.1.25 1000 litres -1
	Non-Domestic	–	Rs.2.50 1000 litres -1
	Commercial	–	Rs.3.75 1000 litres-1
	Industrial	–	Rs.3.75 1000 litres -1
Annual income from water charges		–	Rs. 0.2 millions
Annual maintenance expenditure		–	Rs.0.6 millions

As a part of the augmentation process to the municipality water supply, Thiruvannamalai, a water treatment unit was commissioned with the World Bank aid at the pick up dam near Olgalapadi village about 25 km away from the town. The first scheme comprising a cascade type aerator, clarifloculator and two rapid sand filters having the capacity of treating 55 lakhs litres of water per day, is functioning since 1975 and the second scheme with 6 rapid sand filters with a capacity of treating 140 lakh litres of water per day is also functioning since 1995. At present an average of 90 lakh litres of water is treated every day.

The aerated, clarified and filtered water from the Olgalapadi head works is pumped after primary chlorination and is received in a ground level reservoir (sump capacity 7.0 lakh litres) at Thandarampet road booster station. After secondary chlorination at the booster sump, the water is pumped at various high level reservoirs located at Poonrandakulam (7 lakh litres), Somavarakumal (3 lakh litres), new bus stand (10 lakh litres), old bus stand and Thindivanam (18 lakh litres).